Michael
Schocken

GLOBAL
Awakening

"Take the time to read *Global Awakening* and once you have digested its magnificent context and contents, consider sharing it with your friends and networks and spread the word. The book provides a very real sense of our place in the transformation of culture that holds the promise of reclaiming the world from those who have taken it hostage. The time to act is now, and *Global Awakening* belongs in the library of every awakened activist."

JEFF HUTNER, EDITOR OF *NEW PARADIGM DIGEST*

"Michael Schacker's book reveals the hidden history of great thinkers and change-makers along with progressive solutions for the here and now. This truthful and positive book could have powerful and far-reaching effects and will benefit anyone who reads it."

MAURICE HINCHEY, U.S. CONGRESSMAN FROM NEW YORK

"*Global Awakening* is just what is needed right now: a penetrating analysis of the historical context of our current global crisis, plus the all-important—and hitherto missing—plan of how we get to where we want to be. This is the book the 'organic' movement has been waiting for and, I hope and trust, will embrace as its blueprint for the future. It is as important as *Silent Spring* and as prescient: a must-read for anyone struggling with the question of how to engage with the world in a positive and productive way."

PHILIP CHANDLER, AUTHOR OF *THE BAREFOOT BEEKEEPER*

GLOBAL
Awakening

NEW SCIENCE AND THE
21st-CENTURY ENLIGHTENMENT

Michael Schacker

Park Street Press

Rochester, Vermont • Toronto, Canada

Park Street Press
One Park Street
Rochester, Vermont 05767
www.ParkStPress.com

Park Street Press is a division of Inner Traditions International

Library of Congress Cataloging-in-Publication Data
Schacker, Michael.
  Global awakening : new science and the 21st-century enlightenment / Michael
Schacker.
    p. cm.
  Includes bibliographical references and index.
  ISBN 978-1-59477-482-9 (hardback) — ISBN 978-1-59477-515-4 (e-book)
  1. Science—Philosophy. 2. Science—Social aspects. 3. Science—History.
4. Enlightenment. 5. Science and civilization. 6. Technology and civilization.
7. Social change. I. Title.
  Q175.S305 2012
  501—dc23
                                                                    2012016856

Printed and bound in the United States by Courier Corporation

10 9 8 7 6 5 4 3 2 1

Text design by Jon Desautels; text layout by Virginia Scott Bowman
This book was typeset in Garamond Premier Pro with Arial and Gill Sans used as
display typefaces

The photograph of Werner Heisenberg is used courtesy of Bundesarchiv,
Bild183-R57262 / CC-BY-SA

The images of J. I. Rodale, Robert Rodale, and Maria Rodale are courtesy of the
Rodale Family Archives

To send correspondence to the author of this book, mail a first-class letter to the
author c/o Inner Traditions • Bear and Company, One Park Street, Rochester, VT
05767, and we will forward the communication, or contact the author through his
website: **www.GlobalRegen.net.**

CONTENTS

ACKNOWLEDGMENTS

This book could not have been completed without the help of my wife and partner, Barbara Dean Schacker, who supported my work on this book for over thirty years. After my traumatic brain injury, she updated facts and worked on my behalf with my editor, Anne Dillon. Without Barbara, I would not be here and this book probably would not have made it into print. She dedicated herself to this book, my recovery, and our family in the hardest of times. Barbara finished the book (my speech and writing disabilities prevented me from doing this) and continues to handle the ongoing process of publishing and promoting it as well as directing the Global Regeneration Network (GlobalRegen.net) and all of its related websites.

I thank Stephen Larsen for writing the introduction, helping with the book, and providing LENS neurofeedback treatment, which has made such an enormous difference in my recovery. I would like to thank Anthony Rodale, who helped me develop my knowledge and ideas for global regeneration and for creating brilliant new ways to transition from chemical to organic agriculture and gardening and for paving the way of the future of organic farming.

I thank Dr. John McMillin, regenerative scientist and activist, for his guidance and life experiences and for being the person primarily responsible for the well-known greening of the Antsokia Valley in Ethiopia. McMillin's systematic approach solved the famous African dust bowl of the 1980s, saving more than six hundred thousand people from starvation. He confirmed my experience that for every problem humanity faces there is a solution.

I thank my editor, Anne Dillon at Inner Traditions/Park Street Press, and all the people there who made this book the best it could be.

I would also like to acknowledge all those who provided their love and support: my daughter, Melissa Schacker, for her love and help with the book; my mother, Maxine Schacker, for her brilliance and her tremendous faith in me and my recovery, and for providing financial support in our time of need; my father, Jerome Schacker, for his love and financial support; my sister, Jeannie Schacker, for all she has done; my brother, David Schacker, for his love and support; and my stepdaughter, Jennifer Crosby, for lovingly putting up with her writer stepfather through the many years of research during her childhood.

I extend gratitude and love to my friends Mitch Ditkoff, Evelyne Pouget, Maureen Radl, Michael Latriano, Tad Wise, Marc Miller, and Mae Skidmore; to all my doctors and therapists; and to my caregiver aides for everything they have done. My gratitude goes out to my friends in California, in Woodstock, and all around the world. As this book was thirty years in the making, there is very little doubt that the true length of acknowledgments would shadow what is written here.

INTRODUCTION

REVEALING THE HIDDEN PATH TO THE "POSSIBLE HUMAN"

Stephen Larsen, Ph.D.

Before the stroke that eclipsed half his brain, I knew that my friend was a genius. Whether I was improvising music with him in an impromptu band, attending an opera or a concert he had written, or reading one of his books, there was the sense of a mind on fire with a love for ideas, beauty, and nature. I knew that generally he was engaged in a private and ceaseless act of mega-scholarship. He was attempting to create a global curriculum, an online or computer-based course of study designed to bring new and important knowledge about the world to culturally isolated and historically ignorant Americans. The earlier thousand-page version of the curriculum, I learned, had metamorphosed into a simplified four-hundred-page version for students at American high schools and alternative schools, and for young people being homeschooled. He was also secretly working on another book that was an outgrowth of that research—a history of the current paradigm shift, an organic shift leading to a global awakening where an enlightened civilization would save mankind and the planet from global catastrophe.

Michael was hoping to introduce his learners to a new way of

1

looking at history that didn't shy away from the fact that much of what is conventionally taught history is skewed by ethnocentric values and a dominant Western mechanistic paradigm in which power and the sometimes violent doings of the wealthy elite determine the most important events, almost unnoticeably creating a distorted conception of where we have been and thus where we are going. He held an exciting vision of a different history—one in which, over generations, ordinary men and women have reached out to find the novel perspective, the transformative idea, the discovery or invention that helps to heal the world.

Michael was in search of an alternative history, if you will. I came to understand that his laudable goal was to save future Americans from an even worse naïveté (you could also call it parochial isolation) than current generations enjoy—a naïveté that allows America to throw its weight around like a cultural bull in an international china shop, often lunging destructively among exquisite traditions and cultures that Americans fail to understand.

But this vast labor of learning contained in the global curriculum and this book you are now reading would be put on hold by an even more urgent problem that came to Michael. An amateur but avid naturalist, Michael had learned about a colony collapse disorder among the bees of Europe and North America (primarily). Bees not only produce honey but also pollinate basic food crops, such as the fruits, nuts, and vegetables that are instrumental to our very survival—and apiculturalists were finding their colonies empty. The busy little workers had seemingly gotten lost en route to their hives.

Michael temporarily put the global curriculum aside and for months began a concentrated effort of reading and research that would prove beyond a doubt that neonicotinoid pesticides (and particularly imidicloriprid)—neurotoxic chemicals widely used in commercial agricultural pesticides as well as lawn and garden products—were responsible for the death of billions of bees, thus endangering the entire food chain. I know of no comparable project in modern science research: Michael perused hundreds of monographs and reports, some in other languages, to culminate in an immensely important book, *A Spring without Bees*.

The title echoed and evoked Rachel Carson's *Silent Spring*, the book that showed how the insidious pesticide DDT endangered multiple levels of our ecological food chain, affecting not only insects but the birds that fed on them and, in turn, endangering predators that fed on the birds. That early book was reviled and scorned by the corporate manufacturers of the insecticide and the agribusiness that believed people could not live *without* DDT, but Rachel Carson had sounded the death knell for the deadly pesticide, and by doing so brought the word and the concept of ecology to modern awareness.

When Michael's own book was done, I read it in a few days. A scholar and writer myself, I simply marveled at the scale of the research and the clarity with which he summarized his findings into that important book. To his family and friends, he seemed obsessed, with little time for sleep, let alone rest. In only five months of this frenzied pace he had completed his work. With all the editing finished and the galleys now in process, Michael was finally able to return to a normal pace. But Michael's wife, Barbara, and daughter, Melissa, then twenty, later told me how worried they had become about him during this time. Even while knowing how important this book was for Michael and would be for the world—with a book tour now looming—Melissa had a nameless fear that she couldn't shake. Barbara had frightening and violent dreams.

Then came a day forever imprinted on their minds. On April 2, 2008, only three days after that last edit, Michael was suddenly very pale and complained of terrible pains in his back—pains that suddenly had started to move into his shoulder and chest. Barbara jumped to a computer and Googled in the symptoms of a heart attack and stroke. Knowing there was no time for an ambulance to arrive, she then somehow got Michael into the car for a seventy-mph race from their home near Woodstock, New York, to Kingston Hospital—about twenty-five miles away.

Arriving at the hospital with only minutes to spare before an expert cardiologist would have been unavailable, the doctor confirmed Barbara's worst suspicions. Because the doctor determined that the problem was beyond that hospital's ability to handle, she quickly summoned a Medivac—a helicopter to take Michael to Albany Hospital, one of the few hospitals in New York state that could handle such a high-level emergency. The doctors there confirmed that Michael was in the midst of an aortic dissection, a huge tear in the major artery that feeds the whole body, but especially the brain. Unless the rupture was operated on immediately, he would die, but the rare and relatively new high-tech operation had its own high risks: a statistical probability of causing a stroke in the brain. Thanks to the accurate diagnosis and highly skillful surgery, Michael survived, now with an artificial, reconstructed aorta.

But the expected postoperative stroke *did* happen, and the subsequent swelling in the brain forced yet another operation in which a fist-sized piece of skull plate had to be lifted from the left side of his head to relieve the swelling in the brain. When all was said and done, from the combined effects of the stroke, the brain swelling, and the operation to relieve it, one of the most promising and innovative intellectuals of our time had lost most of the left hemisphere of his brain, including his speech and language centers and the motor center that controls the right side of the body. As is not uncommon in such cases, his right arm and leg were paralyzed and immobile. Barbara was shown the CAT scan of his left brain while the neurologist explained that he would never talk again. When she asked about speech rehabilitation, his only answer was, "There will be no speech."

It was just a few weeks after this that Barbara received the prepublication copy of his book. She rushed to the hospital with book and Melissa in tow. As she placed it in Michael's hands, he wept in gratitude that he had lived to see it published, and Barbara and Melissa wept with him. Carefully, he turned the pages and looked at the text frowning, his eyes slowly, haltingly following the words from left to right. It was then he made an astonishing discovery: he could still *read*. "It looks like you're reading. . . . You can read!" Barbara said. Michael nodded. "He can read!" Melissa and Barbara exclaimed in unison, hugging each other. Barbara then held his left hand and used her Sensory Trigger Method, special techniques and exercises that trigger speech in the *undamaged* side of the brain. She had developed this method years ago to help her father recover *his* speech after his own debilitating stroke, when more conventional measures had failed.

In the hospital, Barbara said to her husband, "Say the word *read*. You can do it, Michael." After forty-five minutes of working with him through countless attempts, he said his first word, *read*, and the tears of joy flowed. "You can talk too, Michael," she said. "You can talk. You will recover."

After some months of not-so-effective rehabilitation efforts, Michael was transferred to the Northeast Center for Special Care, one of the country's premier facilities for head injury located near Kingston, New York. It is a beautiful facility, housing profoundly brain-injured patients, with lots of patient art on the walls and plants everywhere. As a sign that they were in the right place, Barbara was able to arrange for something the previous facility would not allow: she was able to work with Michael using her Sensory Trigger Method, and this brain-plasticity method, along with neurofeedback, put Michael on the road to recovery.

Four years after his brain injury, he speaks in short concise sentences and phrases and is starting to recover his writing. Barbara told me that Michael not only has aphasia (the loss of speech) and apraxia (saying the wrong things) but also agraphia—the inability to write. At this time, his writing lags behind his speech. Having recovered some speech, he is just now commencing intensive work on his writing. Considering that severe adult apraxia is considered "incurable," he has come a long, long way.

On my first visit to Michael, as I threaded the gauntlet of the wheelchair-bound patients in the Northeast Center's halls, I was reminded, again, of just how serious head injuries could be. Michael had his own room on the second floor, and he lit up with recognition when he first saw me—for the first time since his accident. He was wheelchair-bound for the most part, but he could stand to get into bed or to the bathroom. He was wearing a shiny plum-colored football helmet because the piece of skull that had been removed to repair the damage to his left hemisphere was still cryogenically frozen, waiting for restoration on some auspicious future day. Because the living brain was pulsing gently, right under the scalp, any kind of fall, or even a light bump, could be extremely serious. Colorful posters covered the walls, and there was a fairly serious-looking shelf of books (remember, Michael could still read!) as well as the computer—an anomaly, even for this center. This family's considerable cultural and artisitic accoutrements now had to be crowded into this little one-window room in a health-care facility for the brain-injured.

For a long time after the accident, Michael's only spoken language was one word he used for everything: "Read, read, read!" But I gathered immediately that Michael understood so much more than he could say. Often Barbara would have to guess what he was thinking and inter-pret for him; he would listen critically, a smile would light his face, and he would nod enthusiastically or shake his head no. I had the uncanny sense that, even with half his brain missing, in some way Michael was still whole. I could feel his soul present.

As I sat with my wounded friend, part of my own brain was tick-ing away. Over the years I had seen neurofeedback (particularly the kind I practice professionally called the LENS—see *The Healing Power of Neurofeedback*) restore lost memories for people and bring the past back into a much clearer focus. I determined then to do something that might seem to be somewhat ambitious, although in a way it was not so very different from the soul-loss healings I had studied among shamans, where people were thought to be sick because they had lost their soul or a part of it. I wanted to see if a kind of neuro-shamanism could help my friend recover what he had lost. This was not necessarily intended to find his lost soul—I felt like *that* was there all along—but to bring his soul into much sharper focus.

As soon as Michael was able to make day trips out from the Northeast Center, early in the fall of 2009, we began neurofeedback treatments. It wasn't easy in the beginning because Michael was still bound to his wheelchair most of the time. But Barbara was faithful about getting him to weekly treatments, as well as continuing with her Sensory Trigger exercises during the course of the neurofeedback.

Soon real progress became visible.

After a few visits, Michael insisted he wanted to walk into the center—initially with a little support up the wheelchair ramps. Then the day came when he wanted to climb the *steps* to the center. With me supporting his paralyzed side, and Barbara on the stronger side, he was able to do so. Verbally he was improving, and his favorite expression now was "The answer is . . . questions," while pausing meaningfully and raising the index finger of his left hand to emphasize the word *questions*.

One fall day, as the late sun streamed into my therapy office, Michael looked at me intently and asked me for help to "finish book."

"What book?" I asked, in surprise, thinking he thought there was still more work to do on *A Spring without Bees*. It was then that Barbara told me about the manuscript for *Global Awakening*. As Michael had been doing the painstaking long-term research for the global curriculum, she explained, he had been simultaneously writing a book for "grown-ups," if you will. As she talked, Michael nodded excitedly.

It was almost, but not quite, finished. It was his next major project, and it would contain all of the lore and expertise he had accumulated in science and the philosophy of science, history, biology, and the study of cultures in transformation. It is the book you now hold in your hands.

My wife, Robin, and I wrote a biography of Joseph Campbell entitled *A Fire in the Mind*. In it, Joseph Campbell had admitted that during his twenties he had wanted to write "an introduction to everything." This was perhaps a bit grandiose, shall we say, but this unformed youthful impulse would mature into the scholar's encyclopedic grasp of world mythology as a whole, and the defining of a new field: comparative mythology. What saved Campbell's work from superficiality was the depth of his immersion; when asked by Alan Watts what his yoga was, Campbell replied, "My yoga is underlining sentences in books!" This, and a hunch he came to explain only when he was far along on the path of its discovery: the fact that the hero has "a thousand faces" and myths contain the secret DNA of both psyche and culture—a generative wisdom to be *shared and understood* by a world searching for its own soul.

Through his own pursuit of these goals, Campbell brought soul to the modern world.

Michael's hero quest had been an analogous one: He had been sniffing down the corridors of Western and Eastern civilizations looking for something apparently hidden, not really recognized, a seemingly new, but actually quite old, paradigm that was there all the time—caught in the spaces among the official histories of dynasties, conquests, empires, and their wars. As he tells us in the pages that follow, this elusive something is organic, something that functions like ecology does in the biological world but is really cultural and social in its dynamic. It grows out of our own lives, our own vitality, and our own search for meaning. It is something, he explains, that can move us in our development as a species, far beyond the mechanistic myth that has ruled modern civilization for the last three or four centuries.

Michael, like Campbell, was interested in *everything* and, as mentioned previously, his work is saved from superficiality by the depth and breadth of his research. If you look at Michael's bibliography of sources, you will find there what looks to be a lifetime worth of reading. While I had been teaching psychology classes and was engaged in my clinical practice, Michael had been doing quite a "read" indeed. It would not be overstretching to say that his bibliography rivaled Campbell's in scale. (Realistically, there should be no rivalry at all, because each genius is drawn, uniquely compelled, by that which he or she loves.)

As a writer, I always have a slightly pugnacious response to anyone else's prose. I look critically for clarity of thought that builds through each sentence as it is woven into a paragraph and for skillfully argued points that reach a small or great conclusion. It doesn't hurt if the words are evocative or metaphoric, colorful, strange, and inviting, and if the images summoned echo the words in a familiar yet timeless dance of word and image. Eventually, the whole event should be enjoyable—comfortably instructive yet neither glib nor shallow—and lead to a point where there is a deeper resonance in the reader: "Yes, I've glimpsed that or come close to that in my own thinking!"

There is a moment in the film *Amadeus* when the accomplished and successful Italian court composer Salieri looks at a musical score written by Mozart. The film, brilliantly, shows Salieri's nuanced expression as he realizes what he is seeing. Talent contemplates genius!

What you will find in Michael's prose, as it unfolds in clear paragraph after paragraph, is the secret logos that was driving his quest. Now I knew why he said read, read, read!

Some of the places I found Michael going in this book were familiar to me and others totally unfamiliar. I was learning to trust Michael's instincts, his knowing where to go, and how faithfully he recounted what he had discovered. I followed him through the dance of the paradigms in Western civilization, the mechanistic versus the organic, how the latter kept trying to surface under the hobnailed boots of the former—with its empire-building, fascism, and racial and gender dominance. There was no doubt where Michael's sympathies lay—always with the inclusive, the sustainable, the economically and environmentally friendly energies.

In expressing these sympathies, he doesn't hold back from detailed examinations of the robber barons and the hegemony of corporations in all their brutality. For this, I marveled at the evenness of his treatment. He tells it like it is; he never seems to interrupt the informative flow with polemic or ranting.

I told Barbara and Michael that I could not countenance a reality in which such a book would fail to be published or fail to be made widely available. I knew it could change people, for I felt myself changing, as well as learning, from almost every chapter I read. And I suspected that if, in its inquisitive, intelligent way, the book transformed people, ultimately it could help pave the way for a genuine global awakening, one just like its title promised.

This book, as I have come to know it, is itself made of transformational stuff, much like Campbell's own *Myths to Live By*. It is not merely intellectual rhetoric, destined for an elite community, nor mindlessly affirmative mysticism for those preoccupied by the rapture or the culmination of a Mayan prophecy. Michael is much closer to home, giving a factual, reasoned account of our recent history and the stumbling, fitful evolution of a species whose heart is mostly good but whose imagination and capacity for change has not yet even begun to be understood. In short, it leaves room for evolution.

Now Michael can speak in short—often profound—sentences. "The answer is," as Michael kept telling me at our weekly meetings, "questions"—and yet more questions. Who are we? Where are we going, and what can we do now? These are the *great* questions that Michael takes on in this book. Skillfully using hints embedded in the past, he guides us toward the future, never doubting something we all yearn for at the end of our quest: the possible human.

STEPHEN LARSEN, Ph.D., LMHC, is professor emeritus of psychology at SUNY Ulster, board-certified in EEG biofeedback, and the author of several books, including *The Neurofeedback Solution, The Healing Power of Neurofeedback, The Fundamentalist Mind,* and, together with his wife, Robin, *A Fire in the Mind: The Life of Joseph Campbell.* He is the founder and director of Stone Mountain Center, offering biofeedback, neurofeedback, and psychotherapy treatments. He lives in New Paltz, New York.

...

EDITOR'S NOTE: At the time of this writing, Michael Schacker is able to talk in words, phrases, and short clear sentences, and he is now working on recovering his writing as well as his speech. He will continue his recovery with the help of his loving family, friends, and community.

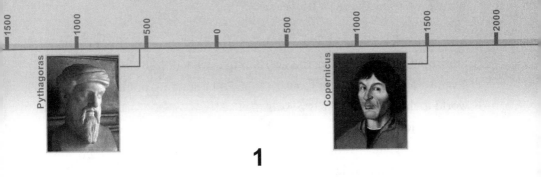

Pythagoras

Copernicus

1

THE COMING CHANGE IN WORLDVIEWS

You can't solve a problem on the same level that it was created. You have to rise above it to the next.

ALBERT EINSTEIN

TRANSCENDING THE MECHANISTIC DILEMMA

Without change, there can be no hope. Our society is headed for a multitude of problems, confronted by the limits of both technology and paradigms. From global warming to toxic pollution, economic inequity, and the spread of nuclear weapons, the world is careening toward a future that simply will not work. Add all these crises together with the end of cheap oil and gas, and the ability of government budgets to handle it all at once becomes impossible. Our direction is clearly unsustainable.

Yet we can shift our current course by charting a new map, one that leads to a workable future, a sustainable future. If we want a better place for our children and grandchildren, many great changes lie ahead of us. Change is not easy. Change is hard. Deep change would involve a lot of work over many years and decades. Yet it is something that we must now do. Just making the decision that we will fight for change can give us hope. As Norman Cousins said, "The starting point for a better world is the belief that it is possible."

Our journey to a better world will mean learning anew the very essence of science. It will mean taking a new look at technology and what its true potential might be. It will require transforming, in no small way, all the institutions of society, including politics. In short, it will mean changing the general way we think. *Global Awakening* explains how we can get from here to there. A new understanding of past, present, and future will show how that sustainable future is now possible.

To arrive where we want to be, we're going to need all the transformation we can muster and not just with technology, such as the new breakthrough wind turbines or the new solar electric generators. Technology is not enough in this case. As Einstein said, you can't solve a problem with the same thinking that created it. You need to have new thinking—and we need some new thinking fast.

Our society, for example, is based on an old scientific worldview that cannot see beyond fossil fuels and nuclear power. Unlike renewable solar and wind, both of these nonrenewable technologies have the potential to devastate the biosphere to the point of no return. Use of fossil fuel for energy and industry, in particular, has created not only global warming but also global dimming, in which 10 percent to 37 percent less sunlight is reaching Earth, reducing crop yields.[1] The proliferation of jet con trails and man-made clouds are contributing to this.

Top-level scientists say that 350 ppm is the safe *upper* limit for carbon dioxide levels in the atmosphere and a UN Environmental Program (UNEP) study from 1999 predicts that when the atmospheric carbon dioxide reaches 450 parts per million (ppm), the global economy will be overwhelmed by climate change, rising sea levels, and superstorms, and an irreversible downward spiral will be unavoidable.

We are only decades away from that level now, at 379 ppm in 2005 and rising rapidly at 3 ppm every single year. In 2004, the Mauna Loa

Observatory, Hawaii, found the CO_2 level in the atmosphere to be a record 379 ppm, compared to 376 ppm a year earlier and 373 ppm in 2002. (This trend apparently slowed for the first time in 2011; data from the National Oceanic and Atmospheric Association [NOAA] at the Mauna Loa Observatory indicated a level of 388.92 ppm, an increase of only 1.77 ppm over 2010.)

However, measurements taken in 2012 indicate that the ppm levels around the world are currently at 395 ppm, and at 400 ppm at the poles.[2]

If China and India fully industrialize on fossil fuel, we would need another two or three Earths to handle all the carbon dioxide. Although corporate media and the so-called global warming deniers try to ignore these facts, 97 of 100 climate scientists think humans cause climate change.[3] Misinformation and out-and-out propaganda paid for by big oil have been spread throughout the media including the Internet, saying the studies the scientists based their findings on were "flawed."

Though scientists are still fine-tuning the testing methods and determining what exactly the data figures will mean in terms of specific long-term effects, this does not invalidate the findings overall. The worldwide scientific community has reached a consensus and is more than convinced we must move quickly to implement change. We, in the United States, should not go to the mainstream media—so prone to corruption—for scientific information but rather rely on the information in the actual scientific reports.

DEAD END ENERGY AND CHEMICAL FARMING

Nuclear power, meanwhile, is only affordable until there is a major meltdown—or damage due to flooding, earthquake, or tsunami, like the disaster that occurred at the Fukushima Daiichi Nuclear Power Plant, Japan, in March 2011.

There are other dangers. The danger of terrorist attack at nuclear plants has been made clear. The 9/11 Commission revealed that

Mohammed Atta himself flew above the nuclear power plants in test flights for the 9/11 attack and contemplated airliner hits on them. Later, he flew his hijacked United Airlines plane right over Indian Point—but kept going to the Twin Towers. Despite the obvious danger, Entergy, the corporation that owns the Indian Point reactors, refuses to even store its spent fuel rods in dry casks, which are much harder to set afire. Even though a fuel rod fire twenty-two miles north of New York City would be the biggest disaster in recorded history, the Nuclear Regulatory Commission and the nuclear power industry fight the hard science of the environmental groups with whitewashed studies and tons of PR.

The environmental contamination from uranium mining and the crisis of growing piles of radioactive waste rule out nuclear for the future. Few people realize that uranium processing burns coal, which releases potent greenhouse gases like chlorofluorocarbons (CFC) into the atmosphere. So each nuclear plant—when construction, mining, ore transportation, ore processing, and endless storage needs are added together—equals one-third the greenhouse gas emissions of an oil plant. To make a dent in global warming, some energy scenarios call for tens of thousands of nukes. Not mentioned is that the world would then run out of high-grade uranium ore after just several years. Since low-grade ore requires far more mining and processing, a nuclear plant running from low-grade ore would emit just as potent a greenhouse-gas mix as an oil plant.

In an effort to offer a "green" energy proposal, former president George W. Bush proposed the production of hydrogen. Hydrogen gas is a potentially green source of energy, created with an electrolyzer, which splits water into hydrogen and oxygen. It's odorless and tasteless and produces no harmful emissions when burned. If the electrolyzer is powered by solar or wind energy, the gas is truly green. But the black hydrogen proposed by Bush would be developed from fossil fuel, which would still pollute in the drilling process.

The actual Bush/Cheney plan was, however, for "pebble" nuclear reactors, without containment buildings, to make hydrogen that would then be injected into coal shale to make—you guessed it—gasoline. It's hard to give up your paradigm. These pebble nukes, combined with new oil refineries, would be constructed at closed military bases to avoid the otherwise impossible permitting process. These are the actual plans put forth by the Bush administration, starting with six pebble nukes at six sites. These plants have never been built because the designs are still being developed. Although touted as inherently safe, the only large-scale pebble power plant in Hamm-Uentrop, Germany, ended up contaminating a two-kilometer area when a radioactive uranium-graphite pebble got stuck in the feeder tube and started a graphite fire in 1986. The German government permanently closed the plant, deeming it unsafe.[4] China has the only operating prototype.

A final worry is that, in the case of Iran and North Korea, certain types of reactors can be used to produce bomb-grade explosives, making the spread of nuclear weapons much easier. So nuclear power can never be an answer to global warming. Both fossil fuel and nuclear energy are economic and technological deadends, supported only by subsidies and an unholy alliance between governments and corporations.

At the end of the day, nuclear power is simply too dangerous, particularly when you take into account the fact that there has never been a safe way to store nuclear waste, which remains deadly for thousands of years. It is imperative that we move to green hydrogen rather than black hydrogen or nuked hydrogen—if we move to hydrogen at all.

Fracking is another bad idea. Fracking, or hydrofracking, is a process of drilling deep wells into rock formations. Highly pressurized water and chemicals, some of them carcinogenic and neurotoxic, are then injected into the man-made cracks to release petroleum, natural gas, or coal seam gas. Many spills and accidents have happened in the United States, and hydrofracking has been suspended and banned in many other countries. Fracking can pollute miles of groundwater and also releases poisonous air pollution. In addition, methane gas, one of the most potent greenhouse gases, can be released from the fissures. The EPA warned of the dangers of groundwater contamination and air pollution, but the study was not thorough and did not take into account all the environmental impacts of the process, and so Congress made the process exempt from the Safe Drinking Water Act in 2004.

Pennsylvania has had the most fracking disasters—the Bradford County blow up and spill of 2010 being one of the worst. The disaster caused tens of thousands of gallons of fracking fluids to leak into the Susquehanna River, which flows into the Chesapeake Bay. In a similar leak elsewhere, eight thousand gallons seeped into a creek near Dimock, Pennsylvania. Wells are poisoned forever, fish and other wildlife are killed, and even forests can die: the damage from fracking spills is irreversible. According to the environmental group, Earth Justice:

> Gas drilling in Pennsylvania has skyrocketed in recent years. And along with the gas rush have come disastrous industrial accidents and poisoned drinking water—earning the state a reputation in the region for gas development gone wrong. Communities fighting to keep their water and air clean and state forests intact have their work cut out for them: Governor Tom Corbett accepted nearly one million dollars from oil and gas companies during his political campaigns and since assuming office in 2011, promptly began repaying his benefactors by cutting down on environmental enforcement and oversight of gas drilling activities.[5]

Oil spills are another serious problem. The Deepwater Horizon Oil Spill in the Gulf of Mexico in 2010 was an enormous environmental disaster. Efforts to stop it were too little and too late, and its cleanup along the coast was nearly impossible and extremely costly. The spill created an eighty-square mile "kill zone" in the ocean surrounding the blown well and there was no way to prevent the ruin of thousands of square miles along the coastline. The damage to marine life and wildlife, as well as fishing and tourism, was immeasurable. Much like the Fukushima nuclear disaster, such an accident was simply "unthought of," and so plans and methods to deal adequately with it were never developed by the mechanistic oil industry.

The proposed Keystone XL pipeline is another disaster in the making. It would transport synthetic crude oil from Canada's tar sands through the heartland of America to the Gulf of Mexico. Alberta has already been privy to a number of oil spills and to pollution from this pipeline, built to transport the dirtiest form of oil there is. The pipelines are prone to leaking and vulnerable to human error, as well as earthquakes and other natural disasters.

The pipeline would endanger the Ogallala aquifer, the largest aquifer in the world. This water feeds the food crops of and supplies water to seven states, from South Dakota to Texas and from Colorado to eastern Nebraska. Contamination of the aquifer would poison a huge region of the United States. The proposed route would cross a wide area as it passes through Nebraska. The tar sands crude oil sludge requires an enormous amount of processing to purify the dirty oil. Most of the oil would be shipped elsewhere in the world.

No one wants the risky pipeline in their state, as the risks are now better understood by the public. The most important reason to reject the plan comes from NASA's Jim Hansen, who said that the approval of Keystone XL would lead to more consumption of tar sands oil, which would mean "it is essentially game over" for the climate. Burning that enormous amount of oil would put so much carbon dioxide into the atmosphere, we would soon reach the point of no return with global warming. After a massive demonstration at the White House and a plea from Nebraska's governor to reject the plan, Obama sent it back to the State Department for further review. At the time this book went to press, the southern portion of the Keystone XL pipeline, from Cushing, Oklahoma, to Nederland, Texas, is being constructed, while the northern portion is still under review, in part because Nebraska is still trying to determine an alternate route around the environmentally sensitive Sand Hill region of the state.

Few jobs would be created, compared to the number of jobs investing in alternative energy sources would generate, which makes it less desirable even with the best-case scenario. If the same amount of money were spent on shifting to an alternative energy infrastructure, it would make the United States the leader in renewable energy and the most energy self-sufficient country in the world.

Another technological dead end is chemical farming, the use of oil-based fertilizers, pesticides, and herbicides in agriculture through-out the world. In the United States, the run off of agricultural and industrial pollutants has created a five-thousand-square-mile dead zone in the Gulf of Mexico, an area devoid of oxygen and fish.[6] This is only one result of the widespread use of chemicals, but it illustrates well the dire nature of the situation. There are many other examples, from Love Canal to Bhopal, of the horrors of toxic pollution and chemical accidents.

CRONY CAPITALISM

Corporations and elites have bought most of the governments of the world, a form of rule known as corporatism, which is crony capital-ism carried out to its ultimate end. This is not to say that capitalism is bad, but that crony capitalism, an unregulated capitalism, is danger-ous to both democracy and the economy. In 2005, it was revealed that corporate and government so-called news releases are mixed into U.S. media reports without letting the public know the source. In an article in the *New York Times,* David Barstow and Robin Stein reported the following:

> Under the Bush administration, the federal government has aggres-sively used a well-established tool of public relations: the prepack-aged, ready-to-serve news report that major corporations have long distributed to TV stations to pitch everything from headache rem-edies to auto insurance. In all, at least 20 federal agencies, including the Defense Department and the Census Bureau, have made and distributed hundreds of television news segments in the past four years, records and interviews show. Many were subsequently broad-cast on local stations across the country without any acknowledge-ment of the government's role in their production.[7]

In 2005, the Government Accountability Office (GAO) ruled the effort illegal, yet the practice persists today in the Obama administration because either no effort has been made to stop it or attempts made to stop it have been obstructed.

This covert propaganda—and the way conservative governments have recently let multinationals write their own legislation—means that the democracy we used to enjoy has been compromised or at least is in jeopardy. At the same time, public relations and the use of phony "independent" institutions have become such an art form, large corporations and the government can outright lie and get away with it nearly all of the time. From misinformation about the safety of prescription drugs to the security at nuclear power plants, PR is used to cover up the bald truth—with the help of the compliant corporate media. Meanwhile, corporate executives become government regulators through a revolving-door employment policy and are supposed to enforce the law against their former companies. This led to the previous FDA failures on Vioxx, Bextra, and Celebrex.

Today, it is being discovered that more and more pharmaceutical drugs have, over the years, been released on the market before they were fully tested or despite cautionary scientific studies on side effects and other dangers. The fox is clearly in charge of the henhouse, and yet people still wonder why those chickens keep disappearing.

The merger of government and big business perpetuates the whole system, even to the point where the Bush administration refused to force the chemical industry to improve security after 9/11. This can also be seen in the threat of genetic engineering (GE), especially of food crops. Beyond the myriad possibilities of environmental damage from wind drift and the elimination of beneficial species, the main goal of genetically modified organisms (GMOs) is to allow corporations to patent basic food crops and *own* them, requiring payment *even if you grow your seed*. The 2005 field experiments testing GMOs in England were a miserable failure. This British study, the world's biggest to date of GMO crops, found that fields planted with biotech sugar beets and rapeseed had fewer butterflies, birds, and bees.[8] Here again, massive PR was deployed to expand GMO use, despite all the problems and failed tests. Like the Mary Shelley novel *Frankenstein,* a monster that science should never have created is now out of control and wreaking havoc—so much so that the environmental groups call them Frankenfoods.

Add to the above crises the spread of nuclear weapons to countries like Pakistan, India, Israel, North Korea, and now possibly Iran. At the same time, the breakup of the Soviet Union made little-known Kazakhstan the fourth largest nuclear power in the world; Kazakhstan has since reduced its arsenal, but the country remains the world's leading producer of uranium to the world.[9] Meanwhile, thousands of weapons and other nuclear material are improperly guarded throughout Russia and Central Asian dictatorships. This is a disaster waiting to happen—and yet the governments of the world do almost nothing to end proliferation, even in the case of Dr. A. Q. Khan, the top Pakistani bomb maker. Although Khan peddled illegal knowledge and machinery to Iran and other countries, he was merely placed under house arrest. We are looking at the total breakdown of the nonproliferation system.

All of these things are connected. The crisis stems from deep within the old worldview, which is based on mechanistic science and technology, seeing reality as the clockwork universe. *We think by analogy,* through models of reality. The model of the machine has been the underlying foundation of most science for the last three hundred years or so—ever since Newton compared the regular movements of the planets to the gears of a clock. Yet machines are not alive, they do not have feelings or reproduce or have their own thoughts. So the machine model is limited to inanimate things. It tells us little about the true nature of life and the environment, or even social justice or what being human means.

The multiple crises of today have developed because this limited science and mind-set has wreaked havoc on just those things: the environment, the quality of life, society, public health, and the state of individuals. We are approaching a true crisis because the old mechanistic science—bought in whole by multinational corporations and now allied with nationalistic governments and the military-industrial complex—has no solutions to the problems it has caused. Nor does it possess any realistic changes for the future. The corporations now in charge of the

Fig. 1.1. Isaac Newton published the Principia Mathematica, *the foundation of classical mechanics, 144 years after Copernicus.*

world simply want to continue their profit points, even though they may attempt to clean up their image now and then with well-publicized "green" or "sustainability" projects. The modern dilemma should therefore more properly be called the mechanistic dilemma, as its root cause is the machine-model scientific paradigm supporting it all.

The circular thinking behind the mechanistic dilemma was described in 1979 in William Barrett's *Illusion of Technique.* Barrett showed how the mechanistic model creates an illusion in the mind of the technician. The smooth operation of the machine becomes everything—even people are mere parts in the machine, mere objects. There is no meaning to be gained from a machine, so the technician sees no purpose to life. Just as the old medieval religious priesthood was replaced by the mechanistic scientific priesthood, the illusion of technique replaced the meaning and purpose faith used to offer.

Our whole mechanistic society now reflects this meaningless and purposeless worldview. And if concern is raised about the future this is all leading to, the answer from the priesthood is "Don't worry— technology will solve all our problems." The *illusion of technique* helps us understand this fatal flaw of mechanistic dogma and how it fails to confront reality. In short, the allure of the machine outweighs the mounds of scientific data showing the fragile interconnections of Earth and its biosphere. Social, environmental, and health concerns are swept under the rug and ignored. The mechanistic paradigm is thus dysfunctional at its core—and so we find ourselves in the mechanistic dilemma.

Extricating our planet from this mess will require more than mere politics and more-than-ordinary thinking and action. For we need to *transcend* the mechanistic dilemma; we need to go beyond the entire mechanistic worldview. We can't just tinker and fix the old world; there must be deep change. We must replace the broken paradigm with a new one. To do that, we must first know how worldviews have changed in the past. Major paradigm shifts are far more than politics alone; they involve philosophy, science, culture—and action. In order to bring about change, we must first learn how change—deep change—occurred in the past. Then we will know how to take action in our own time.

PARADIGM FLIP

Can the current power structure of bought politicians, corrupted science, and suicidal fossil-fuel technology ever be replaced with a sustainable worldview and visionary leadership? If history is any example, the answer is yes. In fact, this book will show that a dramatic shift in worldviews and a social revolution in politics during the twenty-first century is the *most likely future scenario.* Our global society appears to be following a distinct pattern that occurs very rarely in history, one that has led in the past to total reinventions of the world within very short periods of time. *In short, we are in the throes of a classic paradigm shift and are fast approaching the tipping point of the whole process.*

In the 1740s and 1750s, a powerful medieval elite had similar total control over Western society—despite the fact that the medieval worldview had long outlived its usefulness. Tyrannical kings, powerless parliaments, feudal economics, theocratic rule, scientific theories from the Middle Ages—all were failed and obsolete, yet the antique power structure of kings, the aristocracy, and the church still stood. By the 1750s, however, the contradictions and failures built up to a tipping point, and at the same time, inexpensive book printing became possible. A small group of Enlightenment scientists and thinkers such as Voltaire and

Diderot wrote book after book, exposing the medieval world from one corrupt end to the other. A new activist generation with last names like Washington, Jefferson, and Madison grew up on these books, seeing the world through an entirely different scientific and political worldview. Between the years 1755 and 1760, the Enlightenment reached the tipping point and the paradigms flipped, a rapid period of change. Though the flip lasted only a few short years, a large part of the population of the West nonetheless converted to the new worldview of modern science and democracy.

Within a few decades of this momentous paradigm flip, there came not only the American and French revolutions but also the Industrial Revolution. As the modern worldview replaced the medieval paradigm and power structure, society was transformed on nearly every level. Before the paradigm flip, in the early 1750s, such tremendous change in so short a time would have been a laughing matter; it would have looked impossible. The church and the kings had ruled with an iron grip for over a millennia and a half and could ban books or close newspapers with a wave of the hand. How could they ever be rendered powerless? It was merely the Enlightenment philosophers and a young new generation against all the powers that be.

But the new paradigm won. Although it was a long and bitter struggle, the scientific and cultural shift of the Renaissance and the Copernican Revolution finally culminated in the Enlightenment and the overthrow of the old order. In the end, the old worldview simply did not have the arguments on its side. Reality was against the old view and science and democracy took center stage. The Enlightenment philosophers had outlined a future that worked far better than the failed society of the 1750s, and their common agenda for change became a blueprint for action to the young radical democrats.

HOW PARADIGMS SHIFT

The man who made the study of paradigm shifts popular was Thomas Kuhn, who coined the term in his 1962 book *The Structure of Scientific*

Revolutions, defining the paradigm as the framework supporting a whole scientific worldview. The paradigm is a related constellation of ideas, spanning many different disciplines by depending on the same underlying model of the universe. A paradigm shift erupts when the old model becomes obsolete and a more accurate theory of the universe emerges to replace it.[10]

Kuhn revealed the true nature of these epoch-making periods. We have been taught that science is rational and objective, based on observation and data. Kuhn, however, said nothing could be further from the truth. He found that thinking alternated back and forth between times of stable "normal" science and turbulent scientific revolutions caused by "extraordinary" science. The history of science, Kuhn discovered, consisted of a "series of peaceful interludes punctuated by intellectually violent revolutions," in which "one conceptual world view was replaced by another."[11]

Yet rather than being an objective debate about data, these revolutions were actually *ideological struggles* between the old and the new thinking. The struggle was thus not about science and data; it was all about maintaining an ideology and maintaining power. Adherents of the old worldview simply had too much invested in old concepts to accept the truth of the new—even when they saw it with their own eyes. It's basically a war of worldviews.

Kuhn's prime example of a paradigm shift being an ideological battle was the Copernican Revolution of the sixteenth and seventeenth centuries. He told of a seventeenth-century theologian who would not even look through Galileo's telescope. Galileo wanted to show the church scholar that the moons of Jupiter circled the larger planet, just as the small Earth must circle the larger sun. The medieval theologian knew it proved certain lines of the scriptures as well as his revered geocentric thinking incorrect, and so he ridiculed it as an optical trick. As Kuhn points out, this is not science—it's ideology. Despite this resistance by the old worldview, the shift instigates the paradigm flip and the new thinking soon wins the day.

History certainly tells us of other periods when whole societies *did* change relatively quickly. Clearly, one such time occurred when Greek society—within a couple of centuries—went from believing in mythological gods and goddesses to orating about mathematics, philosophy, and the laws of nature. The conception point of that great shift was the learning of astronomy, mathematics, and other sciences from the Egyptians and other Mideast civilizations. In 585 BCE, Thales correctly predicted a solar eclipse, officially beginning the Greek turn to science. Pythagoras, rather like Newton over two millennia later, used mathematics to create a whole new physics when he determined the harmonic ratio in music by measuring the length of strings on a lyre.

"All is number," declared Pythagoras around 525 BCE, and from that point onward, the Greeks measured all of time and space using mathematics, much as the West did over two thousand years later. The golden age of Greece, the fifth century BCE, was the result of the paradigm shift in the sixth century BCE and corresponds to the Enlightenment in many ways. The supernatural power of Zeus was replaced by the science of the laws of nature, as philosophy offered a completely different and far more sophisticated picture of reality—very similar to what happened later in the sixteenth to eighteenth centuries. Herodotus the historian gave Greeks an awareness of other cultures and a "new story," in much the same way as Voltaire did for the Enlightenment. Full democracy came to Athens in the golden age and to France and America in the Enlightenment, to the detriment of monarchs and aristocrats.

We now view both the Greek and Copernican periods as major paradigm shifts, when the underlying model guiding thought was replaced by a entirely new analogy. This change in analogies is most important, for we think only by analogy, by comparing one thing to another, by realizing *this is like that.* The underlying model is the basis for the theorizing of the entire paradigm; it is the first assumed step in any thinking. Change that foundation, and the props are knocked out from beneath society as a whole.

Furious that their model was being challenged, the scholars of the old worldview resisted the new thinking, much as an ardent member of one religion resists conversion to another. In short, the old scientists almost always take their paradigm with them to the grave. A shift's struggle phase thus ends only when the old thinkers die, said Kuhn, for only then can the new science become established. So-called normal science begins again, a centuries-long period of steady research and data collection under the new model. It's a natural pattern of development, alternating between stable and revolutionary periods. Here is a synopsis of Kuhn's main stages of scientific revolution, which, in the case of the Copernican Revolution, took over 150 years to unfold.

- Emergence of anomaly contradicts the old worldview. Nature violates the expectations of normal science and answers must be found outside the paradigm.
- The emergence of a new analogy or paradigm and system of thought. Revolutionary period upsets stability of normal science period.
- Crisis and reconsideration of old paradigm by new thinker to explain anomalies.
- Bitter struggle ensues: paradigm resistance from old scientists; paradigm wars are fought by the new worldview with facts and by the old worldview with ideology and bluster.
- After the old scientists die and the new paradigm wins the struggle, a new normal science period begins, with the new underlying analogy/model, a new scientific method, and a changed set of rules.

In the stages summarized above, it should be noted that a single person can begin a whole paradigm shift. It can start with a thought in the back of someone's mind, a question that cannot be answered using the old worldview. Thomas Kuhn wrote that the "failure of existing rules is the prelude to a search for new ones." When a large enough contradiction cannot be answered without going beyond the usual borders of the old paradigm, then the stage is set for change. The answer to this great anomaly is the conception point of the paradigm shift, the beginning of the unraveling of the old thinking.

Copernicus had wondered why summer was hot and winter was cold, something geocentric astronomy could not explain. The heliocentric theory, however, made sense of it all. In the Copernican Revolution, this one question in the mind of a Polish astronomer thus set in motion the complete reinvention of the world—eventually ending medieval thinking and replacing it with modern, scientific democracy.

The new paradigm had a tremendous power, for it replaced the old *by transcending it*—by making it *irrelevant* and *obsolete*. In Arnold Toynbee's well-known theory of history (*A Study of History*), the "dominant Establishment" is from time to time overthrown by a small "creative minority." The dominant establishment of the eighteenth century—the medieval power structure—in the end had no chance against the new science and the new values of the creative minority, the radical democrats and scientists. A huge change in science thus can transform the whole root value system of society. Table 1 shows the vast differences in beliefs between the medieval and modern worldviews, making clear why each side fought the other so hard. The new values quite simply ripped the power and prestige away from the old structure of theocracy, aristocracy, and monarchy; yet the dominant establishment was not about to relinquish its rule over society without a fight.

This was a huge change in values, politics, and religion, a redefinition of nearly the entire society. As the chart shows, the difference was enormous: we can call it a values gap. This values gap lay beneath the social and democratic transformations of the Enlightenment—a chasm between the two systems that simply could not be allowed to stand. Despite all its power, the end for the medieval dominant establishment was certain—if not in the 1770s and 1780s then a decade or two later.

In the eighteenth century, when the new scientific model entered the mainstream of the social and political realms, it caused cultural, religious, and political battles, creating great turmoil and then a series of rapid, all-encompassing social revolutions. Although it is accepted that the Copernican Revolution led to the Enlightenment, that was never really discussed by Kuhn in his paradigm shift theory. Yet there were distinct stages of development after the scientific shift—and that is what we need to understand for our own time. What happened after the five stages of the Copernican Revolution?

TABLE 1.1. COMPARISON OF MEDIEVAL AND MODERN WORLDVIEWS

MEDIEVAL WORLDVIEW	MODERN OR MECHANISTIC WORLDVIEW
• God is responsible for all events on Earth.	• God or nature merely sets universe in motion, natural law determines the rest; clockwork universe of Newton.
• God's creation only six thousand years old.	• Universe very old, Earth millions to billion years old, formed by natural forces.
• Two sets of laws: one for Earth, one for heaven.	• One set of natural laws governs Earth and the universe.
• Geocentric universe: Earth does not move.	• Heliocentric solar system: Earth orbits the sun.
• King and nobility have divine right to rule.	• The right to govern derives from the people; kings are tyrants.
• Medieval laws and value system designed to protect the lands and power of kings, the aristocracy, and the church.	• Laws and values designed to provide liberty and equality to all men, to protect the pursuit of happiness, and to derive power from the people in a democracy.

The historian Peter Gay put forth the idea that three generations comprised the Enlightenment, but upon further study, the division of the eighteenth century into four stages makes more sense.[12] Going beyond Kuhn's five stages of scientific revolution, the four stages of social transformation shown in table 1.2 best explain how deep change was achieved so quickly.

TABLE 1.2. THE FOUR STAGES OF SOCIAL TRANSFORMATION: THE ENLIGHTENMENT PATTERN

1. **1700–1725 Early enlightenment:** Shocks old worldview with radical ideas and waves of secular change. New science and new philosophical concepts enter public consciousness, but no one realizes society has entered an age of enlightenment. New sexual libertines glorified.

2. **1726–1740 Conservative backlash:** Sets back new paradigm with powerful fundamentalist and political counterreaction lasting decades. First fundamentalist revival gatherings (also called the First Great Awakening).

3. **1741–1760 Intensive phase:** Polarized culture war between the worldviews. New science and inexpensive books thoroughly expose old medieval theories and history as false. Bestsellers and Voltaire popularize new paradigm, while Voltaire's new history radicalizes population and young activist generation, ending in a rapid five-year paradigm flip. Public then recognizes that society has been in an Age of Enlightenment for the last sixty years.

4. **1761–1800 Transformational phase:** Multiple revolutions (American, French, and Industrial) create science-based democracies and institutions to replace old medieval monarchies and theocracies.

Between 1700 and 1800, these four stages of social transformation swept the societies of the West in an all-encompassing Age of Enlightenment. The pattern began with an early enlightenment stage (1700–1725), in which shocking new ideas and sexual mores are introduced yet deep change does not take place. Opposing all the secular change and progress of the early enlightenment, a conservative backlash followed (1726–1740). This fundamentalist counterreaction has also been called the First Great Awakening. There then followed a severe polarization of society, the intensive phase (1741–1760): a bitter struggle between the medieval power structure and the scientific/democratic counterculture of the mechanists and *philosophes,* between the old and the new.

Intensive phase events magnify the difference between the worldviews, the tipping point being reached when a new view of history debunks old beliefs and provides a solid foundation for thinking about a different future. Between 1755 and 1760, as mentioned earlier, the Western world, especially France, England, and America, experienced the rapid change of the paradigm flip. By 1760, the resulting radical- ization of approximately one-quarter of the population led directly to the revolutionary atmosphere of the transformational phase—and the American and French revolutions.[13]

THE CRITICAL CATALYST: A NEW STORY

The crucial factor, the catalyst that finally upends the old worldview and jump-starts the paradigm flip, appears to have been the writings of the author Voltaire. Although known primarily as satirist and essayist, Voltaire was also a historian and scholar. His books on science and the past popularized not only the Copernican Revolution and the signifi- cance of Newton's achievements, but his true history of the French roy- alty also gave a historical center to the new worldview; it gave the new paradigm a story. Every worldview is based on its story of *who we are, how we got here,* and *where we are going*—it must answer those three essential questions.

Voltaire explained to his time a new story, how the king, rather than a divine being, is just a man driven by the lust for power—thus we should have a democracy in which the people have the power. Voltaire declared that laws of nature caused events here on Earth, not God. Science should thus guide us rather than religion. He made the concept of progress a common one, focusing attention on science and the future in a whole new way and giving the multiple revolutions to come a pow- erful image of what they could be.

It was this new story of Voltaire, this reanswering of the three essen- tial questions, that allowed the new Copernican paradigm to finally tran- scend the cultish control the old medieval powers had on the population. Once it was realized that the divine rights of the king and nobility were made up to enrich the few, control evaporated very quickly. By the 1760s, the rulers Louis XV of France and George III of England were suddenly faced with a large part of the population despising them and wanting deep change. Their personal mistakes concerning foreign policy and ruinous tax structures ended up discrediting the whole medieval worldview.

Old paradigms disguise themselves with partial, glorified histories and lots of public relations. As we saw in the eighteenth century, one's view of history guides opinions about everything from politics to religion, so *a history based on a new paradigm changes opinions like nothing else*. In fact, with no new story, there's no paradigm flip and no new future. The retelling of the story is the catalyst to the whole process of the new transcending the old. Once the old is transcended, the new worldview grabs power and reinvents the world.

Although oppression by the monarchy and dogma from the theocracy had always succeeded before, after the paradigm flip, the knowledge of the new history fired up a true level of outrage, generating the courage in an enlightened generation to stand up to the powers that be. The conservative kings did not accept this act of defiance very well, and, with the support of their privileged aristocratic supporters, continued to abuse their subjects. Their ongoing mistakes proved to be most helpful to the young radical democrats. Most European theocratic monarchies of the medieval worldview were replaced by scientific democracies within a few decades of the paradigm flip in the earth-shaking transformational phase (1761–1800). The seemingly invincible power structure was overthrown: 1, 2 3, 4.

FINDING A FUTURE THAT WORKS

Are we in a similar type of enlightenment period today? Although much has changed since the 1700s, the last 250 years seem to be repeating the whole process of scientific and cultural revolution, followed by the stages of social transformation. Certainly, the worlds of modern science, technology, and politics have in their own turn become obsolete and are no longer useful. They are all dead ends. All we have to do is look at the global warming caused by burning a billion barrels of oil every eleven days, the dead zones in the ocean caused by pollution, the increasing wars and violence incited by religious fundamentalism, the accelerating proliferation of nuclear weapons, the radioactive leaks and disasters, the inability to deal with the AIDS pandemic, the pharmaceuticals scandals, and the general corruption of politics, economics, science, medicine, and technology by crony capitalism.

As Einstein explained in the opening quote, we need a new type of thinking. Merely electing a new politician and carrying on with the old science and politics are not nearly enough in this case. Politicians alone cannot make the shift happen. It is unrealistic to think that the election of a progressive president of the United States, though it is a crucial step, will induce the vast changes that are needed. It will take a much larger movement of people thinking in a new way. A new paradigm would change the general way we think, and in so doing, it would change the world. According to the enlightenment pattern, an accepted new paradigm would have clear advantages over the old worldview. In the Enlightenment of the eighteenth century, the truth of the new science trumped the phony universe of the old worldview by exposing it, by making an end run around it—rather than endlessly trying to confront it politically. Such profound truth was *transcendent*—and the old power structure was eventually almost helpless before the changes that would undo it.

Since we cannot return to the medieval worldview (although many may try), is there an alternative science, an alternative politics and economics through which a group of scientists and writers could change the world—as happened in the Enlightenment? Like them, could we find a future that works and create a twenty-first-century transformation—equal in scope to the Democratic and Industrial revolutions?

The first step in our learning process is to deconstruct the world we have been taught as being real by our schools, our parents, and the media. Going beyond the well-known liberal-conservative struggle in politics, the world today initially appears as an ongoing battle between two scientific and cultural worldviews, or paradigms. Leaving aside for the moment the beliefs of the indigenous nations and peoples of

Earth, the worldview of modern science and the medieval fundamentalist worldview are seen as the two main competing mind-sets of today. Whether a conservative, liberal, or even Communist Party runs a government today, all are based on the paradigm of modern science and technology (though not always democracy). And most countries today practice crony capitalism, making the corporations and billionaires of the modern worldview a clear parallel to the aristocracy of the 1750s.

Yet the term *modern* is misleading, merely meaning contemporary, the latest, the most recent. It would be more accurate to call the scientific worldview that overthrew the medieval paradigm as mechanistic, for the models used to create its theories are based on the concept of the machine. Newton's clockwork universe neatly conveys the underlying analogy of the whole worldview: the planetary orbits and revolving galaxies of the universe are like the gears and machinery inside a clock. Clocks and mechanisms are good analogies for inanimate objects, but machine models are too limited to explain human beings, the mind, evolution, ecology, and most branches of biology. Since machines cannot reproduce, think, or evolve of their own accord, a broader model naturally emerged.

THE NEW PARADIGM:
THE ORGANIC WORLDVIEW

In fact, there are not two major paradigms in the world today—there are *three*. All coexist and have their own time lines of development: the medieval, fundamentalist worldview of many religious sects; the ruling mechanistic science and culture; and now a new paradigm, the organic worldview.

This new thinking uses *biological* or *organic models* rather than *machine models,* creating not only a different science but a whole new way to look at science itself. Changing the underlying analogy that all scientific theories rest on, changing the paradigm, as we saw, inevitably leads to a reinvention of nearly all science and then society itself. Thus this *organic shift* is a new Copernican Revolution—the Global Awakening. In his 1982 book *The Turning Point,* Fritjof Capra agrees that the new paradigm is a true scientific revolution on the scale of the Copernican. Capra, moreover, agrees that the old science and power structure will never be able to solve the crises of the modern day; only the radical change of the new worldview—applied to all endeavors of humanity—can succeed in that great task.

Just as the mechanists exposed the contradictions of the astronomy and pure science of the Middle Ages, now the new organic worldview emerges in its turn, based on a devastating critique of the mechanistic worldview. In his book *Person/Planet,* Theodore Roszak showed how ecology is more than just another new science; it changes the very relationship humanity has with nature. Rather than the anthropocentric universe, which places humanity at the center of reality, giving it the right of full exploitation over nature, ecology reveals a nonanthropocentric view. Instead of the typical materialism and rugged individualism of the mechanistic worldview, Roszak said the very concept of the person changes in the new paradigm, becoming "person/planet." We are still individuals, but everyone is now also part of the greater whole, of the biosphere and the planet itself—and that must guide our actions going forward into the future.

Pollution and high-impact technology are then seen as degradations to our being, the opposite of the pride the medievals and the mechanists feel when they devastate the environment with their anthropocentric ways. A six-thousand-square-mile dead zone in the Gulf of Mexico is acceptable to the old anthropocentric worldview, although lip service may be given to the economic problems of fisherman to help offset the bad image this gives to conventional agriculture. To the mechanists, Earth is treated as though it were an inanimate rock, something to be mined, drilled, consumed, exploited, and farmed ad infinitum. In their final analysis, the dead zone is a small price to pay for modern agriculture. To ecologically aware people, however, the dead zone is a living nightmare, a real pain inflicted on their own sense of being.

As in the shift from the medieval to the mechanistic worldview, the new paradigm brings in a whole set of shocking ideas, a change in the very fundamentals of thought and values. Once again, an old set of values that cannot stand drives the new paradigm to reinvent society from top to bottom. The organic worldview is an entirely new set of values, one that is in many ways the opposite of the old thinking.

TABLE 1.3. COMPARISON OF MECHANISTIC AND ORGANIC WORLDVIEWS	
MECHANISTIC WORLDVIEW	**ORGANIC WORLDVIEW**
• Limited mechanistic models underlie traditional science and medicine and cannot explain living systems adequately: ecological, health, economic breakdowns.	• Encompassing organic/biological models underlie new-paradigm sciences from physics to agriculture, medicine, technology, economics, and psychology.
• Clockwork universe, no purpose assigned to humanity or universe; we live in a vast static cosmos.	• Complexity-centered universe and evolution mean we are always evolving to the next level.
• Anthropocentric universe: planet Earth treated as a nonliving thing to be exploited.	• Complexity-centered universe: planet Earth shown to be a living system.
• Newtonian physics limited to macroworld, nonliving things only.	• New physics studies subatomic realm; law of organics and other theories explain living systems.
• Time and space quantified.	• Life, evolution, consciousness quantified and given meaning.
• Studies objects and things as separate parts.	• Studies the relationships between objects and things.
• Old-paradigm culture based on oil, ultranationalism, and militarism; huge military budget, small foreign aid; top 1 percent owns 45 percent of all wealth.	• Counterculture based on transition from oil, world peace, and sustainable development; increase foreign aid to $50 billion to stop terrorism; new economics to eliminate poverty.
• Laws and values designed to protect the rights of men, especially corporations and men with property.	• Laws and values designed to protect the rights of all, from women to blacks, gays, and all minorities, especially the poor and the middle class.
• Belief that war has always been a part of human nature.	• Know that war was invented and that it can be transcended in a future world peace.

From its fundamental rejection of the anthropocentric universe to its respect for ecology, from the new sciences and renewable energy technology to cultural changes such as the suffrage movement, the union movement, and the civil rights movement—the organic world-view has both a broader and more inclusive science and a broader and more inclusive set of values. The organic worldview even transcends the view of present-day nationalism, believing the state of our small planet and the world to be as important as the state of our own country. In the organic worldview, it becomes clear that ultranationalism is in fact the main cause of war. Here again, the new thinking is very far apart from the old mind-set.

THE ORGANIC SHIFT TOWARD THE GLOBAL AWAKENING

As in the eighteenth century, this enormous gap in values and beliefs will likely not last long. Eventually, a tipping point will be reached, and the paradigm flip will then sweep through a portion—but not all—of the population. After the flip, widespread and deep change will finally be possible. To see if this could really happen today, to see if our own time could undergo a global reinvention, we need to compare our recent past and present to the distinct stages of the Copernican Revolution and the Enlightenment.

Is the Organic Shift roughly equivalent to the Copernican Revolution? Does the organic/biological analogy truly provide a new foundation for science and philosophy? If so, who was the organic Copernicus? Who was the organic Galileo and Newton? Further, if we are in the midst of a new enlightenment, how far have we already gone in the enlightenment pattern? How far off, then, is the paradigm flip and the transformational phase? Finally, what would a twenty-first-century transformational phase look like? These are the questions we must answer.

Every major paradigm shift has a profound, epoch-making idea at its beginning—there is a veritable conception point of the new thought.

Who was the originator of this second Copernican Revolution? In looking through the history of science and philosophy, one thinker stands out as an obvious candidate to fulfill the role of the organic Copernicus.

THE ORGANIC COPERNICUS
AND GALILEO

In 1781, the German philosopher Immanuel Kant (1724–1808) published *Critique of Pure Reason,* long accepted as the beginning of modern scientific philosophy and idealism. Yet Kant was much more than that. In his long, tortuous work, Kant basically lays the foundation for a whole new worldview, a new way to think. Rather than a simple mechanistic model of the world, he pointed out that we can only know what we perceive with our senses. Reality is like an iceberg, we can easily see the top but the 90 percent underneath the water line is invisible—and we may only theorize about it. We can never see the whole of reality. Kant's final analysis was that we may only know our own thoughts of reality. It was all rather Buddhist, seeing the world as an illusory appearance, like *maya.* Kant had read Latin translations of Buddhist and Taoist stories and was—in effect—marrying Eastern organic thinking and Western science.[14] In so doing, he created the modern scientific organic worldview.

With this tremendous realization, Kant knocked pure science from its throne and put a balanced philosophy in its place. He himself declared that his discoveries had "many analogies" to those of Copernicus *and actually predicted that a new scientific revolution would come thanks to his momentous discovery about reality.*[15] Since machines do not truly live or evolve or have consciousness, mechanistic models are quite inadequate for fully explaining life, evolution, and the mind. So Kant said, "There will never be a Newton for the blade of grass." By recognizing the usual limits of science, the waterline of the iceberg, he was able to fashion a new science that could more accurately theorize about the whole, extending thought to the "bounds of reason."

Kant was right. A vast scientific and cultural revolution did indeed take place as the result of his advances, and in every way it equaled the Copernican Revolution. Kant taught that the observer was part of the observation, for example, key to the discovery of the new physics, quantum mechanics, and the splitting of the atom—a solid parallel to Isaac Newton and Newtonian mechanics in the Copernican Revolution. Chapter 2 takes a longer look at Kant's many other contributions, including those beyond philosophy and idealism, such as his essay "Perpetual Peace," which began the movement toward world peace and government, and his description of the new awareness in his writings on *homo noumenon* (moral man).

In his *A Brief History of Everything,* Ken Wilber agreed that Kant must be seen as a main founder of the new paradigm. We can therefore safely say that Kant is the prime candidate to be the new Copernicus. Even Kant thought so.

If Kant was the new Copernicus, was there an organic Galileo? Our candidate this time is Johann Wolfgang von Goethe (1749–1832) of the late eighteenth and early nineteenth centuries, typically thought of as a poet and writer. In fact, Goethe was an astonishing scientific genius and researcher. Like Galileo, he not only developed a far superior scientific method, he then used it to uncover new laws of nature. These included the law of organics, as well as the first adequate description of the essence of evolution and nature itself. Goethe included holistic intuition and contemplation into the observation and analysis of science, which turned out to be the only way we can create adequate models of the biology we are attempting to understand. Whereas Newton could never quantify the essence of life, of a blade of grass, with Goethe's participatory method, however, science could at least begin to understand it.

His law of organics explains how nature unfolds through evolution, that by observing the similar structures that exist in plants and animals we can see types, unbroken lines of development and change over millions of years. Through his new method, Goethe was able to make the study of evolution and nature a science, where previously it had been a series of observations and vague reasoning. In many ways, he can be compared favorably to Galileo and the role he played in the Copernican

Revolution, making the new paradigm a hard and practical science. Rudolf Steiner, the Austrian philosopher and social reformer, in fact, declared Goethe "the Galileo of the organic."[16]

THE ORGANIC COUNTERCULTURES

As in the Copernican Revolution, a more open lifestyle was a large part of the paradigm shift. Each change in worldview brings with it new lifestyles and culture, especially in art, literature, music, and theater. Thus the counterculture artists and writers of the nineteenth-century Organic Shift parallel the secular humanists of the sixteenth and early seventeenth centuries. Kant, Goethe, and also early visionary Romantics like William Blake helped create a new view of the artist. Rather than merely glorifying God and the rich and powerful, the artist was now a radical change agent, a bringer of vision, depth, and mystical transformation. Decrying the destruction of nature, Romantic poets and writers also fought against the spread of industrialization and the blight of factories as "satanic mills" (William Blake). Out of all this antimechanist thinking came the bohemian counterculture and the American transcendental movement, which in turn gave rise to the later beatniks and hippies. We can see that the new worldview had a tremendous impact on the nineteenth and twentieth centuries—it was a true scientific and cultural revolution.

THE NEW PARADIGM AND
THE NEW PHYSICS

In *The Tao of Physics,* Fritjof Capra explained how science in the 1920s was able to go beyond the limitations of Newtonian mechanism and comprehend the complexity of the subatomic world. The culmination of this quest was the 1930 uncertainty principle and quantum mechanics, created by a team of physicists led by Werner Heisenberg. It was Heisenberg who used the new paradigm successfully, especially Kant's concept that the observer could not be removed from the act of observation, for we can only look into the atom by destroying it. Heisenberg even invited a Kantian scholar to take part in his brainstorming sessions about the atom.

Fig. 1.2. Werner Heisenberg published the uncertainty principle, the foundation of quantum mechanics, 149 years after Kant.

Newtonian, mechanistic *certainty* must stop at the limits of our senses, but with the new paradigm, you can go beyond and formulate the *probabilistic* mathematics and science of the invisible atom. You had to trade certainty for probability, but then the mathematics was possible. At that moment, the publication of the uncertainty principle in 1930, Newton's clockwork universe was superseded by the new physics of the Organic Shift, which explained both the macroworld and the subatomic realm.

Most interesting in all this is that Newton published his main work in 1687, *144 years after Copernicus,* while Heisenberg published the uncertainty principle in 1930, *149 years after Kant.* This fascinating parallel in years has never been noted before, but it helps to emphasize the striking similarity between the Copernican Revolution and the Organic Shift. Both paradigm shifts took generations to complete and both changed the very foundation of physics about 150 years after the original concept was published. Science was completely reinvented in the process, from physics and philosophy to ecology, psychology, anthropology, medicine, economics, biology, and agriculture, often with competing mechanistic and organic schools of thought coexisting at the same time. Chapters 2 through 7 will explore the role of the thinkers and movements of the Organic Shift.

Following the scientific and cultural revolution phase, the new worldview enters the social and then political mainstream, creating rapid change and great turmoil during the enlightenment part of the shift. Does recent history indicate that we are living in such a time?

A NEW ENLIGHTENMENT

A current-day paradigm flip cannot start until all the essential paradigmatic questions are fully answered with a new story that gives an organic version of our past, present, and future. Once that new story is in place and delivered to the mass of humanity, the deep changes establishing sustainability and true democracy can sweep the world. Today, those aware of the new paradigm understand the new view of evolution and how consciousness has evolved; it's the new answer to "Who are we?" Yet the full new paradigm answer to history and the future has never been fully given. When the other two essential questions of "How did we get here?" and "Where are we going?" are fully addressed, the paradigm flip can finally begin.

Comparison of our own time to the four stages of social transformation, or the enlightenment pattern, reveals many fascinating parallels that shed new light on "how we got here" and "where we are going." We have not thought of our day as an age of enlightenment, but then neither did the public of the 1700s—until about the 1750s. Then, in inexpensive books, Voltaire told them they were indeed living in such a time. The parallels of our own society to the four stages are clear. With the creation of the Internet, we have, today, a similar opportunity for communicating a new way of thinking. We appear to have gone through the first two stages of early enlightenment and conservative backlash and have started the intensive phase.

Following the scientific revolution period, the Organic Shift entered the social and political mainstream in the 1960s and 1970s, upsetting the staid and stable worldview of the 1950s with shocking new ideas like full civil rights, ecology, and the evolution of consciousness. The generation gap of the time was really the chasm between old and new paradigms. This whole period of the 1960s and 1970s was quite similar to the early enlightenment phase of 1700–1726, when the hold of the old medieval worldview on the public was challenged by reason, natural law, and the concepts of liberty and equality.

This would make the 1980s and early 1990s, the Reagan and Thatcher era, very much a conservative backlash phase, trying to reverse the secular change wherever it could, just like the years between 1725 and 1740. Again, the struggle was generational, with the progressive boomers being outvoted as the more conservative and nihilist generation X was born from about 1976–1990.

According to the enlightenment pattern, the struggle between the two cultures was then truly joined in the current intensive phase (starting approximately in 1991), as adherents of the new organic science and culture thoroughly exposed the falsehoods supporting the mechanistic worldview. The last several years closely resemble the intensive phase of the eighteenth century in many ways. There are the new philosophers of the organic worldview, thinkers like Deepak Chopra, Ken Wilber, Andrew Weil, and Thom Hartmann, who have already popularized significant parts of the broader thinking. Similar to the 1750s, they are in the process of together building a common agenda for change. The culture war and the struggle over family values have increased the polarization between these worldviews to an unprecedented level, as the old and new systems confront each other politically and socially. We are in an information war once again, in which the old paradigm tightly controls the traditional media, while philosophers and a new activist generation make an end run around the powers that be through the new blogs and news sites on the Internet.

Today, it is quite clear that the new worldview is beginning to affect politics more and more. There are the new young progressive activists, called the millennial generation by demographers—in contrast to generation X. They are now part of the picture; they have held enormous antiwar and anticorporate globalization demonstrations and, during the 2004 U.S. primaries, became the Internet-savvy Dean generation. Currently, the young progressives and older activists have sparked the Occupy Movement with its 99 percent slogan, which started with the Occupy Wall Street demonstration in New York City in September 2011.

This time the movement for change must be kept alive, for it might take decades to completely undo the damage that regressive leaders have wrought. During the process, there will be steps forward and steps back, and the movement must be strong enough to withstand those temporary, frequent setbacks. This is not something that will be accomplished with one term of a moderately progressive U.S. president like Obama. The ascent of progressive political will in the United States represents a new era. It is likely the first of many deep changes.

Although the new progressive politics is now visible and beginning to flex its mighty muscles, the young activists have learned the answers to only one part of the three essential questions that could change the world. We can thus pinpoint where we are in the enlightenment pattern. As stated earlier, the world is clearly right in the middle of the third stage, the intensive phase. Having gone through the early enlightenment of the 1960s and 1970s, as well as the conservative backlash of the 1980s and 1990s, we have not yet reached the paradigm flip, which will come at the end of the intensive phase.

Although liberals and progressives today may despair, they need only remember that the destructive policies of the conservative kings and aristocrats who ruled throughout the last intensive phase (1741–1760) helped set the stage for *total* change by discrediting and bringing to complete ruin the old system. In a similar way, the neoconservatives, corporatists, 1 percenters, Tea Partiers, obstructionists, and fundamentalist extremists of all creeds—Christian and Muslim—will likely turn out to be their own worst enemies.

While conservatives, Communists, and liberals stand by the mechanistic paradigm and all its machine models and the medieval-paradigm fundamentalists try to beat back nearly *all* the changes of the secular world, the organic worldview offers the alternative politics of progressivism—and a sustainable future based on ecology. In politics, liberals have developed massive credibility problems, and now so have the conservatives. The liberals of today, in fact, correspond closely to the liberal aristocrats of the mid-eighteenth century, who continued to support the old power even though they thought in an enlightened way. As a result, monarchy was free to expand the corrupt management of the treasuries, make worse the unfairness of the tax structure, and commit the folly of starting one unnecessary war after another. Economic and military disaster was the result. If the historical pattern continues, there will soon be one remaining credible worldview: the organic progressives and their new ecological science.

According to the four stages of social transformation during the Enlightenment, we are fast approaching the paradigm flip of the Organic Shift. The last time the world, or the Western world, experienced a paradigm flip was more than 250 years ago. No one alive today has ever experienced such a rapid time of change, but if the historical pattern continues, we are about to see a paradigm flip firsthand. This time the flip will not just be a portion of the population but will be much more widespread. The "united nations"—meaning every country and all peoples everywhere, even tribes of indigenous people—will be in on this change in worldview. It will affirm and embrace their ancient wisdom and—unlike the mechanistic paradigm—will not attempt to change their right to live in native traditional ways or lifestyles.

For many decades now, ever since World War I destroyed the sunny optimism of the nineteenth century, the mechanistic worldview has been suffering from a metaphysics of despair, a lack of hope that humanity will not be able to avoid destroying itself with the machinery of war, pollution, and the effects of corrupt government and science. We have been under the influence of the metaphysics of despair for so long that a future that works can no longer be realistically envisioned. A paradigm flip would change all of that. Similar to the revolutionary atmosphere of the 1760s, 1770s, and 1780s, following the tipping point, society would enter the transformational phase—and deep change will finally be within our grasp. *What was previously impossible would then become capable of manifestation: this would then become the most likely scenario.*

IS IT REGEN OR DEGEN?

Deep change—true transformation of the world—would mean many things. Environmental restoration, the end of the fossil-fuel economy, world peace, and the end of abject poverty would be the first order of business. That is all far easier said than done. Can we really solve such seemingly intractable problems?

If we are smart and go new paradigm, the answer is yes. Yet it means more than moving to a new energy base, phasing out chemical agriculture and industry, or passing stricter corporate governance laws. What will have to change is the very way ordinary people look at the world and their own personal roles in shaping a sustainable future. Though it will be the biggest transformation of all, the science and other solutions needed to shift that many minds have already been discovered. We need only to implement the entire set of solutions on a global scale.

The mechanists have always operated by brute force, by crudely conquering nature with chemicals and oil-powered machinery, rather than trying to work with nature. Yet Goethe discovered that nature has what he called "open secrets." If we uncover these secrets, we could get rid of the need for chemicals to grow our food and oil to drive our society. Furthering Goethe's impulse or initiative, during the 1970s and 1980s, the new-paradigm pioneer Robert Rodale attempted to understand how a chemical-free agriculture is best managed and took it upon himself to research the most minute workings of the soil—and how even dead soil can be regenerated and returned to fertility.

A new order of understanding, the regenerative sciences, emerged from these studies, fulfilling many of Goethe's dreams. Based on the concept that nature has the ability to regenerate itself in many ways and so restore that which had been lost, the regenerative concept reinvented several fields, including agriculture, economics, environmental restoration, regional planning community development, international development, and many others. By reversing the usual way of doing things, and paying attention to the way nature works, Rodale came upon an integrated set of solutions for the workable future. Using regenerative science, now taught in universities around the world, global society can go "green" in a realistic and practical way. Some of today's leading lights in the regenerative sciences—Bill Leibhardt (regenerative grain), Fred Magdoff (organic dairy), and George Bird (soil science and regenerative soil management)—teach at these universities, such as UC Davis, the University of Vermont, and the University of Minnesota.

Through a regenerative index, Rodale tracked the self-sufficiency of energy production, food production, health, transportation, manufacturing, and capital within a regional zone. By creating multiple regenerative sources within a single zone, the restoration of the environment and the community becomes a straightforward matter: it's called regenerative economics. When combined with regional planning, renewable energy technology, and the use of plant and biomass oil for industry instead of crude, along with a new paradigm for education and health, the scientific foundation for a new "industrial revolution," region by region, becomes very real. By using modern soil science and crop research, the regenerative sciences make it possible to design an agricultural and economic system for each region that is sustainable and actually regenerates the environment and restores wildlands that have been degraded or lost.

During the 1980s, in the Ethiopian dust bowl of the Antsokia Valley, hundreds of thousands of people were starving to death. Suddenly, a regenerative specialist, Dr. John McMillin, appeared on the scene and within a relatively short period of time transformed the region—literally. He carried out his own regenerative design for the Antsokia, pumping water from an aquifer and using compost to raise the health of the local soil. He raised tilapia, one of the easiest species of fish to raise domestically, in a combination fish-pond and biodynamic garden, thereby creating an enhanced nitrogen cycle and providing three times as much protein as grain farming. With funding and further help from World Vision, McMillin oversaw the planting of nine million trees and the farming of 250 acres of land, ending the drought by creating a new microclimate.

The Antsokia Valley went from dust bowl to oasis, showing how the regenerative sciences can take apparently hopeless places and barren soils and restore their health from the ground up. Today, the region actually *exports* food. Repeat the Antsokia regeneration ten thousand times, and two billion poor would be lifted out of abject poverty. Later chapters of this book explore these new sciences and how they could change the world in a global regeneration revolution.

When we look at our world today, we can ask ourselves, Is this trend or idea regenerative, or is it degenerative? Does it help restore the environment, economy, or culture, or does it do even more to destroy the workable future the planet needs to survive? Is it regen or degen? An organic farm is regenerative, said Rodale, with no toxic runoff and no need to drill for oil to make pesticides and fertilizers. In contrast, a conventional farm is degenerative, polluting the local streams and air with various types of chemicals. In foreign policy, knee-jerk military reaction to a disagreement is usually degenerative, while a comprehensive regional peace and development plan, fully supported by the world community, is regenerative. Programs to prevent health-care problems and deliver alternative integrative medicine are regen, while getting misprescribed but powerful drugs from a traditional doctor can be very degen.

Here's a test: health food or fast food? Which is regen, which is degen? You get the picture. Eating healthy organic food is not only better for your body and reduces future medical problems dramatically, it also helps shift farmers away from using the chemicals that end up creating the dead zone in the Gulf of Mexico. Our personal actions, accumulated together, have a tremendous impact. Person/planet means understanding that it is all interconnected, that how we live as individuals affects the environment—and the future. In short, we must change ourselves before we can change the world.

We can learn a lot by watching the news and the Internet if we categorize the stories. How many organic shift-related stories are there—

regenerative ones about changing the world? How many reports that are just more public relations for the obsolete and corrupt mechanistic worldview? Is the news report regen or degen? Right now, our thoughts about the paradigm shift are mostly unconscious. We tend to ignore our worldview, much as a fish is unaware of the ocean surrounding it. But at some point before the paradigm flip, the mass media will suddenly become aware of the organic worldview. At the same time, a new progressive media will likely acknowledge the Organic Shift, and the public could then realize society has been in a new Enlightenment since the 1960s. Will the new organic philosophers and the progressive bloggers, so outspoken for the new paradigm and the new politics, then become a parallel to the pamphleteers of the American Revolution and lead the United States and the world through a great series of changes? Could this really come to pass?

We have long been programmed to dismiss all such notions as hopeless utopian pipe dreams—that futures based on such things as solar power, peace, and brotherhood could never work. In the cynical view, the only realistic future scenario is that of the status quo: the corporate-controlled, oil-based world feeding a bloated military will continue indefinitely. Yet is that true, has it ever been true? Given global warming, dead zones, and nuclear proliferation, which future is the hopeless pipe dream? Has a progressive, ecological solution ever been tried on a large scale? The twenty-first-century transformation being described here is not utopia, not the perfect place. But it is a workable future, a sustainable future—a better place.

This is not a religious doctrine; it is sociological prediction based on the study of past paradigm shifts and on the truth of the new regenerative sciences. This book simply says that, at this point, the mechanistic worldview has little hope of creating a workable future, but the odds of success are much higher with the ecology and science of the organic worldview. The industrial mechanists have even ignored the fast-approaching peak of global oil production and so have never come up with a plan B for when the world goes into its inevitable final years of using cheap fuel.

Imagine that, no plan B, despite the fact that a tremendous economic dislocation is the most likely scenario, with the very real possibilities of food shortages and impossibly high fuel and transportation costs. The turn to regenerative agriculture and renewables will then be unavoidable—that is now plan B. It very well may be that the regeneration revolution will have a world economically flat on its back to restore, but it still can do the job.

Poverty and pollution will still be extant in the workable organic future, but they will no longer be threatening the very survival of the race. This grand social transformation does not take place, however, unless enough people rapidly become active and make the changes happen in the midst of almost certain turmoil. That's where the paradigm flip comes in.

A TWENTY-FIRST-CENTURY PARADIGM FLIP

In the early 1750s, similar to today, many ingredients for transformation were present, yet deep change could not have happened without the catalyst of Voltaire's new history. This was the critical catalyst that set the paradigm flip in motion. Voltaire's new answers to the three essential questions of who we are, how we got here, and where we're going gave the eighteenth-century Enlightenment reader a solid center, making the new thinking acceptable. This radical history also debunked the old paradigm's version of the past, changing the opinion of whole segments of the Euro-American population. Higher and higher levels of outrage and desperation, as well as a new image of the future, finally gave the activist generation reading Voltaire's books the courage and ambition to rebel against the monarchy in what became the transformational phase.

So today, although many ingredients are present, deep change awaits its critical catalyst. Without a new history, without a new set of answers to the three essential questions, there will be no paradigm flip, no transformational phase. In the spirit of Voltaire then, let us answer anew these queries of the heart that must be answered before the soul will be pledged. What is our real history—not the glorified, partial history of the mechanists and ultranationalists? How does the Organic Shift transcend and so end the mechanistic dilemma? Are we now approaching the society-wide paradigm flip—the crucial part of the paradigm shift to Global Awakening? And—according to the enlightenment pattern— what does the future hold for us all?

All good stories start at the beginning. So let us go back in time to the conception point of the Organic Shift. We return to that thought in the mind of a single person, the nagging question that led to a whole new model of the universe, to a *global awakening*. How did the new paradigm get its start? What was the contradiction that originally led to the unraveling of the old thinking? Only then can we can understand the second Copernicus—the Copernicus of the organic.

Immanuel Kant

2

KANT

The Organic Copernicus

We have here the same case as with the first thought of Copernicus. . . . In the same manner the laws of gravity . . . would have remained undiscovered, if Copernicus had not dared, by an hypothesis, which though contradicting the senses, was yet true, to seek the observed movements, not in the heavenly bodies, but in the spectator. I also propose in this preface my own view of metaphysics, which has so many analogies with the Copernican hypothesis.

IMMANUEL KANT,
CRITIQUE OF PURE REASON

THE ORGANIC SHIFT BEGINS

Never doubt that a small group of thoughtful, committed citizens can change the world. Indeed, it's the only thing that ever has.

MARGARET MEAD

51

Where did ecology come from? When did the world peace movement start? Why has there been so much interest in Eastern philosophy? Why do some people refuse to eat anything except organically grown food? How could modern subatomic physics be based on something called the uncertainty principle? Why do minorities and women demand their full civil rights? And how in the world did nonconformist beings like the beatniks and hippies come about? All of these seemingly separate trends, movements, and more were actually part of one enormous change, from one paradigm to another: the Organic Shift. It is this change in worldviews that has affected the sciences and society over the last two hundred years, preparing the ground for a new period of enlightenment, which we know as the 1960s and the decades beyond. We cannot understand our own time—and thus ourselves—unless we first study this change in worldviews.

Global Awakening is based on the theory that a new scientific revolution, the Organic Shift, has long been underway, at least from the time of the late 1700s. Proof of this new paradigm as a true scientific revolution would involve showing several things: (1) that the conception point of the Organic Shift be identified, (2) that a new scientific method is shown to accompany the change, (3) that this new worldview affects many of the sciences, and (4) that an equivalent to Isaac Newton's advances in physics be proved. In other words, strong parallels to the Copernican Revolution must exist for the Organic Shift to be accepted as a major change in paradigms. Logically, there should be an organic Copernicus, as well as an organic Galileo, and there should, of course, be a new paradigm counterpart in physics to the great Newton himself.

Finding an equivalent to Copernicus would seem no easy task. The heliocentric theory of Copernicus did nothing less than end a powerful religious worldview by giving rise to another paradigm and a more democratic and scientific society. Any parallel to Copernicus would have to be directly responsible for setting in motion a similar reinvention of the world—from top to bottom.

As mentioned earlier, there is one person in history who stands out as our candidate: Immanuel Kant, chairman of philosophy at the

University of Konigsberg in Prussia. Kant himself explained the parallel between Copernicus and himself. Both expanded our view of our place in the universe. Immanuel Kant was a profound influence on a stellar list of later organic thinkers, including Johann von Goethe; Ralph Waldo Emerson, American essayist and poet; Ernst Haeckel, German scientist and philosopher; Wilhelm Dilthey, German historian and philosopher; Jan Christiaan Smuts, South African statesman and philosopher; Werner Heisenberg, German theoretical physicist; and Pierre Teilhard de Chardin, French philosopher and Jesuit priest. Scholars have long pointed to Kant's idealist philosophy and his concept of world government and world peace as being a great influence on the Romantic period that followed. Yet Kant did much more, for his theories ignited a full-blown scientific and cultural revolution, much as, 250 years earlier, the heliocentric theory of Copernicus proved to be the conception point of the mechanistic world.

Where the astronomy of Copernicus provided the foundation for the mechanistic analogy, for the clockwork universe, Kant's philosophy uncovered a whole new way to think, taking into account the complexity of nature and showing how Newton's formulae cannot measure or understand biology nor the human mind. Mechanistic theories must stop at the "limits of reason."

Yet by using feeling and synthetic thinking, meaning to holistically synthesize and take the broadest of views, science can transcend the limits of reason and go beyond—theorizing about such things as life and consciousness within the bounds of reason. Kant called his system of thought transcendental philosophy, for it transcends the limits and makes possible sciences of reason (sciences of consciousness). His leap in thinking did indeed lead to a scientific revolution, allowing the development of not only significant evolutionary theory and the field of ecology but also electromagnetic wave theory, the new physics, cultural anthropology, and even Jungian psychology, not to mention many branches of the biological and social sciences.

There is little doubt that the best candidate for the role of the organic Copernicus is Professor Immanuel Kant of the University at Konigsberg.

KANT THE MAN

Immanuel Kant is thought to have been such a systematic thinker that his life was as orderly as his philosophy. It has been said by one of his biographers that he lived his life like the most regular of regular verbs. Many writers also noted his Pietist upbringing as influencing his systematic nature—the Pietists being known for their strict set of religious rules. Kant was so punctual in taking his daily walk that the ladies of Konigsberg could set their clocks when they saw his top hat going by their windows at exactly the same time every day.

Kant, however, was far more than a dry, systematic, and stuffy Prussian philosophy professor. His friends knew a man whose luncheons were renowned throughout intellectual Europe for their guests and their deep conversations. One who knew him, his student Johann Herder, and later founder of the field of the philosophy of history, fondly remembered his old mentor:

His forehead, formed for thinking, was the seat of indestructible serenity and peace, the most thought-filled speech flowed from his lips, merriment and wit and humor were at his command, and his lecturing was discourse at its most entertaining. In precisely the spirit with which he examined Leibniz, Wolff, Baumgarten, and Hume and perused the natural laws of the physicists Kepler and Newton, he took up those works of Rousseau which were then appearing, *Émile* and *Héloïse,* just as he did every natural discovery known to him, evaluated them and always came back to unprejudiced knowledge of Nature and the moral worth of mankind. The history of nations and peoples, natural science, mathematics, and experience, were the sources from which he enlivened his lecture and converse; nothing worth knowing was indifferent to him; no cabal, no sect, no prejudice, no ambition for fame had the least seductiveness for him in comparison with furthering and elucidating truth. He encouraged and engagingly fostered thinking for oneself; despotism was foreign to his mind. This man, whom I name with the utmost thankfulness, and respect, was Immanuel Kant.[1]

Kant was no stuffed shirt. He was a fountain of knowledge and humanity, always taking the side of freedom, equality, and the power of ideas to solve the problems of humankind. He was also known as Germany's greatest supporter of the French Revolution and the overthrow of King Louis XVI.[2] Kant was a radical idealist, whose writings would help inspire not only the Romantic era of art and literature but also a new movement for world peace. Beneath his systematic punctuality and proper life, he was a brilliant revolutionary—of the intellectual kind.

KANT'S NEW DESCRIPTION OF REALITY

As mentioned earlier, scientific revolutions start with an anomaly, a contradiction of the known workings of the paradigm, something the old science cannot explain. For Copernicus, the question that ended up unraveling the whole medieval society was "Why is it hotter in the summer and colder in the winter?" The heliocentric theory was his eventual answer.

Immanuel Kant's question was much more complex—yet equal in eventual impact to the one that perplexed Copernicus. How do we know and what can we know? asked Kant. These questions are beyond any machine model. Simply trying to answer these questions meant discovering a whole new out-of-the-box worldview. Kant himself declared, "We have here the same case as with the first thought of Copernicus. . . . In the same manner the laws of gravity . . . would have remained undiscovered, if Copernicus had not dared, by an hypothesis, which though contradicting the senses, was yet true, to seek the observed movements, *not in the heavenly bodies, but in the spectator*"[3] (italics added). As with Copernicus, Kant is saying reality can shift its appearance—depending on where you stand.

THE OBSERVER IS PART
OF THE OBSERVATION

By noting that the observer is part of the observation, Kant showed that so-called objective science is, in fact, subjective, for it does not recognize its own part in the theorizing. In his preface to the second edition of *Critique of Pure Reason* in 1783, Kant explains how both he and Copernicus went beyond the mere appearance of the world, by looking at the role of the observer in the observation. In Copernicus's case, the Polish astronomer realized the observer (Earth) was moving and the observation (the stars and sun) were actually at rest. In the opening quote, Kant advises "to seek the observed movements, not in the heavenly bodies, but in the spectator." That flipping of perspective changed everything, especially the physics needed to explain the movements of the planets.

Kant essentially applies the idea of the observer being part of the observation to the study of thought itself. Using the first moment of time as an example, Kant discovered that thought—by its very organization—creates an appearance that it can absolutely explain the world. As Kant points out, however, no one can say with absolute certainty what occurred at the first moment of time or when it was. Since no one can go back and visit the first moment of time, we can only speculate and theorize; the real reality is unknowable. So where Copernicus exposed the false appearance of the *geo*centric universe, Kant exposed the fundamental flaw of the *ego*centric/anthropocentric universe (including the mechanistic clockwork universe of Newton). Mechanistic science acts as though it directly knows reality, when in fact it only can know its *mental image* of reality.

Kant was the first Western philosopher to understand Taoism and was familiar with the basic thrust of Eastern philosophy. His critique was of both Eastern and Western thought—leading him to consider the good and bad points of each in his own system. Kant's transcendental philosophy is thus shaped by a powerful combination of East and West. In his essay "The End of All Things," he criticized Lao-Tsu's total nihilism as being an escape into the inner world of meditation, an abandonment of using reason to improve the practical world.[4] Kant was familiar with the basic thrust of Eastern philosophy.

How do we know and what can we know? Kant brings the different logic of the East to bear on his question of the ages. Western philosophy had for the most part described the world with "positive logic," or *it is because it is.* In contrast, Taoism and other Eastern philosophies use "negative logic" to describe reality: *it isn't because it isn't.* Positive logic tells us what we can know about the world, while negative logic tells us what we cannot know.

In the construction of his philosophy, Kant thus describes reality with a balance of positive and negative logic. His writing at times seems almost Buddhist, as if he were talking about the illusory material world of *maya.* In *Critique of Pure Reason,* he writes: "[A]ll our intuition is nothing but the representation of appearance. . . . As appearances, they cannot exist in themselves, but only in us. What objects may be in themselves, and apart from all this receptivity of our sensibility, remains completely unknown to us."[5]

His idea that science can never explain the whole of reality, the thing-in-itself, became one of the main concepts of the current organic worldview—a fundamental assumption that leads not only to the environmental ethic but also to the new physics and many other holistic disciplines.

THE ICEBERG OF REALITY: PHENOMENA AND NOUMENA

Kant divides reality into *phenomena* and *noumena.* Like the visible part of an iceberg, phenomena are the appearance of things we can see and sense with the instruments of science, while the noumena lies beyond the reach of our senses; it is the mysterious and unknowable thing-in-itself—the invisible part of the iceberg. Yes, instruments like the telescope will grow more and more powerful, yet we will never be able to say that we can know, see, and measure anything in its totality. Knowing that we cannot know everything tempers the bragging rights of science, balancing its power with the skepticism of nonabsolutism.

It is said that absolute power corrupts absolutely. In science and philosophy, that is doubly so. Without the balance of the noumena, science has free rein, allowing absolutist theorizing of many kinds, leading directly to dogma, undue arrogance, and false science. This can be dangerous, as the citizens of Chernobyl and Japan found out.

Mechanists believe that the measurement of anything can be so all-encompassing that eventually we will know the thing in all its complexity. If you have enough data, they say, one can predict the future with such accuracy that it is, for all practical purposes, infallible. Kant essentially asked, "How can you ever be sure you have enough data or that you are even asking the right scientific questions?" The possibility of a mechanistic "philosophy" that could pass Kant's test of nonabsolutism is therefore zero.

KANT'S COMBINATION OF
POSITIVE AND NEGATIVE LOGIC

By combining positive science with negative logic, West with East, Kant changed the equation in philosophy for succeeding centuries. The Chinese historian of philosophy Fung Yu-Lan (1895–1990), who originated the description of positive and negative logic,[6] agrees that Immanuel Kant, by being the first Westerner to use negative logic, made a significant contribution to the scientific philosophy of the future: "In the West, Kant may be said to use the negative method of metaphysics. In his *Critique of Pure Reason*, he found the unknowable, the noumenon. . . . When one knows that the unknowable is unknowable, one does know, after all, something about it. On this point Kant did a great deal."[7]

Fung goes on to explain that Kant was on the right track. The combination of positive and negative logic is the foundation of a future scientific philosophy:

[T]he two methods do not contradict but rather complement one another. A perfect metaphysical system should start with the positive method and end with the negative one. If it does not end with the negative method, it fails to reach the final climax of philosophy. But if it does not start with the positive method, it lacks the clear thinking essential for philosophy. . . . In the history of Chinese philosophy, the positive method was never fully developed; in fact it was much neglected. Therefore, Chinese philosophy has lacked clear thinking. . . . On the other hand the history of Western philosophy has not seen a full development of the negative method. *It is the combination of the two that will produce the philosophy of the future*[8] (italics added).

To us today, "the philosophy of the future" in Fung's quote above fits the description of the maturing organic worldview—and it began with the professor from Konigsberg. By combining positive and negative logic together along the lines of Kant's description of phenomena and noumena, new-paradigm science has been able to go beyond the limits of reason and theorize within the bounds of reason. Within the bounds of reason, however, science becomes more of an art. Although workable theories may be created, exact measurement is not always possible.

Yet, said Kant, it *is* possible to gain a genuine insight about a subject if the broadest perspective is taken, if feeling and the heart are used in an attempt to gain a sense of the admittedly unknowable whole. Mechanists, in relieving God of the responsibility of causing the daily events on Earth, threw out something essential: purpose. The scholastics had assumed a God-given purpose for each living being and part of nature. Kant felt that if science wants to truly understand and be able to make theories about nature, biology, and the human condition, it must theorize as though nature itself gives each living thing a purpose.[9]

Scientists must look at the big picture and feel what the purpose of the living thing might be. Without understanding that each organism has a purpose that drives it (finding food and reproducing, for example), or how a society may be driven by collective purpose, science will not be up to the task of studying biology and social change. If the complex set of causes behind their actions and characteristics are to be known, one must assume purpose *in all that lives.* This is the scientific leap that later made possible the fields of evolutionary science, modern biology, ecology, cultural anthropology, and psychology—sciences of organic, living things, of life, society, and the mind.

KANT DECLARES A
NEW SCIENTIFIC REVOLUTION

Kant boldly predicted that new sciences of reason would develop out of his philosophy. Just as Francis Bacon called his philosophic work that opened the door to a society based on mechanistic science *Novum Organum Scientiarum* (new order of sciences), Kant called his broader realm of science "the new organum." Kant himself states that *Critique of Pure Reason* sets the stage for the next scientific revolution, as science studies *thought itself:* "my own view of metaphysics . . . has so many analogies with the Copernican hypothesis."[10]

Indeed it did, as his new vision of science evolved into the Organic Shift over the following two hundred years, right up into the twentieth century. Just as Copernicus had led to the quantification of time and space, Kant's discoveries made possible the quantification of life, culture, consciousness, and subatomic space. Kant fully understood that he was starting a whole new intellectual revolution, and he compared his discovery of a new way to think not only to Copernicus but also to Thales, the Greek mathematician and astronomer who started a similar "intellectual revolution" in the sixth century BCE. These changes in worldviews are, of course, the same three paradigm shifts we have already compared. In the following passage, Kant explains how Thales's paradigm shift in the early sixth century BCE transformed math into a true science: "[T]he change is to be ascribed to a *revolution,* produced by the happy thought of a single man, whose experiment pointed unmistakably to the path that had to be followed, and opened and traced out for the most distant times the safe way of a science.

"The examples of mathematics and natural science, which by one revolution have become what they are now, seem to me sufficiently remarkable to induce us to consider . . . as sciences of reason, of imitating them."[11]

Just as Copernicus disproved the medieval worldview by showing that Earth was not the center of the universe, Kant undid the mechanistic paradigm by proving that human consciousness, or reason, was not the center of all reality. The unknowable thing-in-itself, the noumena, is the center of Kant's universe. Copernicus disproved the geocentric universe, and Kant the egocentric universe. Geo-ego—both words have the same letters, so it's easy to remember.

THE NOUMENA-CENTERED UNIVERSE

Rather than the ego-centered anthropocentric worldview of the medieval or the mechanistic worldview, Kant proposes a *noumena-centered universe*. That is, our first thought must be that reality is fundamentally unknowable, and this must be accounted for in each subsequent thought. As scientists are so quick to confuse their theories for reality, this is critical. For without the check and balance of the noumena, they may try to make reality fit the theory, rather than the other way around!

Kant's noumena, moreover, obliterated the anthropocentric foundation of the medieval and mechanistic worldviews, which simply assumes that human beings are the center of reality (giving our species the right to exploit nature without respect or prudence). Like egocentric five-year-olds who think the whole world revolves around them, it is merely appearance that nature seems placed here for humanity to exploit without limit. The actual reality is something far different, said the German philosopher. Kant showed that the ego-centered anthropocentric universe is merely the illusion of appearance, similar to the illusion of the geocentric universe. But what does that mean to us in the everyday world? It made possible the new scientific revolution, but what does it mean in personal terms, in how we live our lives?

The noumena is a "check and balance" to the absolutist tendencies of science and society. It is the dogma of absolutism that also supports the various chauvinist dreams of history that each culture creates for itself, leading to history's endless series of wars. Understanding what Kant discovered means taking a personal step toward deprogramming yourself, toward awakening, toward enlightenment.

Realizing the noumena changes your entire outlook: it makes you a different person. You might see old absolutist concepts that ruled your mind for what they are. You suddenly know that everything has an inner essence, an essence that could be totally different from the appearance of the object. Your personal thinking may still be limited in many ways, as you are imprinted with and brought up on many absolutisms. Yet understanding the noumena makes it possible to free yourself from these dogmas and untruths. This is the crucial first step you must take.

THE TWO TYPES OF AWARENESS

Kant said there were two kinds of people: *homo phenomenon,* or natural man, and *homo noumenon,* or moral man (Kant of course uses the word *man* for humanity, so these terms include women).[12] *Homo phenomenon* is the initial, natural state of human awareness. He described natural man as being surrounded by a series of mental concentric circles—circles that increasingly limit thought. *Homo phenomenon,* said Kant, thinks first of his family, his job, his house, his horse, and then in a vague sense his neighborhood, town, and nation—and in almost no sense does he consider the world.

Homo phenomenon, spellbound by these circles of phenomena, spends most of the time dealing with and thinking about his immediate surroundings. Easily swayed by appearance, he is prey to the allure of consumerism and the drumbeat of nationalism and war. Thinking for himself is almost never attempted. Nor does *homo phenomenon* want to change. For example, *homo phenomenon* may give lip service to whole concepts, the ideas of God and morality. Yet there is an aversion to really exploring the true nature of anything. It's part of the spell spun by the material things around us, like a circle of enchantment.

Homo phenomenon cannot get past the circles and so does not know the wholes that exist in any but a vague sense.

In contrast, *homo noumenon* has broken through the magic circle of material things and lives a life full of meaning and purpose. Rather than looking from the "inside-out," he thinks first of the whole and works his way back to the level of the individual. He not only understands the concept of noumena; he can experience the unknowability of it all in a mystical state, a union with the whole. According to Kant, *homo noumenon* is enlightened, realizing that although a person can never know the whole directly, he can feel it through his heart. The head is, after all, little use in the case of the unknowable. Becoming aware of the noumena is, therefore, the foundation of the broader awareness of the organic worldview.

Kant's categorization of people into *homo phenomenon* or *homo noumenon* is not elitist, for the broader, progressive awareness is the truly inclusive mind-set, one that wants to lift up the rest of the world through regenerative science and grassroots community development. It is the conservatives and mechanists that are elitist and wish to maintain the status quo and keep control in the hands of the few. That is why they often accuse those who challenge them of being elitist. It's an old political trick: accuse your opponents of being what you are yourself.

A natural step for the next paradigm to take is to possess the global view of the *homo noumenon*. Yet how does one become so enlightened? How can we, according to the old professor, develop our own understanding and reach this state?

INDIVIDUAL ENLIGHTENMENT EXPLAINED

In one essay ("What Is Enlightenment?"), Kant explains that the way from the state of *homo phenomenon* outward to the enlightened state of *homo noumenon* means daring to think for yourself, to be brave enough to throw off all the programming, all the education and tutelage of institutions.

Enlightenment is man's release from his self-incurred tutelage. Tutelage is man's inability to make use of his understanding without direction from another. Self-incurred is this tutelage when its cause lies not in lack of reason but in lack of resolution and courage to use it without direction from another. Sapere aude! "Have courage to use your own reason!"—that is the motto of enlightenment.

Laziness and cowardice are the reasons why so great a portion of mankind, after nature has long since discharged them from external direction nevertheless remains under lifelong tutelage, and why it is so easy for others to set themselves up as their guardians. It is so easy not to be of age. If I have a book that understands for me, a pastor who has a conscience for me, a physician who decides my diet, and so forth, I need not trouble myself. I need not think, if I can only pay—others will easily undertake the irksome work for me.

That the step to competence is held to be very dangerous by the far greater portion of mankind (and by the entire fair sex)—quite apart from its being arduous is seen to by those guardians who have so kindly assumed superintendence over them. After the guardians have first made their domestic cattle dumb and have made sure that these placid creatures will not dare take a single step without the harness of the cart to which they are tethered, the guardians then show them the danger which threatens if they try to go alone. Actually, however, this danger is not so great, for by falling a few times they would finally learn to walk alone. But an example of this failure makes them timid and ordinarily frightens them away from all further trials.

For any single individual to work himself out of the life under tutelage which has become almost his nature is very difficult. He has come to be fond of his state, and he is for the present really incapable of making use of his reason, for no one has ever let him try it out. Statutes and formulas, those mechanical tools of the rational employment or rather misemployment of his natural gifts, are the fetters of an everlasting tutelage. Whoever throws them off makes only an uncertain leap over the narrowest ditch because he is not accustomed to that kind of free motion. Therefore, there are few who have succeeded by their own exercise of mind both in freeing themselves from incompetence and in achieving a steady pace.

But that the public should enlighten itself is more possible, indeed, if only freedom is granted, enlightenment is almost sure to follow.... For this enlightenment ... nothing is required but freedom ... the freedom to make public use of one's reason at every point.[13]

Most people think of enlightenment as something one would find in a far-off Asiatic land from a guru or Zen master. Yet to be enlightened, said Kant, we must actually learn to think for ourselves. We must free ourselves from the grip of the shallow absolutisms that pass for science, philosophy, and politics. Then, with your own deep look at the world, you can walk on your own and make your own map of the world. That is true awakening.

Kant himself had his own map. He said that every living thing has a purpose—that trying to understand the purpose must be the first goal when studying something biological. What does being an enlightened *homo noumenon* mean, then, in terms of purpose? What is our individual purpose? He explains that, in the end, our individual purpose is tied to our collective purpose. Rather than wasting time and precious resources in the unconscious pursuit of war, we are here to attain a state of world peace, so that humanity can develop to its full potential. Making the change from *homo phenomenon* to *homo noumenon* is key to making it all happen.

Kant next told how to nurture that world change.

PEACE THROUGHOUT HISTORY

Looking at the evolution of thought from the early myths of the Greeks to the scientific view, in which democracy and the laws of nature replaced the power of kings and the Greek gods and goddesses, as well as more recent advances of his own time, Kant concludes that human history is primarily the movement from *homo phenomenon* to *homo noumenon*.[14] In other words, our awareness as a species is increasing . . . we are becoming more moral, more evolved, as time goes on. The ultimate end of this evolution of morality, the obvious purpose driving this movement is world peace, decided Kant, the end of war.

> [O]ur world leaders at present have no money left over for public education and for anything that concerns what is best in the world, since all they have is already committed to future wars. . . . Through wasting the powers of the commonwealths in armaments to be used against each other, through devastation brought on by war, and even more by the necessity of holding themselves in constant readiness for war, they stunt the full development of human nature. . . . In the end, war itself will be seen as not only so artificial, in outcome so uncertain for both sides, in aftereffects so painful in the form of an ever-growing war debt that cannot be met, that it will be regarded as a most dubious undertaking.[15]

Kant's ultimate plea to make world peace come about was for philosophers to write a universal history to teach and guide future generations. Rather than the usual history of wars and politics, of "dubious undertakings" with no overarching "purpose" in mind and so no real meaning, this was to be a history of all the good that has been done. The universal history would describe the cultural and scientific advances of humanity that reveal the true purpose of it all—to finally reach a democratic and lasting world peace. This universal history or global history, taught as the center of a new public education, is what will break the final enchantment of *homo phenomenon* while at the same time empower *homo noumenon* and allow him to have not only the courage but also the tools to end all war.

Kant complained mightily about the damage done by histories that glorify war and treat it as the central fact of human existence: "[W]hat is the good of esteeming the realm of brute nature and of recommending that we contemplate it. . . . If we are forced to turn our eyes from it in disgust, doubting that we can ever find a perfectly rational purpose in it and hoping for that only in another world?"[16]

Here he is not saying that these histories of war and politics should be censored but rather that the central history taught to the young should be a universal history written by a philosopher. Endless stories of war and national chauvinism only perpetuate the culture of war; what we must do is study the stories that nurture the healing culture of peace. Kant goes on to warn that the need for a universal history designed to reveal the grand purpose of humanity would only grow; that without it, people would be prey to war, expansionism, and the various absolutisms of immoral dictators and the greedy. So no universal history, no world peace; without universal history, the nations will never be able to find their way out of the dilemma of war. Without the universal history, the young are not taught the purpose of humanity and so most graduate purposeless—nihilistic in one way or another. You cannot learn purpose from the history of greed and evil. Kant worried that the lack of social purpose in such histories would lead to more centuries of war:

"Otherwise the notorious complexity of a history of our time must naturally lead to serious doubts as to how our descendants will begin to grasp the burden of the history we shall leave to them after a few centuries. . . . So long as states waste their forces in vain and violent self-expansion, and thereby constantly thwart the slow efforts to improve the minds of their citizens . . . nothing in the way of a moral order is to be expected."[17]

Finally, reasons Kant, the universal history needs to be learned by generations of young people to prepare the ground for a lasting peace. It is difficult for nations to give up their pride and take the chance for peace; it will be a long and arduous process. Yet after being educated in the universal history and understanding what the most meaningful purpose of the human race and thus each individual is, we will have an environment in which world peace can finally be established. Then, with true peace, humanity will at last have the time and the resources for progress, art, culture, and science—what the magnificent mind of our species should be doing, instead of thinking up new ways to kill ourselves. With the universal history, "the human race finally achieves the condition in which all the seeds planted in it by Nature can fully develop and in which the destiny of the race can be fulfilled here on Earth."[18]

The universal history is thus required to set the mental environment for peace. A new legal environment, however, is also necessary. So Kant—never one to let others do the dirty work for him—adds to his résumé and becomes the first thinker to systematically propose the legal framework for a league of nations, the first world government.

THE NEED FOR A WORLD GOVERNMENT

In his essay "Perpetual Peace," Kant thinks out the structure that must exist in order for war to be abolished. He knew that world peace first required a federation of free states, pledged not to attack each other in the usual wars of aggression. These were the definitive articles, the principles that must guide law and government, national and international:

1. The civil constitution of every state should be republican.
2. The law of nations shall be *founded on* a federation of free states.
3. The law of world citizenship shall be limited to conditions of universal hospitality.

Under the third principle, universal hospitality, he outlines the modern attitude of tolerance on a global level of other people's culture and religion. World citizens, especially diplomats, must be free to travel and not be persecuted because of their race or land of origin—a new idea in his time. In the following quote, Kant explains how to ultimately attain perpetual peace:

The state of peace among men living side by side is not the natural state; the natural state is one of war. This does not always mean open hostilities, but at least an unceasing threat of war. A state of peace, therefore, must be *established*, for in order to be secured against hostility it is not sufficient that hostilities simply be not committed; and, unless this security is pledged to each by his neighbor (a thing that can occur only in a civil state), each may treat his neighbor, from whom he demands this security, as an enemy.[19]

Kant's ideal of a state of peace burgeoned into a powerful pacifist movement, especially after the end of the Napoleonic Wars in 1815. Apart from the short war between France and Germany in 1870, the new diplomacy and the desire to end war helped Europe enjoy a century of peace—until 1914. Kant's concept evolved into the government-supported peace movement of the late nineteenth century and the *establishment of the League of Nations after World War I.* There is no debate as to Kant's importance in inspiring the world to strive for an end to war. His influence on the preamble to the UN Charter in 1945 is often noted as are his other ideas, which have been put into action by the United Nations in a myriad of different ways.*

KANT HELPS TO INSPIRE THE ROMANTIC MOVEMENT

Kant's discovery of the significance of thought, feeling, and the power of the idea elevated the intellectual, literary, and artistic world of his day—and led directly to the Romantic era and the idealist movement in philosophy. Kant opened up the inner world of the mind, where artists and thinkers can be free in their imagination, no matter how little freedom their political system may allow. Enlightenment to Kant meant freeing yourself from the limited mind-set of books and from the institutions of the world.

He said a person learning how to think for him- or herself is like learning how to walk. After falling down a few times, the individual finally learns to walk alone. He implored everyone to think for him- or

---

*Concepts from Kant's essay on peace can be found in the general UN Charter (1945), as well as the UN Declaration of Human Rights (1948).

herself—to go beyond the mechanical laws of nature and feel what's in the heart. All you can know with certainty is what is in your heart.

The power of getting in touch with your feelings was clear to the Romantic. Nature and love's beauty could now be celebrated in a much deeper way. The creative imagination set free, poets and artists celebrated the newfound freedom of being able to express pure feeling in their work. Meanwhile, philosophers and reformers took up the mantle of Kant's idealism and the task of creating a society based on great ideas. Kant was thus a tremendous influence in his own day and in the decades that immediately followed, as the German idealist movement took shape.

KANT'S RELEVANCE TO TODAY

How is all this philosophy relevant to us today? Although philosophers don't do anything except plant ideas, those ideas may soon take on a life of their own among the general culture. The concept of the two types of awareness is critical to understanding ourselves, crucial to seeing where we are today. Kant's broader, transcendental awareness was adopted by the Romantic artists and the countercultures that followed. His expanded worldview was also reflected in various ways by the science, technology, medicine, art, and writings of organic-paradigm thinkers, as well as the lifestyles of the later bohemians, beats, and hippies.

As mentioned earlier, the organic countercultures in the nineteenth and twentieth centuries are a clear parallel to the more secular lifestyles lived by the humanists and early mechanistic scientists two centuries earlier. In both the mechanistic shift and the Organic Shift, a small creative minority slowly developed the new thinking until a comprehensive scientific and social revolution was underway, determined to eventually replace the old establishment, the old paradigm. The emergence of Kant's idea of *homo noumenon,* or transcendental awareness, as a mass movement helps us understand the 1960s and the changes that led up to that momentous decade.

That was the real generation gap back in the 1960s. Most of the parents of the baby boomers were fundamentally a *homo phenonmenon* generation, while almost all the boomers lived with the much broader awareness. Although not totally aware of it, the boomers were thinking in a completely different way from all previous generations. According to Kant, they were thinking from the outside in, rather than from the inside out. It was the first mass emergence of the organic paradigm and *homo noumenon,* a huge leap in the complexity of worldviews.

Even though he lived over two hundred years ago, Kant is thus *very* relevant to understanding our own time. From transcendental philosophy and idealism to world peace, *homo noumenon* and the exploration of feeling and the inner world, Kant's ideas have transformed our understanding in an astonishing number of ways. Most of all, his discovery of the noumena revealed the foundation of the modern organic worldview, leading directly to a whole new scientific revolution, one that quantified life, culture, and consciousness itself—as Kant had predicted.

For too long after their introduction, many of Kant's ideas were delayed, typical for a truly innovative thinker. The basic concepts of Copernicus had suffered the same fate. Kant's perpetual peace and the balancing of scientific technology are *still* opposed by various absolutisms.

As in the Copernican Revolution, the new worldview nevertheless began a long, two-hundred-year struggle to wrest away control of science and society from the mechanists. Others got involved in the struggle, as we shall soon see. Yet so much of the new paradigm started within the mind of one person. There can be little argument that Immanuel Kant is indeed the organic Copernicus, the conception point of a whole new scientific revolution.

Just as Kant can be held to be the organic Copernicus, to maintain the parallel we have established between the Copernican revolution and the latter one of the 1700s, we must determine the identity of the latter-day equivalents to Francis Bacon and Galileo in the Organic Shift—for these two created the scientific method. The significance of the invention of the method of induction and deduction, of hypothesis and experiment, cannot be emphasized enough. Sir Francis Bacon and

Galileo provided a system for theorizing through which all of time and space could be examined and measured. Without the scientific method, the Industrial Revolution would not have taken place, nor would there have been the rise of the mechanistic priesthood, which rules today.

Who created a similar "organic" scientific method, a method that could crack the deeper mysteries of nature and the world beyond the visible? The next organic pioneer was more than a philosopher or researcher, he was also a writer and poet famous throughout the Western world. The combination of deep contemplation and scientific discovery were his greatest contributions, and this is what we need today: the wide use of an organic scientific method from which to fashion a new world and a new future.

Johann von Goethe

3

GOETHE

The Organic Galileo

Goethe was to me the founder of a law of organics. . . . When I looked back to Galileo in the history of modern spiritual life, I was forced to remark how he, by the shaping of ideas from the inorganic, had given to the new natural science its present form. What he had introduced for the inorganic Goethe had striven to attain for the organic. Goethe became for me the Galileo of the organic.

RUDOLF STEINER,
THE STORY OF MY LIFE

WHY GOETHE IS THE KEY
TO A BETTER FUTURE

As Francis Bacon and Galileo stood on the shoulders of Copernicus and earlier scientists, so Goethe stood on the shoulders of Immanuel Kant. Like Kant, Johann Wolfgang von Goethe was a supergenius, named in one study as the highest IQ tabulated.*

Known mostly for his drama *Faust* and his solidification of the German language, Goethe was much more than that. As a poet and an artist of nature, Goethe went beyond others and strove to see the essence of the scene before him. If he was studying botany or writing a poem about a flower, he would envision not only the plant but the archetype of that plant; he could understand the nature of it in a deep new way by immersing himself and contemplating the organism under study from within. He "became the plant" by comprehending its essence through his own human essence.

By relating what you feel of your human essence, says Goethe, you can see the essence of the phenomena before you; you can envision the archetype. To do that, however, you must first immerse yourself in the object of your study—you must lose yourself, merge the two essences together, *and try to view the world as the plant sees it.* This was a powerful new scientific *and* artistic awareness, and it became a major factor in the era that immediately follows. Goethe has been hailed as the primary inspiration for the German Romantics.

Although mechanists disqualify him as a Romantic writer who merely dabbled in the sciences, we shall see that he was, in fact, the organic Galileo. Goethe not only envisioned the seminal law of organics, but he also, along the way, invented an entirely new scientific method, an organic method.

THE ORGANIC WAY TO THEORIZE: THE PARTICIPATORY METHOD

Goethe's scientific method is of utmost importance for us to understand and use, for it can be used in many ways to transcend the mecha-

---

*Average range of intelligence quotient (IQ) is 85 to 115 (Stanford-Binet scale). One percent have an IQ of 135 or over. Psychologist Dr. Catherine Morris Cox, and Dr. Lewis M. Terman, Dr. Florence L. Goodenaugh, and Dr. Kate Gordon, published a study in 1926 of the most eminent men and women between 1450 and 1850 and projected their IQ based on their writings and art before age seventeen. Goethe came in first with an IQ of 205; Kant, by the way, came in at 175.

nistic Dilemma. His "participatory method,"[1] once understood and used by the institutions of the world, has the potential to transform global science and complete the paradigm shift. It is a way to transcend the multiple deadends of so many branches of learning and put science and technology on a forward track again. Today's struggle between mechanistic and organic science would then change dramatically, as the mechanists will be forced into a long retreat, forever discredited and disqualified as merely objective science. That's how important Goethe is. He is literally the hidden key to a better future, the way to defeat the mechanists forever.

A core problem of mechanistic science can be seen in the writings of Francis Bacon. In the early 1600s, Bacon described his new scientific method, explaining the need to test theories with mathematics and a series of experiments. He was not only a scientist, he also served as attorney general for King James I of England and participated in many trials of witches, which often wrested satanic secrets from the unfortunate women by means of mechanical torture. The language of his other occupation seems to have carried over into Bacon's writings on science, for he declared that nature should be "put in constraint," and the purpose of all the experimenting was to "torture nature's secrets from her."[2] Nature must be "hounded in her wanderings," "bound into service," and become a "slave" to man. Before Bacon, the goal of science was primarily to gain wisdom by understanding and living in harmony with nature. Now science was to dominate, torture, and then control nature for the good of humanity.

From the beginning, Bacon's direction and method, although a basic advance at the time, set the pursuit of knowledge off down a strictly mechanistic path. In the 1630s and 1640s, Galileo completed Bacon's initial realization of the scientific method, allowing him to make his basic discoveries on astronomy, motion, and gravity. Newton in 1687 then provided all the mathematical formulae for the new physics, equations that can also be used to quantify most of time and space. Galileo's work on the scientific method and astronomy thus laid the foundation of the clockwork universe.

What was so different between Goethe's method and the traditional approach of Bacon, Galileo, and the mechanists? In short, with the mechanist method one basically only uses analysis. It is the science of quantity. A phenomenon is taken apart to study its mechanics. Then a series of questions to be answered is developed—and a theory based on those answers is created. Experiments testing the theory are then carried out, involving still more data collection and measurement. This analysis is then presented as supporting or disproving the theory. Yet such reductionism allows all sorts of mechanistic dogma to be "proved."

In contrast, as Henri Bortoft, a researcher and writer on physics and the philosophy of science, explained, Goethe's way is the science of quality.[3] While mechanists assume there is only one kind of seeing, observing the phenomena with the senses, Goethe showed how there was also "imaginative seeing." As philosopher of science Norwood Hanson explained, "There is more to seeing than meets the eye."[4] This calls to mind Einstein's famous quote: "Imagination is more important than knowledge." In addition to the sensory seeing, Goethe implored scientists to also use their inner eye, their imagination. Before theorizing on merely the sensory data, one should seek out the meaning of the phenomenon and use the mind to *feel* the archetype behind the thing being studied. That is when true insight into the subject will emerge.

This is not to say that quantity, measurement, and data collection are not performed in the participatory method. The organic method of Goethe goes back and forth between analysis and holistic synthesis, proposing a theory only after immersion into the thing being studied. Goethe said true science forces the theorist to become one with his or her subject, to discern the subject's purpose. After studying the phenomena in all its detail, the theorist must immerse him- or herself in the whole and try to reach no conclusions at all—let the phenomena tell the observer what the theory should be.

Goethe explained, "Seek nothing behind the phenomena, they themselves are the theory."[5] In order to avoid the mind's inclination to make snap judgments based on mere perceptions, Goethe wants us to slow the entire process down to a crawl, for the observer to mentally become the observation, to consciously participate in whatever nature is showing us. In this way, intuition is allowed to rise to the surface. That is not to say that you lose yourself to the thing being studied. You are still a conscious observer; you are only mentally looking at the whole, by understanding its essence through your own personal essence and wholeness.

Yet Goethe says, even with all that, it's *still* too soon to theorize. Think about it all once more; immerse yourself again and again. When you feel you are absolutely sure of your intuition, *then* you can finally, safely ask nature the right set of questions and create your theory and experiments. This was the opposite of the mechanistic method, which "tortures nature," researching and analyzing just the parts of problems as fast as it possibly can, and then comes the quick theory dump and experiment check. Goethe saw this as *unevolved science.* As William Barrett showed in *The Illusion of Technique,* the scientist becomes enchanted with the technique and the data itself, never seeing the forest for the trees. The time was never taken to *first ask the right question,* the result being you always end up with the wrong or incomplete answer. As in the parable of the seven blind men and the elephant, where each touched a different part of the creature and each instantly came up with a different theory of what it was, machine-model science cannot see the big picture, and so it does not ask the right set of questions. It is the right question that leads to the underlying truth, the all-encompassing theory that ties together a slew of previously unrelated facts. It is only through the participatory method that the true nature of life and humanity can be realized.

The Swiss ecologist Adolf Portmann (1897–1982) studied Goethe's method and found a very useful analogy to explain his new approach. Imagine nature as a theatrical play. The mechanistic method looks behind the scenes, studying the way the lighting works and how it changes the colors on the stage or seeing how the curtains close and hide the set change. The focus is on the mechanical nature of the parts of the play being presented to find out how nature works.

In contrast to this reductionism, Goethe essentially said understanding the parts is, of course, necessary; but we should also sit in the audience and see the whole of the play being performed. Only then can we grasp the overall meaning of what we are looking at. It's like trying to figure out who committed the murder in a murder mystery play by seeing how the lighting works. The mechanists never sit in the audience, and so they are at a loss to describe the whole of nature's mysteries and riddles. You must let yourself become absorbed in the plot and submerge your behind-the-scenes knowledge so you can focus on the mystery at hand and then find an answer to the mystery.

Goethe realized that nature loved to pose riddles; what may seem to be the answer is something else entirely, which may in fact be hiding the truth. Going back to the analogy of the theater, the mechanists don't even know it is a mystery that is being performed for them, as they are not sitting in the audience and watching the show. Goethe not only knows nature is a mystery drama, he also knows that the hidden truth would not be revealed in the beginning of the play or even in the middle. A scientist must sit and watch *all* of nature's show, for it is only at the end of the piece that the secret is revealed. One must therefore immerse oneself in the whole and abstain from reaching any conclusions at all. Only after a long period of immersion in the subject can one even *begin* to properly envision the underlying, controlling archetype, the secret interconnection that solves the mystery.

In a murder mystery, the writer and director may trick the audience and have them suspect a slew of possible culprits. Even halfway through the drama, the viewer may think he or she knows the ending yet is often surprised to discover who really did it. Goethe says sit back, enjoy the show, and then when it's all over, cogitate on it some more, and *then, if your intuition feels right,* theorize on the underlying archetype.

Again, the real science lies in not what answers you come up with but simply to figure out *what the right questions are.* What should we be asking nature to reveal? What are the questions whose answers will lead us to a better future? If one is but patient and waits for the mystery to reveal itself, one can discover many things—things that will always be beyond the limits of the mechanistic paradigm, so obsessed with analysis. Goethe saw analysis and holistic synthesis like breathing in and out—analysis and synthesis should alternate.

Without ever writing it down per se, Goethe had, in effect, created a new scientific method through his poetry, plays, essays, and letters. In particular, his poem "The Metamorphosis of Plants" says directly that if we just observe the way nature works—rather than trying to trick or conquer her—she will eventually give up her most precious secrets. Goethe's thoughts on science can most clearly be found in the following works:

"The Content Prefaced"
"The Enterprise Justified"
"The Formative Impulse"
"A Fortunate Encounter"
"History of My Botanical Studies"
"Judgment through Intuitive Perception"
"The Metamorphosis of Plants"
"The Purpose Set Forth"
"Reconsideration and Resignation"

Just as Bacon's scientific method of the seventeenth century was the foundation of our current society, Goethe's holistic process could be the bedrock upon which a new and workable future can be built. We must come to know all of it.

THE FOUR TYPES OF SCIENTISTS

Fleshing out his view of the new science, Goethe believed there were four types of scientists who study botany, categorized by the kinds of questions that they pose. Mechanists occupy the bottom two rungs of this intellectual hierarchy, which in fact can be applied to all of science. There are, moving up the ladder, users, knowers, beholders, and encompassers:

1. Users. These innovators only ask enough to make practical use out of the thing or organism they are studying. Having gained a level of expertise and knowledge about a subject, they pursue it no further.

2. Knowers. This next level of scientist wants to know more of the details, studying the parts within the parts. They only work with the users and with what is already known about a subject. No use of imagination is attempted in trying to see the whole of a problem—as their goal is simply to increase the number of practical applications for the users to take advantage of.

3. Beholders. Beholders will discount the scientific value of the human imagination, yet seek to behold the whole of a thing or organism. To do this, they must use the imagination, many times in a subconscious way. Beholders are far more productive than knowers or users in making scientific breakthroughs.

4. Encompassers. The highest level of scientist is the encompasser. Understanding how nature poses riddles and how one must employ imagination and intuition to see the whole, these thinkers follow the participatory method and immerse themselves in the whole of a phenomena. They become the phenomena; they become the plant, animal, or culture they are studying and ignore their own opinions, knowledge, and past thinking about the subject. Once encompassers become what they are studying, they may be intuitive enough to uncover the archetype, the essential question that has been hidden for so long. This is the question that leads to the underlying truth, the all-encompassing theory that ties together so many previously unrelated facts.

Obviously, the world today is still being run by the users and knowers, a hazardous situation, as a little knowledge—especially in the hands of governments and industrial scientists—is a very dangerous thing. By learning the holistic participatory method, however, science and society can be brought back into harmony and balance as they will then understand what the right questions are to ask of nature—and ourselves. This would be the first step to the creation of a better world in the transformational phase.

NOUMENA OR BEING?

Remember that Kant began a new Copernican Revolution when he tried to answer the question, "What can we know?" Just as Galileo stood on the shoulders of Copernicus, Goethe took Kant's noumena and developed it into a practical tool to guide science. Kant had said the noumena was the unknowable whole, that reason could never know it directly. Goethe agreed with Kant's basic concept of phenomena, which said that the whole of reality was not directly knowable with the senses.

Yet Goethe realized that since *human beings have within themselves the same essence that makes up the universal being and nature, human intuition has the ability to envision that essence in its mind.* You can know through your own being that the universe is here, that it too is in a state of *beingness*. If it is found that this beingness explains and relates to a wide range of phenomena, it is more than a mere idea or concept, it is for all practical purposes an objective reality not a subjective one—as Kant would say it is. Goethe had in fact taken scientific philosophy and the scientific method one step further, *making holistic thinking a practical tool for the real world.*

He immediately used it to make one significant organic discovery after another.

THE GREAT UNFOLDING

Goethe used his participatory method to decipher several mysteries of nature and evolution. For example, Kant had said that in order to understand the biological, we must assume there is an underlying purpose to nature and to humanity. It is around this purpose that we can *feel out* our theories.

Looking consciously and intuitively for nature's teleological purpose, Goethe was struck by a great realization: all of evolution is an unfolding, a journey by nature to explore all possibilities and forms of beauty and function. Humanity, by extension, is the crown of the unfolding of nature, so that nature may enjoy the pleasure of love or a beautiful sunset and then write soaring poetry about itself. Humanity, in short, is nature enjoying itself. The great unfolding, moreover, means we live in a nature-centered world, not a human-centered one. Despite Goethe seeing humanity as the crown of creation, we are here for nature to enjoy itself, rather than humanity having the right, as in anthropocentric worldviews, to exploit nature to the end. Nature and humanity are one, not two opposing forces.

Goethe's writings profoundly influenced many thinkers and artists, as they developed the organic worldview over the next two centuries. Besides shaping the thought of Jung, Dilthey, Haeckel, and Steiner, he directly influenced the naturalists of his time on the matter of evolution and human origins. His discovery of the type and his comparison of embryo development to evolution blazed a path in naturalism that Charles Darwin would later take in his theory of natural selection. Darwin also credited Goethe for discovering that the intermaxillary bone—thought to only exist in animals—is also found in humans. This proved that humans evolved out of animals and helped set Darwin off on his journey of thought about the origin of the human species.

Goethe also directly influenced his friend Johann Gottfried von Herder (1744–1803), a student of Kant and a philosopher of history. Herder developed and popularized the idea that history evolves, that one era transforms itself into the next in a continual process of change. Correspondence between the two thinkers reveals it was Goethe who led him to focus on the concept of evolution. Herder outlined the concept of historical and cultural evolution first in *Auch eine Philosophie der Gerschichte zur Bildung der Menschheit* (1774)—yet another philosophy of history for the education of mankind.

Goethe was shaped by the books and science of his time and, like Kant, was further enlightened by reading recently translated epics and Buddhist writings from India. Taking holistic Eastern thought and synthesizing it with Kant and mechanistic science, he discovered several fundamental principles underlying evolution and all organic life.

THE TYPE

In trying to understand how plants grew from seed to flower, Goethe made his first major organic discovery: the *type*. Up until the discovery of the type, botanists had solely classified and detailed the various parts of plants and animals, not fully realizing the many connections between seemingly unrelated species.

Seeing many similarities among all plants in their metamorphosis from the seed, Goethe suddenly understood the oneness of all plants and envisioned in his mind the archetypal plant. Using this as an organizing image, he was able to see where each species of plant had evolved its own unique features, its own type, yet still retained most of the attributes of the archetypal plant. With this one universal discovery, this scientist/poet had taken naturalism into the modern era. The type became the foundation for understanding and classifying the many species of the plant and animal world.

He knew he had come upon a great advance and wrote in Naples on May 17, 1787, that "[it] will be possible to apply the same law to all living things."[6] This was a scientific breakthrough on the highest level. Just as Galileo had shown that a law of gravity forces the smaller moons of Jupiter to circle the larger planet—and thus Earth must likewise revolve around the sun—Goethe had discovered a universal scientific principle, the law of organics.

Within a short few years, his holistic study of comparative anatomy led him to another major discovery. With the help of other scientists, he was able to prove that the embryos of all animal species follow the same pattern in their structural development. Organs develop from the inside out. For example, the end of the embryonic spinal column develops into the brain and the sensory organs. What was first there to sense the outside world solely through crude feelings of pain and pleasure, unfolded and—striving to sense more of the outside world—created the brain, eyes, ears, nose, and taste receptors.

Goethe's inside-out theory also explained the march of evolution. First primitive creatures with no sense organs evolved, then these developed into higher levels of organisms with far greater abilities. Human beings are the highest form of life, to Goethe, the crown of creation, a reflection of evolution's drive to develop a being with all its abilities unfolded. Later on, these thoughts from the poet/playwright/novelist/ scientist helped inspire future discoveries about the evolution of consciousness.

Goethe saw all of this change unfolding over the course of millions of years, in the same way an embryo develops from a few cells to a fetus in a matter of months. This biological comparison was very instructive to early thought on evolution. The field of naturalism and Darwin, himself, soon followed his lead and fleshed out the details of his law of unfolding.

HUMANITY IS NATURE ENJOYING ITSELF

One of these principles explained one of the key purposes of the human race. As mentioned earlier, to Goethe, humanity is nature trying to understand itself, to enjoy itself. The human eyeball, for example, was created by nature to take in the wonder of light. Light was very meaningful to him as it was the visual reflection of the beingness of the universe. How light breaks down into the various colors of the spectrum was, for Goethe, both a physical and a spiritual experience.

Each person is moreover a reflection of the universal being of reality. Humanity was the flower, the blossom of all evolution. From his essay on Winckelmann:

"When the healthy nature of man functions as a totality, when he feels himself in the world as in a vast, beautiful, worthy, and valued whole, when a harmonious sense of well-being affords him pure and free delight—then the universe, if it were capable of sensation, would exult at having reached its own goal and would marvel at the culmination of its own development and being."[7]

Going further, Goethe believed that humanity as a whole was evolving and had potential far beyond those suspected by mechanistic science and its reductionism. Each of us has the potential to mentally transform ourselves and evolve into the next stage of being human, of creating a better world. That is our archetype, the purpose of our unfolding. The Organic Shift would continue to explore this concept, which grew in importance over the centuries to come.

NATURE'S OPEN SECRETS

For all these reasons and many more, Rudolf Steiner, in his comprehensive review of Goethe's work, declared Goethe to be the organic Galileo. As the founder of the law of organics and the creator of a whole new scientific method, he gave the new paradigm the tools with which it could make the leap to a whole new understanding of nature and humanity itself.[8]

This was a crucial step for the new worldview, as important as Galileo's astronomical discoveries were to the mechanistic shift. Once Galileo had proved that Earth does indeed *move,* the eventual end of the medieval worldview's oppressive control of thought and society was inevitable. To end mechanistic control of society and create a "healing culture," as educator John Michael Barnes talks about, we therefore need to understand how we can apply the participatory scientific method to the problems of our own time.[9]

Goethe said nature had "open secrets," secrets that were hiding in plain sight. If we just observe the entire play until the last scene and realize the meaning, we can understand these hidden answers to the mystery of nature. Encompassers, using the new science, can thus discover many things about the world, create the healing culture, and transcend the seemingly insoluble problems of the mechanistic dilemma.

Since Goethe, many thinkers after him have done just that,

founding such sciences as ecology, cultural anthropology, Jungian psychology, and organic agriculture, discovering many of nature's open secrets. When put together in a future regeneration revolution—equal in scope to the Industrial Revolution—these discoveries and many more can transform society, away from our dangerous, mechanistic science to a healing culture, creating organic solutions that work *with* nature instead of against it.

The book you are now reading is part of that process. Beyond the mystery of nature's open secrets, what is the mystery of history? What are history's open secrets? What can be discovered if we just stop focusing on the parts and look at the whole of history? What has been hidden from us by the incessant reductionism of the mechanist historians and the usual focus on the history of war? What has been right under our noses the entire time? What is the key to the mystery of history?

As Goethe said in his dying words: "More light!"

Werner Heisenberg

4

THE ORGANIC SHIFT
IN SCIENCE AND
TECHNOLOGY

The ecocentric argument is grounded in the belief that, compared to the undoubted importance of the human part, the whole ecosphere is even more significant and consequential: more inclusive, more complex, more integrated, more creative, more beautiful, more mysterious, and older than time. The "environment" that anthropocentrism misperceives as materials designed to be used exclusively by humans, to serve the needs of humanity, is in the profoundest sense humanity's source and support: its ingenious, inventive life-giving matrix . . . in the ecocentric view people are inseparable from the inorganic/organic nature that encapsulates them. They are particles and waves, body and spirit, in the context of Earth's ambient energy.

STAN J. ROWE

87

THE SECOND COPERNICAN
REVOLUTION

Just as Kant was in many ways the new paradigm equal of Copernicus, as Goethe was of Galileo, so the development of the new science in the Organic Shift equaled the scope and significance of the Copernican Revolution. Although the science of the new paradigm studied life, evolution, and consciousness, rather than measuring time and space, it was a major paradigm shift—a true scientific revolution. Starting slowly at first, with each succeeding decade the organic worldview had an increasing impact on the sciences. At the same time, the new thinking began to influence art, politics, and people's personal lifestyles.

By studying the work of many different individuals and schools of thought over the last two hundred years, we can see which of them saw the world through the organic model or through the mechanistic. This distinction gives a more accurate history of science and technology over the last two centuries, clearly showing the shift to the new paradigm. At the same time, by knowing the stories of each of the organic pioneers, we shall indeed learn a lot about the world—things we were never taught in school. We shall let the original thinkers explain how the new worldview and its values are different from the old. If the mechanistic dilemma is to be transcended, it is essential that we learn the teachings of these new-paradigm pioneers.

THINKING WITH THE NEW ANALOGY

The mechanistic worldview has been economically powered by mining coal and iron and cutting timber ever since its inception. This worldview continues to depend on petroleum, metals, uranium, and old-growth timber. Mechanists, including the liberal capitalists who created the Industrial Revolution, primarily see short-term cash to be made by taking natural resources in the most inexpensive, unsustainable way. In the new organic paradigm, however, the way nature inspires the imagination and spirit is just as important as the bounty it provides. Nature is seen as a living but fragile wonder to be respected and used only in a low-impact, sustainable way. In the eyes of the English artist and poet William Blake, the Industrial Revolution was wrongheaded from the beginning, and he hoped for a return to an agrarian-based society. This was how great the difference was between the two worldviews—right from the start. The organic way of looking at the universe was truly a revolution in thought for the West.

Kant himself recognized that he had started a second Copernican Revolution, a new way to think, one beyond the two paradigms of the time. This third paradigm was neither the fundamentalist medieval worldview nor the mechanistic scientific thinking of the Enlightenment. As developed by those coming after Kant and Goethe, the third, organic way was a more sophisticated worldview, one that looked at almost everything in a new light. Even though many in the general population may not have directly read Kant's work, by the beginning of the nineteenth century his influence was immense. Many heard about Kant's breakthroughs from others. They were excited that metaphysics had taken a great leap, that a new idealist philosopher from Germany had solved many previous conundrums. Kant was so significant to his time that he has often been compared to Socrates.

At the same time, Goethe was the most famous German author in the world. Yet he was known for his poetry, drama, and writing, not his science. Those who knew him say that, for most of his life, he primarily considered himself a scientist and spent most of his time on research. Having been borne up on Kant's shoulders with a new worldview, Goethe had an even more sophisticated organic lens. Goethe's science was buried, however, among his many essays and his poetry. Only those who found and understood it were able to make use of it—until the young Rudolf Steiner explained it all in the late nineteenth and early twentieth centuries. Still, between Kant and Goethe, the new paradigm had been perceived and established—at least among the very intelligent.

Aided by Goethe's new scientific method and his law of organics, over the century and a half from 1800 to 1950, the organic paradigm developed the new order of sciences Kant had predicted would arise.

Those thinking with the new analogy focused on the very subjects the mechanistic model could not explain: life, evolution, consciousness, and the subatomic world. As we said earlier, this led directly to the development of evolution theory, ecology, cultural anthropology, Jungian psychology, and the new physics. Each one of these new-paradigm sciences has had an enormous impact on the world already, especially evolution.

It was a new world after 1790. While the American Revolution was establishing a true democracy, the French Revolution had just swept away the French monarchy, church, and aristocracy. The Industrial Revolution was just getting started in England. The understanding of the past was also in a state of revolution. Geology had disproved the idea that creation was only six thousand years old and had revealed earlier epochs of time stretching back hundreds of millions of years. Naturalists were, at the same time, systematizing the study of fossils. Through observing the obvious similarities between skeletons, a new idea was born: evolution. Erasmus Darwin, Charles Darwin's grandfather, wrote about evolution in the late 1700s. And, as noted in the previous chapter, Goethe had proved how humans and animals were directly related through his study of the intermaxillary bone and had postulated that human beings were the result of the unfolding of evolution through many different forms of life.

Evolution has affected nearly every single branch of science. Evolution is a paradigm-shattering concept in that it tells us that we live not in a static world but in one that is ever being reborn, one that is in the process of *becoming*. On the suggestion of Goethe, Johann Gottfried van Herder—a philosopher of history and a former student of Kant— integrated the concept of evolution into history. Herder, in *Outlines of the Philosophy of Man,* showed how each society dynamically changes and advances to the next step, how seeds of change gestate within the previous culture and then emerge, evolving into mass movements and social transformation. He shifted the focus of history from the mere chronicling of great events and leaders to attempting to understand why new trends and movements began and how they developed.

Unlike previous views of the past, which presented each period as part of a static succession of ages, Herder said that *social change is not impossible but inevitable*. The old view, of course, makes change look hopeless, conveniently supporting the continuation of the status quo. Herder's work caused a frenzy of interest in history—called the "history revolution"—which lasted all throughout the nineteenth century.

The field of ecology was another major contribution of the Organic Shift and came directly from Darwin's recognition of the importance the environment plays in natural selection. After reading Kant, Goethe, and Darwin, the German evolutionist Ernst Haeckel envisioned the new wholistic science, naming it oecology in 1866. Over the next century, ecology developed as a field science, eventually inspiring a new land ethic and the political environmental movement. It is highly likely that future historians will mark the emergence of environmental awareness as the most important step in the survival of civilization—for it will someday make possible the reconciliation of science and technology with the reality of a fragile biosphere.

Given the significance of both evolution and ecology, there can be little doubt that the advent of the organic worldview in science, and then society, was as significant as the Copernican Revolution and the Enlightenment were centuries ago. A short history of the main contributions to science and technology by the Organic Shift follows. It is a broad paradigm shift, affecting fields as diverse as psychology, physics, medicine, anthropology, agriculture philosophy, and many others.

THE EVOLUTION REVOLUTION

During the later nineteenth century, the organic analogy began to have a big impact on the sciences through several significant theories. In the first great leap in understanding, Charles Darwin offered his theory of evolution and natural selection, countering the story of Adam and Eve by showing how we, ourselves, are related to apes. In a still deeply religious age, this theory was so controversial that Darwin sat on his epoch-making work for seventeen years—and only published it when he heard that a similar thesis was about to be released.

A raging debate began, with the dogmatic arguments of creationism on one side and the facts of long-term evolutionary and geological change on the other. From the first "young Earth" geologists of the seventeenth and eighteenth centuries to the Princeton theology (which gave an intellectual foundation to today's fundamentalism), scientific creationism, and now intelligent design, the old medieval worldview fought, and continues to fight, the concept of a godless evolution. Darwin had to finally endure the religious criticism he had dreaded all those seventeen years.

Darwin's theory made evolution *the* concept for the rest of the century, one that soon led to a great leap in fundamental awareness. Evolution, most of all, means that we are not in a static world. We are in a world that is always being born again in new divergent forms. This is significant in that it shows the status quo never lasts; things *constantly* change. Natural selection is nevertheless at its core a mechanistic theory and was later used by mechanists to support anthropocentrism and such reprehensible theories as social Darwinism—and was even mentioned as a rationale for eugenics programs. (Darwin's own cousin, Francis Galton, coined the word *eugenics* in 1883.[1])

Yet despite all the bad ideas evolution has spawned and beyond the details of the many different forms life has taken, the realization of evolution as a whole and later theories that *some kind of force is driving all the change* must be seen as part of the Organic Shift. What is that force? Why has life become increasingly complex through its many species, and why have these species become increasingly conscious? These questions, rather than merely supporting the hierarchy of the anthropocentric universe with the human race at the top, suggest a radical and revolutionary vision of evolution.

HENRI BERGSON AND THE ÉLAN VITAL

The concept of evolution affected nearly all the other sciences over the next one hundred years, becoming a powerful driver of the Organic

Shift. Early in the twentieth century, evolution science developed in a radical direction—challenging the credibility of mechanistic theory in several ways.

Fig. 4.1. In 1907, Henri Bergson talked of a "life force" that evolved more complex organisms with greater levels of consciousness.

In 1907, Henri Bergson, in his book *Creative Evolution,* talked of a life force that caused organisms to evolve into more complex forms with greater levels of consciousness. Bergson disputed Darwin's idea that evolution was the survival of the fittest and that adaptation is the force driving the overall path of evolution. According to Bergson, Darwin confused the incidental for the grand, primary driving impulse. This point of contention between these two schools of thought served to divide the mechanists and the new thinkers.

The truth is that adaptation explains the sinuosities of the movement of Evolution, but not the general directions of the movement, still less the movement itself. The road which leads to the town is obliged to follow the ups and downs of the hills; it adapts itself to the accidents of the ground, but the accidents of the ground are not the cause of the road nor have they given it its direction.[2]

To properly explain the direction of evolution, Bergson speculates about the existence of an *élan vital,* a vital life force that drives creativity and evolution. Bergson wrote eloquently about the élan vital, explaining how it is a single original impulse, "an internal push that has carried life, by more and more complex forms, to higher and higher destinies."[3]

As the smallest grain of dust is bound up with our entire solar system, drawn along with it in that undivided movement of descent which is materiality itself, so all organized beings, from the humblest to the highest, from the first origins of life to the time in which we are, and in all places as in all times, do but evidence a single impulsion, the inverse of the movement of matter, and in itself indivisible. All the living hold together and all yield to the same tremendous push. The animal takes its stand on the plant, man bestrides animality, and the whole of humanity, in space and in time, is one immense army galloping beside and before and behind each of us, in an overwhelming charge, able to beat down every resistance and clear the most formidable obstacles, perhaps even death.[4]

Vitalism, as it was called, greatly aroused the ire of the mechanists, who derided it as pure poppycock. The famous philosopher and organic thinker William James became America's greatest champion of Bergson. After the translation of *Creative Evolution* into English in 1911, James made Bergson's concept of a transformational force a subject of much discussion in the United States, empowering the nascent American bohemian and modern art movement with the idea of the élan vital. Bergson's élan vital not only drives evolution, it also powers our own personal development, causing us to constantly create ourselves over and over, ever changing and growing in knowledge and wisdom. Artists and writers could be seen everywhere with copies of *Creative Evolution,* and Bergson's book on awareness, *The Creative Mind,* tucked under their arms. Bergson instigated a true consciousness raising among evolutionists, philosophers, artists, and bohemians.

Another science coming out of the Organic Shift would also prove to be popular with the counterculture, and it too would have a great impact.

ERNST HAECKEL: THE MAN WHO DISCOVERED ECOLOGY

After evolution, the second great leap of the Organic Shift came with the discovery of the significance of the environment and the interrelations of the ecosystem. As stated earlier, to explain Darwin's idea of the environment shaping natural selection, Ernst Haeckel, a biologist, evolutionist, artist, and freethinker, invented the science of oecology in 1866.[5] The importance and interconnectedness of life and the environment was, in many ways, forgotten wisdom in the West while Native Americans, indigenous peoples, and other groups of people retained an intimate connection to nature. The advent of the ecological movement differs from this ancient wisdom in that this was a scientific discovery linking nature with science in a new way. The new science was named ecology, and it dealt not only with the organic qualities and systems of the natural world but also with human interactions and inventions that impact the natural environment. Haeckel defined *ecology* as "the relation of the animal both to its organic as well as to its inorganic environment," combining the Greek words *oikos,* meaning "household or habitat," and *logos,* meaning "reason"—the *reason that brings order to nature.* Like a modern-day ecologist, he even argued against mechanistic anthropocentrism, denying that humanity is the goal of the universe.

The discovery of the ecosystem would prove to be most significant, as understanding ecology meant that people had to think with the holism of the organic worldview and learn the myriad interconnections

Fig. 4.2. Ernst Haeckel coined the word "oecology" in 1866.

that make up a living system. As the concept and study of ecology developed over the next one hundred years, the forces of change gathered for the leap to mass environmental consciousness, which began in the 1960s. Moving from a human-centered society to an eco-centered society is the great transformation we are undergoing, clearly necessary for planetary survival. Haeckel's discovery of ecology thus played an enormous role in the Organic Shift.

Haeckel was also the first to write directly about the evolution of consciousness, seeing that a "long scale of psychic development ran unbroken from the lowest, unicellular forms of life up to the mammals, and to man at their head."[6]

He openly hoped for a future science that would study the psychology and development of consciousness, an idea that later inspired Carl Jung to develop his theory of the collective unconscious.

And where Darwin declared that God created the first organism in a warm rocky pool, Haeckel was brave enough to say that life developed out of nonliving matter without divine intervention—a very difficult thing to say in his day. Haeckel was a freethinker, giving him the temerity to tell his true views. Darwin, afraid of the religious response to his ideas and religious himself, gave God the role of creating the first organism.

It was common among the evolutionists of the day to write about the white race being the most evolved. Darwin put forward the idea that some races are "favored" and suggested that the white race was further up the evolutionary scale. Darwin's subtitle for *The Origin of the Species* was actually *On the origins of the species by means of natural selection or the preservation of favoured races in the struggle for life.* It was this unfortunate supremacist theory that Haeckel continued, which the Nazis later exploited to support their Aryan race theories. Haeckel said Germans were racially superior and that they "deviated furthest from the common primary form of ape-like men."[7] Haeckel even recommended the breeding of people and euthanasia programs, which the Nazis later put into practice. As a result, Haeckel's contributions on ecology are often ignored. Yet there is no doubt that the German evolutionist and monadist Ernst Haeckel was the man who put ecology, as we know it today, on the map.

THE FIRST FIELD ECOLOGISTS

Following Haeckel, around the turn of the century, other ecologists rediscovered concepts that Goethe had preached—to immerse yourself in nature, to use your intuition to see the whole of the thing you are studying, to use the natural phenomenon itself as the basis for a theory. Some of the first ecologists, such as Charles Adams, left the lab and began spending most of their time in the field during the first two decades of the twentieth century—in an effort to let nature teach them.

The next step came when Ed Ricketts's work on tide pools in the 1930s combined field ecology with holistic philosophy. Ricketts immersed himself—literally—in the tide pools of Monterey Bay, studying all of the interactions between species and environment. Ricketts was seeing every day, in front of his own eyes, that the whole is much more than the simple sum of its parts. He realized that each organism in its own way contributes to the overall health of the whole and the loss of any life-form threatens the future existence of the entire system.[8] *Between Pacific Tides,* written by Ricketts and Jack Calvin in 1939, became a classic, a textbook still used today.

Ricketts had his laboratory in Monterey, California, on Cannery Row. He was the inspiration for John Steinbeck's character Doc in his barely fictional novel, *Cannery Row.* Others who knew Ricketts and swapped stories and ideas with him included the young mythologist Joseph Campbell and the writer Henry Miller. They all admired his holistic worldview and his live-in-the-moment philosophy of life—which saw the interrelated parts of everything in the world as one glorious whole. A moving piece of music, a meaningful poem, or a sea urchin—all had equal significance in Ricketts's eyes, just as every denizen of an ecosystem had equal status.

ALDO LEOPOLD,
THE FIRST DEEP ECOLOGIST

A thing is right when it tends to preserve the integrity, stability and beauty of a biotic community. It is wrong when it tends otherwise.

ALDO LEOPOLD, "THE LAND ETHIC,"
A SAND COUNTY ALMANAC

Ecology became a universal science with Ed Ricketts's study of the Monterey Bay tide pools in the 1930s and with the publication of *Our Plundered Planet* by Fairfield Osborn in 1948—which explained in detail the depletion of natural resources on a finite Earth. Yet the deeper philosophical questions raised by ecology had never been discussed. That ended in 1949 when Aldo Leopold's *A Sand County Almanac* was released, the first book on what would later be known as deep ecology.

Leopold debunked the idea that man can ultimately tame nature.

Fig. 4.3. The ecologist Aldo Leopold taught that the anthropocentric worldview must be replaced with one that respects nature.

He noted there would always be unexpected repercussions, with the solution being worse than the problem. He pointed out the history of the Army Corps of Engineers and their many failed attempts at flood control. *A Sand County Almanac* directly challenged the old paradigm and its simplistic economics, which constantly ignores the cost to the environment. In his conclusion on controlling rivers through dikes and dams, Leopold showed how it was a losing battle, as walls upriver create higher floods downstream. Instead of the anthropocentric drive to conquer nature, Leopold insisted that people must learn to live in harmony with nature; there must instead be a land ethic—based on his principle of sustainability. It is sustainability that must guide our actions and our economics: "In short, a land ethic changes the role of *Homo sapiens* from conqueror of the land-community to plain member and citizen of it. It implies respect for his fellow-members, and also respect for the community as such."[9] Society should find low-impact management systems for the environment and try to have little or no management for wilderness areas—and there must be a new respect for nature and all species. Aldo Leopold's work led directly to the deep ecology of the later 1970s and 1980s when Norwegian philosopher Arne Naess, poet and environmental activist Gary Snyder, and American philosopher

George Sessions agreed with Leopold in rejecting the anthropocentric basis of the last few worldviews. Nature was not made for humanity to ruin through greed. Humanity was born from nature and must protect and respect it—as you would your own mother and your own family. This attitude is similar to the worldview of many indigenous peoples, including the Native Americans.

Leopold's essay "The Land Ethic," from *A Sand County Almanac,* was a seminal blueprint for the environmental movement, creating a whole new set of parameters to live by. Leopold explained how *valuing the parts of an ecosystem by economics alone is a real mistake.* In contrast, preserving nature as a whole gains an entire set of intrinsic values—and it was essential that this be recognized in the near future. Since only ecologists understand the bioticvalue of each part of an ecosystem, the piece-by-piece destruction of the environment by the old worldview forces ecologists to suffer "alone in a world of wounds." To Leopold, there are two separate value systems in the world, and the one in control has but a shallow understanding of the ecological havoc it creates on a daily basis.

In the most interesting passage from "The Land Ethic," Leopold actually described the difference between the mechanistic worldview, transfixed by money, and the organic paradigm, devoted to the biotic value system and all life within it. He called the paradigm struggle the A-B cleavage, with the mechanists being the group A mind-set and group B expressing the organic point of view:

> Conservationists are notorious for their dissensions. Superficially these seem to add up to mere confusion, but a more careful scrutiny reveals a single plane of cleavage common to many specialized fields. In each field one group (A) regards the land as soil, and its function as commodity-production; another group (B) regards the land as a biota, and its function as something broader. How much broader is admittedly in a state of doubt and confusion.
>
> In my own field, forestry, group A is quite content to grow trees like cabbages, with cellulose as the basic forest commodity. It feels no inhibition against violence; its ideology is agronomic. Group B,

on the other hand, sees forestry as fundamentally different from agronomy because it employs natural species, and manages a natural environment rather than creating an artificial one. Group B prefers natural reproduction on principle. It worries on biotic as well as economic grounds about the loss of species like chestnut, and the threatened loss of the white pines. It worries about whole series of secondary forest functions: wildlife, recreation, watersheds, wilderness areas. To my mind, Group B feels the stirrings of an ecological conscience.

In the wildlife field, a parallel cleavage exists. For Group A the basic commodities are sport and meat; the yardstick of production are ciphers of take in pheasants and trout. Artificial propagation is acceptable as a permanent as well as a temporary recourse—if its unit costs permit. Group B, on the other hand, worries about a whole series of biotic side-issues. What is the cost in predators of producing a game crop? Should we have further recourse to exotics? How can management restore the shrinking species, like prairie grouse, already hopeless as shootable game? How can management restore the threatened rarities, like trumpeter swan and whooping crane? Can management principles be extended to wildflowers? Here again it is clear to me that we have the same A-B cleavage as in forestry.[10]

Aldo Leopold's land ethic and his recognition of the paradigm shift in the A-B cleavage section made clear the difference between the mechanistic and the organic analogies. Where the mechanists (Leopold's group A) see nature primarily as a money machine to harvest, the organic scientists and farmers primarily view nature as a living integrated being that must be respected first if an income is to be made off the land. Leopold's land ethic became the new bedrock of paradigm B—or the organic worldview.

THE ORGANIC SHIFT IN PHYSICS

In the 1870s and 1880s, at the same time Kant was revived in philosophy and education, a systematic investigation beyond classical physics began, an attempt to discover the workings of the invisible—of light, electromagnetism, and the atom. Science had finally encountered a realm where instruments, such as the telescope in astronomy, do not help. It all brought back Kant's question: What can we know? And so there was a return to the old philosopher, a revival of the teachings and writings of the organic Copernicus. The impact on the rest of the nineteenth century was profound.

In 1873, James Clerk Maxwell put forth the wave theory, a credible hypothesis on electromagnetism based on mathematics. But how could one confirm it? How *do* you measure the invisible? What can we know about the world? Thanks to electromagnetism and the centennial of *Critique of Pure Reason* in 1881, Germany enjoyed a full neo-Kantian movement in philosophy and science.

Kant's philosophy led one physicist, Heinrich Hertz, on a metaphysical quest to make the wave theory a tangible reality. Hertz believed there really is a world beyond the visible. In 1888, Hertz proved Maxwell's wave theory correct when—without the use of any wire—he successfully sent a signal from a transmitting device through the air to a receiver. It was the first radio wave. When a curious pupil asked whether these waves might someday have a practical use, Hertz unprophetically answered: "None whatsoever. I don't see any useful purpose for this mysterious, invisible electromagnetic energy."[11] He was simply interested in proving that electromagnetic waves existed.

Naturally, Hertz could not know that others would soon take his work and invent radio and television. Nor could he see that his proof of Maxwell's wave theory would eventually lead to the new physics—and the unlocking of the atom. Others quickly went beyond Hertz and the merely invisible—and explored the mysterious nature of the subatomic world. In 1895, Wilhelm Roentgen discovered X-rays, and science was able for the first time to see inside an object, even inside the body. Then, during 1896, Antoine Henri Becquerel explained radioactivity, which led to the Marie Curie's discovery two years later of entirely new elements: radium and polonium. Without stopping to ask if this was a good idea or not, physicists set about to carefully analyze and fully understand the new elements and their puzzling property of radioactivity.

Nikola Tesla, perhaps the greatest mechanical genius of all time, was opposed to the study of the atom, which he knew would eventually be used as a power source. He felt that it was against nature to use radioactivity—that the world would come to regret it. Worried that unlocking the atom would open a Pandora's box of problems and horrors, Tesla throughout his life showed there are many other ways to make power—the harnessing of Niagara Falls in 1895 being only one of his amazing accomplishments. Despite Tesla's warnings, the path started down by Maxwell and Hertz would inevitably lead to a true understanding of the world beyond classical physics: the realm of the subatomic.

The scientific realization of the whole was about to go far beyond the merely invisible. The new paradigm now transformed the world—irrevocably—by allowing humanity to finally begin understanding the subatomic realm in a practical way. Even today, we have not yet dealt with the consequences of that great change—for it made real the danger of nuclear power and the horror of nuclear weapons.

NEWTONIAN PHYSICS AND THE NEW PHYSICS: PARALLELS IN TIME

Newtonian mechanics and quantum mechanics each appeared about 150 years after the conception point of their worldviews—a thought-provoking similarity in time. A strong parallel to Newton is needed to show that the Organic Shift was indeed a major change on the level of the Copernican Revolution. The obvious counterpart to Newton, Werner Heisenberg, was the physicist who led in the formulation of quantum mechanics, helped by a team of scientists. Given that Newton was the basis of classical physics, which made possible industrial society, and Heisenberg helped provide the foundation of the new physics—which gave us the nuclear bomb, nuclear energy, and our current understanding of subatomic particles—this parallel in time goes a long way toward making the case that the Organic Shift is in fact a classic revolution in paradigms, fulfilling Kuhn's description of an innovative extraordinary science.

As Kuhn himself noted, the organic paradigm parallel to Isaac Newton is most revealing, with both being preceded by new philosophical descriptions of reality. From Kuhn's *Structure of Scientific Revolutions:* "It is no accident that the emergence of Newtonian physics in the seventeenth century and of relativity and quantum mechanics in the twentieth should have been both preceded and accompanied by fundamental philosophical analyses of the contemporary research tradition."[12] Copernicus, Bacon, Descartes, and Galileo preceded Newton, while Kant, Goethe, Hertz, Einstein, and Bohr preceded Heisenberg.

The world's greatest minds came together in the 1920s in a great collective effort to unlock the mystery of the subatomic world. In the attempt to decipher the atom, science directly confronted something that could not be known—at least not in the way classical physics understood the world. The puzzle was that in some experiments light acted as a wave and in others as a particle. Physics had reached the point where the instruments peering down into the depths of subatomic phenomena altered the phenomena themselves. The very act of observation changed the phenomena being studied—*talk about the observer being part of the observation!* The subatomic world was simply unknowable in the old sense. Kant was thus proven very correct. The observer in this case was *directly* affecting the observation, making the ultimate reality unknowable. Saying that something was *either* this *or* that was simply arbitrary, as reality might not follow such rigid concepts. A new physics had to be invented.

To assist in unraveling the mystery, Heisenberg actually brought in Kantian scholars to help guide the discussion. The team ended up developing a both/and solution—light is *both* a wave *and* a particle. Rather than the either/or rigidity of the mechanists, physics was forced to accept the holistic both/and reality. Heisenberg found that by adding a mathematical formula to handle the resulting uncertainty, physics could once again conduct its subatomic exploration—although it was all probabilities now instead of good old Newtonian certainty.

Heisenberg did far more, however, than fix a thorny mathematical problem for physicists. The new physics represented a leap in thought, one very different from the classical physics that came before it. Atoms, which had been thought to be hard, solid particles, were instead found to be a maze of interrelated fields, forces, and electrons—not hard or solid at all. Rather than being indivisible, atoms could be split, releasing incredible amounts of energy. Atoms are mostly space—subatomic "solar systems" in which electrons and other particles orbit the nucleus with charges and pure energy instead of gravity and matter.

In the final analysis, subatomic particles have no real meaning as separate individual things. They can only be understood in terms of how they relate to the whole swirling atom—we cannot even know their precise location. Only how they interrelate within the whole can be observed; only the resulting interconnections can be observed by the observer. Another physicist, Neils Bohr, explained that the concept of the subatomic particle itself is just that, an idea. It is not the thing in itself. "Isolated material particles are abstractions," said the Danish scientist, "their properties being definable and observable only through their interaction with other systems."[13] Here one cannot even pin down the individual object; one can only talk about the relationships between objects. This discovery had many implications for the rest of science—and for philosophy.

JAN CHRISTIAAN SMUTS AND HOLISM

One of the more interesting characters in the Organic Shift was the retired prime minister of South Africa, Jan Christiaan Smuts. Not just a political leader but also a great philosopher, in his 1926 book *Holism and Evolution,* Smuts made several original discoveries while studying the wholes that appear in our universe. He defined holism as "[t]he tendency in nature to form wholes, that are greater than the sum of the parts, through creative evolution."[14] Coming from the Greek word *holos,* meaning all and everything, Smuts explains that nature has no rigid parts as a machine does; there are only interacting wholes. Not only are there wholes in nature, these systems evolve, through Bergson's creative evolution.

Thus we cannot understand nature through simple analysis of these alleged parts—as mechanistic science attempts to do. Smuts said mechanists overspecialize, focusing on some part and then declaring it to be the whole. We must instead study the whole in all its complexities, as something continually evolving to the next level of wholeness. For example, consciousness is a whole greater than the mere sum of all the brain's neurons, synapses, and so on. A mind functions only through a synergistic effect of all the different areas and parts of the brain. Another example would be ecosystems. These are wholes made up of many elements and life-forms and cannot be understood by separating the parts rigidly from one another. The boundaries between the different things in an ecosystem are fuzzy; they're flexible. Each part in the whole influences many other parts, and they in turn affect even more parts, which affects the ecosystem as a whole. By fragmenting information into artificial isolated units, mechanistic science oversimplifies and overlooks the wealth of discoveries visible when we step back and consider the whole. So we must look at the whole *first* and then see the next level of wholes and the next level after that. Only *then* can we approach the granular, individual level with a well-researched view. Only then can we begin to grasp the vast complexity of reality.

This new realization had a profound impact on research methods in biology and medicine and in all later systems theories. Smuts saw that nature was made up of wholes within wholes and that these wholes were evolving into greater wholes—the process of creative evolution. Every succeeding level of evolution is broader than the last, as what was once a whole becomes part of a still greater whole.

Smuts wrote *Holism and Evolution* as the League of Nations was trying to form a greater whole, a league of peace, seeing the movement toward world government and law as an example of the evolution process. He also noted how whole species and types of animals and plants would evolve into new forms en masse. The old whole would become part of the new whole, which is swallowed up in turn by a greater whole later on.

It is through the imperfections or inefficiencies in each whole, Smuts says, that the impetus for change begins, building until there is a transformation from the old to the new. Through evolution, organisms and systems are increasingly more efficient in their use of energy and their ability to harness energy from the environment. Without the imperfections, there would be no creative evolution. Perfection is sterile: there is no need to change anything. Holism was thus a basically optimistic philosophy and held out hope for an imperfect and inefficient world. According to Smuts's concept, we will soon be creatively evolving our way out of the mechanistic dilemma, and its very imperfections will be the impetus that drives the great change along. (We should thus thank the oil companies, Bush, and Cheney for their misbegotten ways and mistakes.)

Evolution, ecology, and holistic philosophy explain a lot about life, about who we are and how we got here as a species. Now other new sciences, born of the biological model, would emerge, transforming the sciences that study society and consciousness itself. Although mechanistic science has its own schools of thought in these areas, we shall here survey the landscape on the organic side, which has been increasingly replacing the old paradigm.

THE ORGANIC SHIFT IN THE SOCIAL SCIENCES

The new holistic understanding of the world eventually transformed many other fields, setting the stage for the Organic Shift to enter mainstream society. In the 1920s, an antimechanist movement emerged nearly simultaneously in physics, biology, and psychology, all based on finding the holistic patterns within systems.

Jean Piaget, the famous developmental psychologist and philosopher, explained how children process and acquire knowledge through natural stages of development. Benjamin Whorf, an American linguist, gave a radical new view of language and worldviews, and Alexandre Koyré, the French philosopher, had seminal breakthroughs in the history of science. The German philosopher Christian von Ehrenfels meanwhile coined the term Gestalt, meaning organic form, or the irreducible pattern. This concept soon inspired the field of Gestalt psychology, which in turn transformed many other fields later on. "The whole is more than the sum of its parts" was one of the primary understandings of this school.

By demonstrating the importance of seeing the whole and understanding how the observer is part of any observation, the new physics served as a lesson for all of science. It is not only individual things that should be studied, we must also understand the relationship of things to everything else. In the end, that quest is far more significant, for it is by understanding the interconnectedness of things that we arrive at true wisdom—and the right path to a workable future. This is the science we must teach our children. A thing should not be defined for what it is by itself but by its relation to the systems that inanimate objects and living beings form. By proving the truth of the unknowable whole and how the observer is part of the observation, the Kantian logic of the new physics began to affect many other branches of study, especially the life and social sciences.

WILHELM DILTHEY AND THE STUDY OF WORLDVIEWS

In the 1880s, Wilhelm Dilthey (1833–1911), the great neo-Kantian philosopher of history, envisioned a new worldview, which he termed *Weltanschauungslehre*. It was based on Kant's *weltanschauung,* defined as "general view of the universe and the place of human beings in it, especially as this view affects conduct."[15] Dilthey made a fundamental realization about history and culture: every society is convinced that its worldview and culture is superior, that it is *the ultimate culture of destiny*.

Yet the reality is that no one worldview has cornered the market on truth, for every paradigm is merely one step in a long and continuing

*Fig. 4.4. Wilhelm Dilthey discovered the "worldview" in
1882, founding cultural anthropology.*

evolution of worldviews. Moreover, in order to take into account the fact that the observer is clearly part of this observation, *we must imagine ourselves outside our own worldview.* We must try to attain cultural relativity.

Such a perspective naturally leads us to question our own worldview and where it comes from—a jumping-off point for a total transformation of any system of thought. This had tremendous implications. For many, the realization of the worldview allows them to understand their own culture—and the cultures of all of the world—from the outside. This provided a new appreciation of how we are consciously and subconsciously programmed to think by our worldview.

As Dilthey's teachings spread to America, primarily by his student Franz Boas, who started the Cultural Anthropology Department at Columbia University, they became known in the New York bohemian counterculture. Many artists immediately adopted these teachings, making them a part of their inner selves. Stepping outside the old paradigm and seeing it as an obsolete worldview became a necessary first step to being a true bohemian. Margaret Mead came from this school, giving her the outside-the-paradigm thinking so visible in her writings and quotations.

Dilthey proclaimed that the highest form of study is the study of *weltanschauungs,* the study of the history of worldviews. How did one worldview transform itself into another? Here Dilthey was asking the same question Kuhn asked eighty years later in his paradigm shift theory. This study of worldviews, said Dilthey, should be the intellectual foundation of any well-thought-out educational curriculum. Like Haeckel, Dilthey's contributions to the Organic Shift were profound, yet few today even know his name.

GREAT STRIDES IN KNOWING OURSELVES: JAMES AND JUNG

In 1890, the philosopher William James (1842–1910) made the study of the mind a science with his *Principles of Psychology,* which removed the investigation of behavior from the grip of philosophers, designating it as a medical art. This brought the inner world into the real world, allowing many other thinkers, such as Sigmund Freud (1856–1939) and Carl Jung (1875–1961), to map out the influences psychology has on our lives. Those who understood James saw that many previously hidden layers of thought could now be studied, and study of them produced practical clinical benefits. With *The Principles of Psychology,* the inner world was suddenly exposed to the light of day.

Shortly after James wrote his brilliant theories, Jung made his great contribution to psychology. Most important, he taught that mental health comes from creating and believing in a story of who we are and what we want to do with our lives. Inspired by Haeckel's explanation of how consciousness evolved from the lowest forms to the consciousness of humanity, Jung, in his *The Archetypes and the Collective Unconscious,* wrote of a collective unconscious, a set of archetypal memories that inspire our inner story—and so our feelings.

*Fig. 4.5. Carl Jung discovered the collective
unconscious and said mental health depends on
having a personal "story."*

Understanding your archetype and story can be crucial to happiness and well-being. Jung's psychology from the organic perspective is far more complex than the simple stimulus/response of Freud and the behaviorists. Jung explains how—after studying sixty-seven thousand dreams—the use of the biological model led him to the concept of the psychological archetype. The organic analogy made it possible for him to make psychology more of a science: "My scientific methodology is nothing out of the ordinary, it proceeds exactly like comparative anatomy, only it describes and compares psychic figures."[16] And "Psychic events are observable facts and can be dealt with in a 'scientific' way. . . . I observe, I classify, I establish relations and sequences between the observed data, and I even show the possibility of prediction."[17]

In short, Jung realized that the concept of the type in biology could be applied to psychological archetypes, to certain qualities people were subconsciously trying to emulate. There is not enough room in this book to do Jung justice, so the reader is simply implored to explore the subject further, perhaps by discovering your own archetype and understanding your own personal story.

In the 1920s, 1930s, and 1940s, Jung also did much to meld together Eastern beliefs and Western philosophy and was one of the participants in the 1933 Eranos seminar, held to find common ground between Eastern and Western beliefs. Later on, he explored the strange phenomena he calls synchronicity, or meaningful coincidence. He concluded that each moment has its own special quality, and that synchronicity is likely very common—but we often miss it because strict mechanistic analysis and causality ignores it. Jung's work would later inspire many in the 1960s and beyond to delve deeper into the inner realms he discovered.

THE ORGANIC SHIFT
IN MEDICINE

Just as the mechanistic shift—or Copernican Revolution—had led to the development of a new medicine, the Organic Shift in the late 1700s led to a new organic and natural approach: homeopathy, invented in the 1790s by Samuel Hahnemann. Based on the idea that disease could be cured by a law of similars, Hahnemann discovered that specific preparations made from herbs and other substances—if given in minute doses—can mimic the targeted disease, activating the body's natural immune defenses. It was the same idea as Jenner's discovery of smallpox vaccination, where microdoses of the actual disease created long-term immunity. Just recently, a consortium of scientists was able to duplicate the main point of Hahnemann, that preparations—so diluted not even a molecule of the original substance remains—still have an effect on organisms.[18]

Able to read and write in seven languages by the time he was twenty-four, the life of Hahnemann shows that he was indeed a supergenius. He translated over twenty major scientific and medical texts as well as researching and writing *The Pharmaceutical Lexicon,* used as a reference book for the pharmacists of his time. This strong grounding in all medical and chemical knowledge allowed Hahnemann to add it all up in his mind—then make one significant discovery after another.

*Fig. 4.6. Hahnemann excelled in so many areas,
he was called the supergenius.*

As both a brilliant experimenter and a respected chemist, he was not afraid to criticize traditional medicine and suggest alternatives. His new medicine was systematized into the founding text of homeopathy: *The Organon*.

The Organon was an exposition of remedies and the bible of homeopathy, establishing its organic medical principles and goals. The book's influence was so great that a whole new school of medicine quickly developed in Europe, challenging the allopaths (the traditional doctors), who at the time were still using leeches, arsenic, and mercury. In 1825, the Dutch homeopath Hans Gram brought the new medicine to the United States. Homeopathy expanded so rapidly that the first national medical society in the country was formed in 1844: the American Institute of Homeopathy.

Abolitionists, including William Lloyd Garrison and Zabina Eastman, told all who listened about the benefits of homeopathy.[19] As today, homeopaths were often progressive, and the new medicine in the United States became identified with the radical causes of black and female liberation. While this might have helped in the North, it no doubt slowed its advance in the South. Women especially flocked to homeopathy, and many became homeopaths themselves. In fact, the world's first women's medical college was opened in 1848: the Boston Female Medical College, a homeopathic school.

In the 1831 Austrian cholera epidemic, homeopathic microdoses cured the disease with only a 21 percent death rate—versus over a 50 percent death rate with allopathy. Despite statistics and case studies such as this, traditional doctors and pharmacists were so alarmed at the direction medicine had taken they started antihomeopathy magazines, with one actually called *The Anti-Organon*. As Thomas Kuhn found, scientific data means nothing in the ideological struggle between paradigms. Science be damned, all data showing the truth of the new worldview must be fought with *whatever means available.*

Despite all the opposition to come, homeopathy somehow survived and is today stronger than ever. Hahnemann realized the good he had done by discovering a new version of medicine and asked that his tombstone read: I did not live in vain.

THE MECHANISTS STRIKE BACK WITH THE AMERICAN MEDICAL ASSOCIATION

The ideological nature of the struggle between the mechanistic and organic paradigms was perhaps most clear in the field of medicine. Traditional allopathy, after all, was still using harmful preparations, many containing mercury, lead, and even arsenic. Nor were patients wild about the age-old practice of using leeches. In 1833, France alone imported forty-one million leeches. One U.S. firm imported five hundred thousand leeches in 1856. Additionally, the education of doctors then was simply ludicrous. After some apprentice work, traditional medical college could be completed after only sixteen weeks and some morgue work. Even doctors of today would agree that, during this period, traditional medicine no doubt did more harm than good.

The successful outcomes of homeopathic practices made traditional medical practices look inferior in comparison. Additionally, homeopathic treatments were self-delivered and cost a mere fraction of the usual preparations and potions. Homepathy was based on inexpensive regimens of tiny, daily microdoses taken by the patients themselves. But doctors could not make the same kind of money from homeopathy that pricey leech treatments and strong preparations garnered, so traditional doctors had no financial incentive to switch to homeopathy. Today, the situation remains much the same, although the leeches and strong preparations have been replaced by expensive radiation treatments and chemically based drugs.

The struggle between these two medical disciplines had little to do with scientific data or theory: this was clearly an ideological war. Even more than ideology, homeopathy quite simply threatened the income of traditional doctors, and so paradigm war was declared. Pharmacists and conventional doctors everywhere saw the new medicine, growing in popularity every single day, as a direct threat to their reputations and their incomes. Unfortunately for the alternative doctors, they were typically progressive pacifist types and no match for the vicious blackball campaign soon launched against them.

Oppression of homeopathy had begun with the persecution of Hahnemann himself in 1820, when he was forced to flee his home in Germany. In 1830, homeopathy was outlawed in Austria and oppressed in many other places. The American Medical Association (AMA) was then formed in 1846 and 1847 with the clear intent to suppress its direct organic competitors and to deal with widespread abuse in medical education. Johns Hopkins University of Medicine was considered to be the first modern medical school, requiring a Liberal Arts degree, three to four years of medical courses, and a one-year internship. It opened in 1893.[20]

In 1850, the AMA established a policy that excommunicated traditional doctors if they so much as even consulted with a homeopath. In 1856, the ban was expanded to *discussion of homeopathic theory*. In that same year, *the AMA even forbade marriage to a homeopath*. These rules were enforced with utter ruthlessness. In the 1860s, modern pharmaceutical companies were formed and traditional doctors gained a powerful new ally in the fight against the little pills and tinctures of the homeopath. Soon, conventional medicine would have its own little pills and serums.

Despite this war of the doctors, homeopathy expanded rapidly around the world and became standard practice among progressives in the North, with people such as Frederick Douglass using homeopathy. During the course of the nineteenth century, many colleges, books, and research on cures were undertaken, and often homeopathic women doctors led the way. In 1890, Mark Twain wrote about the alternative treatments in *Harper's Weekly,* explaining that "[t]he introduction of homeopathy forced the old school doctor to stir around and learn something of a rational nature about his business."[21] He also intoned, "You may honestly feel grateful that homeopathy survived the attempts of the allopathists [orthodox physicians] to destroy it." Yet the AMA ignored Twain and redoubled its efforts to get rid of its pesky competitor.

By 1900, allopathic medicine had reached such a miserable state of affairs that few doctors belonged to the AMA and even fewer bothered with what its policies were. Medical colleges were poorly funded and did not teach a standard curriculum. One did not even have to attend a medical school to practice—just hang out a shingle and start seeing patients.

Quackery was everywhere, along with fake cancer cures and scores of snake oil remedies. Meanwhile, there were some fifteen thousand homeopathic doctors in the United States and twenty-two homeopathic colleges.

Then things started to turn around for the AMA. The homeopath's archrival succeeded in convincing philanthropists to stop funding the homeopathic schools, and by 1904 the flow of money to homeopathic colleges was shut off. This was easy to do as the rich were now owners of the new pharmaceutical companies. Despite the ability of alternative medicine to cure many diseases with simple remedies, the loss of the colleges was catastrophic, and the practice began to collapse in America.

Nevertheless, in 1910 the AMA found itself on the verge of bankruptcy—and then the Carnegie Foundation and Rockefeller stepped in: Henry Prichard, president of the foundation, bought control of the AMA in 1910 for a mere $10,000. From this point on, the AMA was used by the industrialists to sell expensive pharmaceuticals. The organization issued the *Flexner Report,* which "reformed" the whole field of medicine. Medical colleges were hereafter funded by the robber barons, but only if they embraced the drug-centered curriculum recommended by the report. The AMA was basically for rent to the highest bidder. The organization even let itself be used for decades by the tobacco industry to convince everyone that smoking was not bad for your health, *with some doctors even prescribing it for nerves*. Many, many doctors calmed their own nerves with it, and ad after ad in the magazines of the 1940s and 1950s showed an AMA doctor puffing on a cigarette and smiling. The rate of smoking in the general population soared over 40 percent in the 1940s, killing millions of people. It was the utter corruption of science by industry.

Despite this, many good changes took place, and medicine did actually improve for a while, especially with the introduction of penicillin. As the pharmaceutical-centered curriculum grew increasingly more drug oriented, however, lifestyle suggestions, nutrition, bedside manner, and the art of listening to the patient were lost. So it was that traditional medicine reached the sad state of affairs that we know today.

Currently, according to recent research published in the *Journal of the American Medical Association,* 106,000 a year die from adverse reactions to drugs.[22]

The number of people having in-hospital, adverse reactions to prescribed drugs is now approximately 2.2 million annually.[23] Doctors continually make the fatal error of not reading all the warnings and then prescribing the drugs anyway. Unnecessary surgeries cause some 12,000 deaths per year.[24] The total number of medically induced deaths in the United States is nearly 800,000 every year.[25] Meanwhile, these drugs were designed to treat symptoms, rarely addressing the underlying cause of a disease. This approach keeps the doctor and the pharmaceutical companies conveniently rich but the patient continually sick and paying indefinitely. Even today, the medical and pharmaceutical priesthood keeps these facts buried as deeply as it can. The media, paid by pharmaceutical advertising, helps to conceal the fact that certain drugs are often given in a way that quickly kills or injures the patient—just because the doctor misses a warning or there is an unknown synergy with another drug. In addition, there is the problem of fudged and false data in the company-bought studies written by allegedly "independent" institutes. And then there is the sad fact that often the drug just doesn't work at all.

The AMA, the now-corrupt FDA, and the drug companies have ruined the healing science to a large degree, making the sale of chemicals more important than the health and even the lives of the patients.

In a *New York Times* article, Dr. Sandra Kweder, a top official at the FDA, acknowledged before a Senate panel that there were "lapses" in the agency's oversight of the pain pill Vioxx, which was withdrawn in 2005. Kweder noted that "the agency took too long to get information about Vioxx's heart risks into the prescribing label that is provided to physicians."[26] Other articles have exposed the revolving door of pharmaceutical employees and FDA regulators.[27] Plus the drug companies *now pay much of the cost of the FDA's budget.*

Since the 1992 passage of the Prescription Drug User Fee Act, pharmaceutical companies can pay the FDA to review their drugs. The FDA began looking at the drug companies as their clients, instead of the public interest. The review process is a major portion of the entire FDA budget. Congressional oversight has been drastically reduced compared to the frequency of hearings prior to the early 1990s. In the twelve years prior to 2003, only one or two days of oversight hearings occurred.[28]

Although homeopathy had never lost favor in Europe, by the 1930s the last homeopathic school closed its doors in the United States. This field, and other alternative forms of medicine, came back strongly in the 1970s, however, and once again they present a challenge to the drugs and surgery of traditional Western medicine. Many diseases and conditions today are helped by, and in some cases treated effectively with, acupuncture, herbs, vitamins, and homeopathic remedies. These alternatives and Integrative Medicine (the combination of alternative and Western medicine) as well as preventative medicine and lifestyle changes have made great strides toward healing—although the AMA would still disagree.

THE ORGANIC SHIFT IN TECHNOLOGY: THE SOLAR POWER PIONEERS

Today the world must choose: continue with fossil fuel and gravely disrupt Earth's climate, or begin a rapid transition to clean renewable energy. The world will never transcend the mechanistic dilemma until society's addiction to oil is ended.

The story of the renewable-energy pioneers of the nineteenth century and early twentieth is not taught in schools, but these technologists of the Organic Shift must be counted as some of the greatest inventors of their time. We are indebted to engineers like Augustin Mouchot, John Ericsson, and Frank Shuman, who explored renewable energy production beyond coal-fired power plants and gasoline-powered engines. Seeing their goal as essential to the future of industry and agriculture, these solar pioneers often pledged their lives and fortunes to the mission of making solar power a commercial reality. These new-paradigm thinkers were prescient enough to foresee the end of the fossil-fuel era and the economic pain and pollution that would occur as the coal and oil ran out. One can only imagine what our world would be like today if the solar path blazed by these pioneers had been taken rather than the road to cheap oil.

Solar power was seen as the answer, even though the technology of photovoltaics was not yet fully understood. The photovoltaic effect had first been discovered by Becquerel in 1839 and the great Heinrich Hertz himself created photovoltaic cells in the 1870s. However, his devices were only 2 percent efficient at directly converting sunlight into electrical energy. Rather than using photovoltaic energy, these great solar inventors made power the old-fashioned way—with steam engines.

The first solar-powered motor was created by the French inventor Mouchot, who worried about the finite nature of fossil fuel way back in 1860: "Eventually industry will no longer find in Europe the resources to satisfy its prodigious expansion. . . . Coal will undoubtedly be used up. What will industry do then?"[29] Using a curved silvered mirror, tilted to redirect and focus the sun's heat, Mouchot invented the solar oven in 1860, which cooked up a "fine pot roast" in four hours. With the same boiler he next powered a still. Being French, he was soon sipping solar brandy made from two quarts of wine. He then attached his contraption to a siphon and was able to pump water in a ten-foot stream for thirty minutes before the pressure, having been exhausted, had to rebuild again.

Mouchot soon had funding for a large solar boiler to run a still in Endre-et-Loire, the wine region where he lived. Even in the northern climes of France, the eight and a half foot wide solar reflector made enough steam to generate one-half horsepower, which vaporized the wine at the rate of five gallons a minute. He went on to make portable solar ovens, used by the French military for decades, as well as a solar ice maker. The ice maker used Ferdinand Carré's discovery that an ammonia water solution when heated will produce the refrigeration effect, as the vaporized ammonia condenses strongly upon cooling. When hooked to a water pump, the ice machine could deliver over eight gallons of ice a minute.

Yet the cost of silvered mirrors and keeping them polished for maximum heat made these devices too expensive. So Mouchot, just before ending his two decades of solar research, started down the road of producing electricity directly from the sun's heat. By heating plates of copper and iron soldered together, an electric current was produced strong enough to separate ordinary water into its components of oxygen and hydrogen. When the gases were recombined, very high temperatures were generated. It was a primitive fuel cell technology.[30]

Mouchot wrote in 1879 that he had "already made a few experiments which bode well for this procedure. . . . Some very primitive devices have given me significant amounts of electricity." If one could release hydrogen from water with solar power, Mouchot knew that would be a "reserve of fuel as precious as it is abundant."[31] Unfortunately, soon thereafter, Mouchot left his solar research to return to his original career of mathematics.

Despite his inability to start a "sun age" in France and its colonies, Mouchot was the first modern solar-power experimenter and showed how simple sunbeams could create a new future for the world. His research blazed a path for other great solar inventors, including:

- 1870 John Ericsson, the designer of the ironclad battleship *The Monitor,* built a secret solar-powered steam engine
- 1872 Ericsson built a solar-powered *air-piston* engine. No water

was required; the sunbeams heated the air in the piston, which pushed down from the expansion, expelling the hot air and sucking cold air to repeat the cycle.

- 1884 Ericsson improved reflectors by using a cheaper and lower maintenance mirror made out of glass silvered on the back; this prevented tarnish.
- 1888 Ericsson announced that he had perfected the solar engine for irrigation but died months before he could sell the first one; his secret design goes with him to the grave.
- 1890 *Pacific and African Conquest by the Sun,* a book by Charles Tellier, the father of refrigeration, described his experiments with low-temperature steam engines running on ammonia hydrate, which boils at minus 28 degrees Fahrenheit, and sulfur dioxide, which boils at 14 degrees Fahrenheit. This lower temperature engine can then run on mere hot water, without the 1,000 degrees Fahrenheit required for a water steam boiler. A simple "hot box" collector heated the pipe without the use of any expensive mirrors. Tellier claimed that his low-temperature solar engine was able to pump five gallons a minute, with a potential of ten gallons per minute in the tropics.
- 1891 American Clarence Kemp patented a solar hot-water heater and began selling them; 1,600 were installed by 1900; many others followed.

Although these were great advances in harnessing the sun, the problem remained that on cloudy days, or when the sun went down, the steam engines would stop and the hot water in the heaters would grow cold. The era of great inventions was not over, however, and even this obstacle was soon hurdled in the new century to come.

SOLAR POWER BUSINESSES

During the early years of the century, solar power and electric cars became real, profitable businesses. The story begins in 1901 when

Aubrey Eneas, an English inventor living in Massachusetts, erected his latest solar pump design to prove the usefulness of his patent to the farmers of the Southwest. Heating a regular water steam boiler hot enough to produce 15 horsepower, the Ostrich Farm solar engine could pump 1,400 gallons per minute from a well sixteen feet deep—enough to irrigate an entire three-hundred-acre citrus orchard.[32]

The sight of the machine operating was so fascinating, people came from miles around just to wonder at it. More than a dozen articles were written about it: "Sun Power Is Now at Hand!" being a typical headline. By 1903, Eneas opened his doors, offering the first pump from the Solar Motor Company for $2,160. Yet the cost and the fragile nature of the equipment doomed the company from the start. After making only a couple of sales, Eneas faced failure.

Fig. 4.7. The solar pump designed by Aubrey Aneas in 1901

Yet another inventor soon succeeded in building a viable solar motor: Frank Shuman, of Tacony, Pennsylvania. Shuman first understood that to improve the efficiency, the design had to be based on a low-temperature steam engine. Rather than using ammonia hydrate or sulfur dioxide, however, Shuman invented his own low-temperature, low-pressure engine so that ordinary water could run it at a lower tem-

Fig. 4.8. The solar power company designed and owned
by Frank Shuman in 1907

perature of 200 degrees Fahrenheit. This motor allowed him to use the collector box design rather than the expensive mirrors, as Tellier had discovered before him.

Shuman, an experienced entrepreneur, soon raised the funds necessary to start the Sun Power Company in 1907. He built a sophisticated plant in Egypt, which was 40 percent efficient at absorbing the sun's heat. With five 200-foot-long solar troughs arranged to face the south, the Cairo plant opened in 1913 to an enormous crowd and the entire colonial ruling class. Dignitaries were amazed to see the plant run a fifty-five-horsepower engine and pump an incredible *six thousand gallons per minute.*[33] Not only that, excess hot water was diverted into a series of holding tanks, allowing the motor to operate during the night and on cloudy days. Although it was two hundred square feet of collector for one unit of horsepower, Shuman had nevertheless broken through the cost-effectiveness barrier. Shuman's machine, needing little maintenance and possessing the ability to withstand a windstorm, sold for $8,200.

The Sun Power Company soon had many orders, including big ones from different national governments. But then World War I began, and Shuman dove into defense inventions and business, putting the Sun Power Company on hold. Unfortunately for the world, Shuman did not live through the war years. The first practical solar engine was lost in the era of cheap oil and ignored by a corporatism led by mine owners and oilmen thoroughly opposed to the free distribution of power from the sun.

ELECTRIC CARS

The history of the early electric car also illustrates how cheap oil destroyed the potential of another nonpolluting technology. By 1905, 7 percent of all cars sold were electrics, and electric vehicles *held the land speed record from 1898 to 1902*. In 1899, a Jenatzky electric racecar set a speed record of 66 mph, and a B.G.S. electric could go 180 miles on a single charge. A gasoline-powered 1912 Model T went 35 mph, the same as an electric road car of the day. By 1930, over 130 electric car companies had started up around the world. Unfortunately, the heaviness of the batteries, the lack of an electrical infrastructure to charge the batteries, the extra speed of gas cars, and, especially, the availability of cheap oil put them all out of business.

The Gen 1 or General Motors EV1 was an electric car produced and leased by General Motors Corporation (GM) from 1996 to 1999. The design was born out of the favorable reception of the 1990 Impact electric car concept. The California Air Resources Board (CARB), impressed with the potential of the electric car, passed a state mandate that made the production and sale of zero-emission vehicles a requirement for the top seven U.S. auto manufacturers selling cars in the United States.

GM complied with lease-only agreements as part of a marketing study that was limited to Los Angeles, California, as well as Phoenix and Tucson in Arizona. Though the leases were brisk, GM claimed that the cars were unprofitable and literally crushed them out of existence, despite the pleas and quite visible protests of the customers who loved them.

It seemed GM had a hidden agenda to sabotage the cars because they began putting out strange auto commercials and ads complete with dark lighting and eerie music—obviously designed to leave a disturbing and unpleasant image of the car in the viewer's mind. When interest in the new leases predictably lessened, GM then claimed the cars were unprofitable and used their clout to force CARB to abandon its mandate. In the end, most of the 660 cars were repossessed and destroyed.

It is useful to understand this history of solar power and electric cars, for despite the initial blocking of solar energy and the electric car, these promising technologies will, in the near future, be part of the regeneration revolution. It is obvious that these sustainable technologies were opposed by the energy industry and the car industry, even though they worked and were efficient. And they were opposed for reasons of greed and avarice, as oil and gas production are very profitable, while selling solar engines running on free sunbeams is not.

Another thinker who showed how oil could be replaced was a botanist, philosopher, and humanitarian. His life's work is another critical piece of the workable future that awaits full-scale implementation.

GEORGE WASHINGTON CARVER, THE MAN WHO TALKED TO PLANTS

I found that when I talk to the little flower or to the little peanut they will give up their secrets.

GEORGE WASHINGTON CARVER

The great African American scientist George Washington Carver was one of the most important contributors to the Organic Shift, but not for his famous work on making the peanut a food staple, although that is certainly appreciated. Besides being a riveting orator, Carver was also known as a healer, artist, and pianist. In addition and perhaps more significantly, Carver blazed a path that, if followed, could help end civilization's dependence on crude oil. We must understand the true contribution of his genius and become familiar with his discoveries.

Technology and agriculture must work with nature, rather than against it, Carver said. So one of the first things Carver did was to research and describe how to make a superior compost to restore the fertility of southern soil and make chemical fertilizer unnecessary. He even called the farmer who merely exploited his land a soil robber, while one who improved his land was a progressive farmer. "The farmer whose soil produces less every year, is unkind to it in some way; that is, he is not doing by it what he should; he is robbing it of some substance it must have, and he becomes, therefore, a soil robber rather than a progressive farmer."[34]

He was the first great proponent of organic farming and soil management.

Carver then became the leading light of the chemurgy movement, whose proponents believed that plants, grown on farms, could produce organic alternatives to most of the chemical needs of industry. All in all, Carver developed 118 products for industry from twenty-eight plants. These products included adhesives, axle grease, bleach, creosote, dyes, fuel briquettes, ink, insulating board, linoleum, metal polish, paper, rubbing oils, salve, soil, conditioner, shampoo, shoe polish, shaving cream, synthetic marble, synthetic rubber, talcum powder, vanishing cream, wood stains, and wood filler.[35] In 1927, he invented a process that could turn soy oil into paints and stains, for which he was awarded three patents.

Carver must have been familiar with Goethe and his poem "The Metamorphosis of Plants," for he certainly seems to have followed the participatory scientific method. "When I touch a flower, I am not merely touching that flower, I am touching infinity. . . . You have to love it enough. Anything will give up its secrets if you love it enough. Not only have I found that when I talk to the little flower or to the peanut they will give up their secrets, but I have found that when I silently commune with people they give up their secrets also if you love them enough."[36] Carver had his own spiritual view of the universe, calling it "the great whole." There is an organic unity to the great whole, he said, which humanity must understand in order to live in harmony with nature. For Carver, the solution to everything was love. Love allows you to get in touch with God or nature, which then permits you to go deeper into discovery, helping you to find answers about the world or yourself.

To Carver, three steps allowed one to reach the answer to difficult questions:

1. Love your subject with the compassion of a mother.
2. Be humble before nature, lose your self and relax; commune with nature through love. Know that everything you seek is all part of creation and is good.
3. Expect a miracle in understanding, knowing that the answer is just beyond your grasp. It is then that you must use your intuition and make a mental grab for the solution to the problem.

This was all similar to Goethe's organic approach to science and can be applied universally in many areas. For his work in organic agriculture and his teaching of the benefits of composting, for his innovations in chemurgy, and for his holistic, scientific spirituality, George Washington Carver deserves much appreciation.

One person who appreciated Carver a great deal was the industrialist and agrarian Henry Ford, who in 1928 had begun independent research into chemurgy. By 1931, Ford became obsessed with the potential of the soy bean and started using it in the making of automobiles, his second goal always being to create a better income for farmers. He and Carver became partners. They made much progress and in 1941 actually produced a complete car body made out of plastic derived from soy oil.

Ford and Carver reasoned that once bioreplacements were made, farmers would not only be producing food but also much of the energy and chemicals needed to power civilization and they would grow rich in the process—rather than all of the profits going to corporations. Agricultural woes would then be a thing of the past, and farmers would prosper—an agrarian near-utopia. The *San Diego Union* agreed, editorializing that the use of plastics made from plants may "well bring about something in the nature of a highly desirable and peaceful agricultural revolution."[37]

While the serious media declared it a revolution in the auto industry and agriculture, the mass media—dependent on advertising from oil and car companies—ridiculed the soy car with jokes. "A man could eat his car and have it too" went one. The soy car was "part salad and part automobile."[38] One wag noted it was the triumph of the vegetable over the steel industry. Once again, however, cheap oil and opposition by the corporate establishment smothered the true potential of chemurgy, although soy production today is a major industry and much of the soy that is produced is used in chemical applications. Yet there is so much more that could be replaced and thereby help bring income to today's family farm. Most countries have far more in agricultural potential than they do in fossil-fuel resources. It only makes sense to switch to Carver's chermurgy and give incentives to industry to do so too.

Unfortunately, big oil used public relations to ridicule renewable energy and the research of pioneers like Carver and Shuman. There is no hope from solar power, said these PR campaigns, or from wind power, chemurgy, electric cars, or compost to replace chemical fertilizers. Dupont's winning slogan was "Better living through chemistry." On that last point, our next organic pioneer would disagree.

SIR ALBERT HOWARD AND SOIL HEALTH

[T]here are many discontents in agriculture which seem to add up to a new vision of "biotic farming.". . . The discontent that labels itself "organic farming," while bearing some of the earmarks of a cult, is nevertheless biotic in its direction, particularly in its insistence on the importance of soil flora and fauna.

ALDO LEOPOLD, *A SAND COUNTY ALMANAC*

During the 1940s, a very small organic agriculture movement began. Just as scientific agriculture was part of the mechanistic shift in the seventeenth and eighteenth centuries so also organic agriculture was part of the Organic Shift. Europeans, Scott and Helen Nearing in Maine, and lesser-known farmers were pioneers, showing the world how chemical pesticides and fertilizers were no longer needed and how the farmer could be reconnected to the earth. In 1940, Lord Northbourne's *Look to the Land* was the first book to propose a modern return to organic agriculture. Although out of print and rare, *Look to the Land* should be required reading in every economics 101 class, as it clearly shows the link between losing soil fertility and the downfall of economies and whole civilizations and how we are now repeating that terrible mistake.

Also in 1940 came the classic study on soil fertility by Sir Albert Howard, *An Agricultural Testament*. Although this book had a grandiose title, it did indeed become the bible of the new organic science. Throughout the 1940s, a series of key articles on organic farming by Howard laid the foundation of modern organic agriculture. Howard eventually spent twenty-five years in India researching methods such as composting.

Fig. 4.9. Sir Albert Howard developed the scientific foundation of organic agriculture by explaining soil science and fertility.

In 1942, inspired by Howard, J. I. Rodale first used the term *organic gardening* in his magazine *Organic Gardening and Farming*. Primarily through the efforts of his son Robert Rodale, and then the Rodale Institute, organic agriculture would become a science in the 1960s and 1970s. In 2010, Maria Rodale, a third generation Rodale, published her powerful book, *Organic Manifesto, How Organic Farming Can Heal Our Planet, Feed the World and Keep Us Safe*. Here she brilliantly details how regenerative agriculture and science is the core solution to *all* our interconnected crises—including the climate crisis. These are the critical steps we must take to insure a workable future, for we no longer can afford to use toxic pesticides and herbicides on our farms and lawns. We are all indebted to these pioneers who defied the mechanist school and used science to discover the truth about the soil and how to use it wisely.

While organic agriculture was being rediscovered and the beginnings of soil science were researched in the 1940s and early 1950s, the chemical and oil industries were hard at work marketing chemical fertilizers, herbicides, and pesticides to the world's farmers. Unfortunately, this mechanistic system eventually depletes the soil of its humus and fertility, besides introducing toxic chemicals into the food chain, the environment, and the farmer himself.[39] Thousands of chemical salesmen combed the countryside, persuading growers with slick pitches like "Just listen to me and I'll tell you how to increase your yield." Farmers eventually forgot many of the old tried-and-true natural techniques of their grandfathers.

By the 1950s, herbicides began to be sprayed on suburban lawns and gardens—chemicals now known to cause cancer, learning disabilities in children, and other health problems. This crisis has invisibly grown to this day. We now know, for example, that eliminating lawn chemicals lowers blood levels of pesticides and herbicides in learning disabled children, quickly raising their test scores in school. The pollution points of every community were then multiplied as tens of thousands of chemical products were introduced to the market with little or no testing for health or environmental effects. The synergistic effects of those chemicals mixing together was completely ignored for decades, a terrible crisis that is gravely affecting the quality of our water and environment.

DEEPER CHANGE IS NEEDED

Humanity has come to a crossroads in the history of science and technology. There are two paths to the future. One is to continue down the road of oil, gas, and nuclear power, a race to the bottom of the resource end game. As regards nuclear power—even if the construction of plants were geared up to have an impact on global warming (and by some new invention we could deal with the long-term problem of nuclear waste)—the high-grade uranium that nuclear reactors require is limited. The known high-grade uranium resources of the world would only last seven years, which is totally inadequate for the long-term. After that, burning and mining the low-grade uranium would create just as many greenhouse gases as oil does today, and pile up mountains of radioactive waste that remain a danger for hundreds of years.

The other path is organic science and low-impact renewable-energy technology, and yet corporations and governments dominated by the mechanistic paradigm and the profit motive fight the alternative solution wherever possible. Paradigm resistance and the current power structure support a fossil-fuel and nuclear future.

In the twentieth century, the tide of history slowly started to turn against the now corporate-ruled mechanists, who nonetheless continued to fight for every penny of profit and control they could get away with—no matter how much carbon was emitted and how much blood was shed. The world became even more dependent on drilling and mining as pollution leapt to today's outrageous levels. As long as the old power structure and thinking maintains its control, the ends justify any means.

That's why new technology and new science by themselves are not enough to change the world. The current structure of information dissemination—closely held by the corporate media and old-paradigm governments around the world—must be somehow transcended. Deeper change is needed. There must be some kind of end run around ordinary politics and institutions. Mechanists are not about to change of their own accord. The ordinary people of the world, the general public, must become more aware of how they are being had and make the change happen themselves.

A broad and comprehensive social and political transformation will redefine how we use energy and supply industry, how corporations and governments relate to people and international problems, and how we bring about peace, which in the end is the ultimate goal of the world. Deep change means not only learning what is regen and what is degen, it also means social and personal transformation in the Kantian sense of evolving from *homo phenomenon* to *homo noumenon,* of learning to see the big picture and of knowing reality from the outside in. It means changing lifestyles from the conspicuous consumption and exploitation of nature to living in harmony with Earth and ourselves.

Hand in hand with the new scientific revolution from the mechanistic to the organic paradigm, the Organic Shift, as we have seen, sparked many new social and political movements in the nineteenth and twentieth centuries, movements that eventually began to influence the mainstream. All these movements had several things in common: They were all based on the broader definition of personal and spiritual freedom of the new Kantian paradigm and on the more compassionate morality of *homo noumenon.* They were based on new politics, new beliefs, and new lifestyles. And they were the beginning of the end run around the mechanists that we so sorely need to understand today.

Although little progress was made at first, in the second half of the nineteenth century, the fight for rights and social fairness accelerated, becoming rapid in the twentieth century to follow. There was so much to accomplish by this new social revolution, so many parts of society that were disenfranchised or actually enslaved, and there was so much ideological resistance by the old worldview and power structure. It would take over two centuries for the new social and civil rights revolution to succeed, and its struggle for freedom and equality still continues today.

Little known, however, is how this new definition of freedom began—and how it led to the changes of our own time. It is a fascinating story, another part of the hidden history of the new paradigm.

Mary Wollstonecraft

5

A SOCIAL/CIVIL RIGHTS REVOLUTION

If the abstract rights of man will bear discussion and explanation, those of women, by a parity of reasoning, will not shrink from the same test.

MARY WOLLSTONECRAFT,
A VINDICATION OF THE RIGHTS OF WOMAN, 1792

FREEDOM FOR ALL

Hand in hand with the Scientific Revolution, the shift to a broader worldview meant a much wider definition of freedom and equality, just as it had in the Enlightenment. The mechanists of the American and French revolutions fought solely for the freedom of white males with property, while continuing to disenfranchise minorities and women. In the organic worldview, all should be free and equal and have the vote— minorities, women, and gays and lesbians. Workers should also get their rights and always be free to unionize, strike, and negotiate contracts. Industry must also cease its polluting ways. Communities should be free to organize and protect the very air and water needed to live, to protect nature itself. Although these were unimaginable demands in 1781 when Kant began the Organic Shift, after two centuries, most of those rights have been won, although temporarily threatened by the conservative backlash and regressive politics.

Full-blown movements to free the slaves, give suffrage and liberation to women, save the environment, and create a more equitable distribution of wealth all emerged during the nineteenth and twentieth centuries and succeeded to some degree or another. This broader definition of equality in the new worldview paralleled the "liberty, equality, and fraternity" championed in the shift from the medieval world to the mechanistic. Because new politics are always an integral part of a paradigm shift, so it was that the many progressive trends and movements of the nineteenth and twentieth centuries accompanied the Organic Shift in science as the new definition of freedom for all—even slaves and women—took shape.

THE FOUNDING MOTHERS OF FEMINISM

As Kant was laying down the philosophical foundation of the Organic Shift, a small group of free-thinking writers and radicals gathered together in America and Paris in the 1780s and 1790s. It was this small circle of thoughtful but committed citizens who eventually changed the world centuries later. Among them were the very first feminists.

In 1790 Judith Sargent Murray's essay "On the Equality of the Sexes" was widely read in *Massachusetts Magazine*. In it, she pointed out that the basic equality in mental capacity between men and women is made unequal by the educational paths into which the different genders are channeled:

> Will it be said that the judgment of a male of two years old, is more sage than that of a female's of the same age? I believe the reverse is generally observed to be true. But from that period what partiality! How is the one exalted and the other depressed, by the contrary modes of education which are adopted! The one is taught

*Fig. 5.1. Judith Sargent Murray was an early feminist
who argued that women's intellectual capabilities
were equal to those possessed by men.*

to aspire, and the other is early confined and limited. As their years increase, the sister must be wholly domesticated, while the brother is led by the hand through all the flowery paths of science. . . . At length arrived at womanhood, the uncultivated fair one feels a void, which the employments allotted her are by no means capable of filling.[1]

Another fiery feminist, Olympe de Gouges, lived in Paris during the French Revolution and wrote "The Declaration of the Rights of Woman" in 1791. Later beheaded in the Reign of Terror for hanging around with the wrong crowd, de Gouges pulled no punches in expressing a more inclusive definition of "freedom and equality":

"Man, are you capable of being just? It is a woman who poses the question; you will not deprive her of that right at least. Tell me, what gives you sovereign empire to oppress my sex? . . . Woman, wake up; the tocsin of reason is being heard throughout the whole universe; discover your rights. . . . Oh, women, women! When will you cease to be blind?"[2]

She went on in her declaration to spell out those rights, demanding a national assembly of women, as well as a new concept of marriage that did not include the wife surrendering all her property and her person to her husband. De Gouges made a great contribution to women's rights, yet her efforts, and those of Judith Sargent Murray, were overshadowed by still another founding mother.

MARY WOLLSTONECRAFT

Men submit everywhere to oppression, when they have only to lift up their heads to throw off the yoke; yet, instead of asserting their birthright, they quietly lick the dust, and say, let us eat and drink, for to-morrow we die. Women, I argue from analogy, are degraded by the same propensity to enjoy the present moment; and, at last, despise the freedom which they have not sufficient virtue to struggle to attain.

MARY WOLLENSTONECRAFT,
A VINDICATION OF THE RIGHTS OF WOMAN, 1792

Like America and France, England had a small circle of radical thinkers who were busily reinventing religion, art, philosophy, and politics. Mary Wollstonecraft (1759–1797) joined the radicals in the mid-1780s, after striking out on her own and starting a school for girls. She met the publisher Joseph Johnson, who commissioned her to write *Thoughts on the Education of Daughters* in 1786—predating Judith Murray by four years—in which many of her thoughts took form. She and Johnson started the radical magazine *Analytical Review* in 1788, which closely followed the thinking and the events of the French Revolution.

In 1789, Wollstonecraft wrote a spirited defense of the French Revolution in response to conservative criticism: *A Vindication of the Rights of Man.* Here, she not only stood up for democracy and the sharing of prosperity, she also slammed the slave trade as immoral, along with the English game laws (which outlawed hunting by commoners) and the general treatment of the poor. This book put Wollstonecraft on the map, and she was soon praised by the likes of Thomas Paine, Joseph Priestly, and William Godwin. A woman had written *A Vindication of the Rights of Man*! The first part of Paine's *Rights of Man,* which caused such a radical stir that Paine had to flee England, was not written until 1791.

Wollstonecraft then did for women what she had done for men, when she wrote *A Vindication of the Rights of Woman* in 1792. Realizing that all the talk of freedom and equality was for men only, she laid out the reality of a woman's life for everyone to read. The true precursor to modern feminism, Wollstonecraft correctly pointed out that it must be men who change society and allow "true freedom" to flourish, by not only freeing the slaves but also the women.

As women could not attend school at the time, she explained that little or no education for women leaves them in a state of "ignorance and slavish dependence." So women are made to be "docile and attentive to their looks to the exclusion of all else." Rejecting the marriage laws of the eighteenth century, which turned all property of the wife over to the husband, she declared that women "may be convenient slaves, but slavery will have its constant effect, degrading the master and the abject dependent."[3] The rights of woman and the rights of man were in fact one and the same thing, explained Wollstonecraft. She argued for the female right to vote—for a women's congress— over a century before 1920, when U.S. women finally won their full suffrage. Moreover, she cried, true equality could only be won when the male hierarchies of the monarchy, the church, and the military were done away with. Wollstonecraft, like the strong-willed progressive women who followed her, was not one to pull any punches.

As one can imagine, a book like this—in 1792 London—caused a huge controversy. She was vilified as "a hyena in petticoats." In 1793, the conservative Edmund Burke, worried that the Reign of Terror of the French Revolution would cross the English Channel, declared the London radicals "loathsome insects that might, if they were allowed, grow into giant spiders as large as oxen."[4] A proclamation forbidding all seditious writings and meetings was quickly issued by King George III. Like Thomas Paine before her, Wollstonecraft decided it was time to flee, and she left for France.

Sadly, Wollstonecraft died of blood poisoning when she gave birth to her daughter Mary in 1797. Yet that daughter grew up to become Mary Shelley, the writer of the ultimate antimechanist novel *The Story of Frankenstein*. After a long period of great influence in the nineteenth century, Wollstonecraft's personal life and psyche were savaged in a relentless series of conservative attacks, covering up her contribution to the new worldview's expanded concept of true freedom.

The founding mothers were all consigned to obscurity, another hidden history never taught. Nevertheless, their influence in their own time was significant. Like Mary Wollstonecraft, most feminists were also against slavery and helped to energize the second movement for freedom for all: abolition.

We will never know where women's rights would be today without these founding mothers of feminism.

FEMINISM AND ABOLITION BECOME MASS MOVEMENTS

In the early nineteenth century, the embryonic forms of the new freedom movements were considered highly radical: the abolitionists and the feminists. In the new worldview, slavery *must* be outlawed for there to be true freedom, with both blacks and women getting full and equal rights. Pacifism, feminism, and abolition all became strong mass movements in the first half of the nineteenth century. It was strongly interconnected, with feminists packing abolitionist meetings and vice versa.

The antislavery movement in the United States entered a new era in 1829 when a freed black slave, David Walker, printed and distributed the "Appeal," an inflammatory pamphlet distributed to five hundred thousand people. Walker was one of the first militants to go national, as the "Appeal" was read to crowds of free and enslaved blacks all over the young country, demanding that slavery be ended—by rebellion and violence if necessary. He wrote: "[T]hey want us for their slaves, and think nothing of murdering us . . . therefore, if there is an *attempt* made by us, kill or be killed."[5] Although he called for African Americans to throw off their chains and fight for their freedom, he predicted that in the end, after emancipation and full suffrage, both races would live in friendship.

> I declare to you, while you keep us and our children in bondage, and treat us like brutes, to make us support you and your families, we cannot be your friends. . . . Treat us then like men, and we will be your friends. And there is not a doubt in my mind, but that the whole of the past will be sunk into oblivion, and we yet, under God, will become a united and happy people. The whites may say it is impossible, but remember that nothing is impossible with God.[6]

This new militancy of freed blacks led many, especially northern women, to organize and demand the end of slavery immediately rather than gradually, and that made abolition a powerful mass movement. William Lloyd Garrison's fiery antislavery newspaper *The Liberator* then appeared in 1831, ratcheting up the debate. A number of slave rebellions broke out in the South, with at least fifty-five whites being killed in Nat Turner's rebellion, which erupted in Virginia during August 1981. Many white and black antislavery societies were founded, and these soon coalesced into mass meetings around the North. A national meeting was held in Boston in 1835, and a world conference followed in 1840 in London. All of it was radical—and hated by the slavers, who themselves organized mob attacks on the abolitionist newspaper offices.

In 1841, at a meeting of the Anti-Slavery Society in New Bedford, Massachusetts, the speaker heard that an escaped slave was actually in the room and asked him to come up and tell of his escape. The young twenty-three-year-old stepped to the front of the crowd and then held them spellbound for two hours with his powerful plea for the end of slavery. His name was Frederick Douglass. He went on to energize the whole abolitionist movement, giving inspiring, searing speeches in a tour of one hundred regional conventions in 1842.

Frederick Douglass later became the most famous and radical African American in the country, writing articles, advising Lincoln, and recruiting black regiments to defeat the Confederacy. After the Civil War and after his wife died, Douglass shocked society by marrying a pretty white woman twenty years his junior. In his final years, he strongly advocated equal rights for women, repaying the feminists for their support of abolition. By the time he died at seventy-eight in 1895, he was a senior statesman, despite his interracial marriage.

While the big fight was over slavery, in 1848 women held the world's first feminist convention in Seneca Falls, New York, which was the first large assembly of women devoted to finding their own freedom. Although the idea of feminine equality had been part of an emerging counterculture since 1792, Mary Wollstonecraft's *A Vindication of the Rights of Woman* helped tip the balance even further in the direction of women's rights. Ever since the patriarchy took over in the fifth and fourth millennia BCE and made male-oriented myths a central theme of religion, men had never had a serious challenge to their social and political power. That free ride was suddenly over. Men now had to deal with activists like Elizabeth Cady Stanton and Lucretia Mott.

Using the Declaration of Independence as a model, Elizabeth Cady Stanton led the Seneca Falls convention by reading her declaration of female independence. "The Declaration of Sentiments" proclaims that "all men *and* women are created equal." Just as the founding fathers had

*Fig. 5.2. Elizabeth Cady Stanton was
another early, outspoken feminist.*

presented a list of complaints to King George, the feminists listed their grievances with the opposite sex and their laws.

> The history of mankind is a history of repeated injuries and usurpations on the part of man toward woman, having in direct object the establishment of an absolute tyranny over her. To prove this, let facts be submitted to a candid world.
>
> He has never permitted her to exercise her inalienable right to the elective franchise.
> He has compelled her to submit to laws, in the formation of which she had no voice.
> He has withheld from her rights, which are given to the most ignorant and degraded men—both natives and foreigners. Having deprived her of this first right of a citizen, the elec-

tive franchise, thereby leaving her without representation in the halls of legislation, he has oppressed her on all sides.

He has made her, if married, in the eye of the law, civilly dead.

He has so framed the laws of divorce, as to what shall be the proper causes, and in case of separation, to whom the guardianship of the children shall be given, as to be wholly regardless of the happiness of women—the law, in all cases, going upon a false supposition of the supremacy of man and giving all power into his hands.

After depriving her of all rights as a married woman, if single, and the owner of property, he has taxed her to support a government which recognizes her only when her property can be made profitable to it.

He has monopolized nearly all the profitable employments, and from those she is permitted to follow, she receives but a scanty remuneration.

He closes against her all the avenues to wealth and distinction, which he considers most honorable to himself. As a teacher of theology, medicine, or law, she is not known.

He has denied her the facilities for obtaining a thorough education, all colleges being closed against her.

He allows her in Church, as well as State, but a subordinate position, claiming Apostolic authority for her exclusion from the ministry, and, with some exceptions, from any public participation in the affairs of the Church.[7]

The declaration fell, for the most part, on deaf ears as the white male-dominated mechanistic civilization rolled on—even though slavery was ended in the Civil War. As most of the feminists were also strong abolitionists, they rejoiced at the end of slavery. Yet when black males were enfranchised after the Civil War and females of any color could still not vote, the women organizers took deep offense. It took another fifty-five years—a tough half century of noisy protests, jail time, and recrimination—to finally win universal suffrage in 1920. It ended up being the longest and most successful sustained civil rights movement in history, enfranchising half the population on the same day. The feminist pioneer Elizabeth Cady Stanton and her fellow organizers deserve much of the credit.

THE SUFFRAGETTES

Following "The Declaration of Sentiments" in 1848, women began their fight for the vote. After visible protests, dynamic parades, much organizing, and letter writing, the feminists began to see results in 1861, when the new state of Kansas let women vote in school-board elections. It was a small first step. Between 1868 and 1870, Elizabeth Cady Stanton published a newspaper, *The Revolution,* with a young Susan B. Anthony, which advocated equal pay for women, an outlandish idea at the time. They also insisted that women be emancipated along with the slaves. They were, for the most part, ignored as much as possible.

So Anthony took action, leading a group of women in Rochester to the polls in 1872, demanding that the poll workers give them their rights. Two weeks later, she was arrested. She next set out on well-publicized tours, telling halls full of women her views on the matter. After being convicted on voting law violations, Anthony refused to pay the fine—and got away with it. She went on to work for female suffrage full-time until her death in 1904, establishing organizations and councils that carried on the fight for a full half century.

Winning state suffrage was far easier than getting the right to vote in federal elections. That required a constitutional amendment needing ratification by two-thirds of Congress and three-fourths of the state legislatures. In 1869, the Wyoming Territory granted the vote in local and state elections to women and even the right to hold public office—an attempt to lure more people to their region, as well as a way to support the pioneer women already there. Women were civilizing the place with schools and community building, and the need was there for them to sit on school boards. Wyoming, in short, needed the talent. The Utah Territory followed suit in 1870, along with Colorado and Idaho in the

1890s. Not only were the farmers in the streets in the 1890s, so were the unions and now the suffragettes. It was a time of protest and a broader awareness. Despite the death of Elizabeth Cady Stanton in 1902 and Susan B. Anthony in 1906, the women's suffrage movement nevertheless fared far better than the Union Movement during these decades.

Australia gave women the vote in 1902, and Finland in 1906. Meanwhile, in England, there were many suffragette associations, which all united in 1897 under the banner of the National Union of Women's Suffrage Societies. The agitation was ignored as much as possible by Parliament. Then the situation became far more tense. Emmeline Pankhurst founded the radical Women's Social and Political Union (WSPU) in 1903. Their motto "Deeds, Not Words" reflected a new conviction to take action and hold protests to achieve their single-minded goal. Pankhurst's daughter was the first to physically fight back and be arrested after crying out at a 1905 public meeting, "Will the liberal government give the vote to women?" From that point on, militant actions were the strategy for the WSPU, which included the 1908 stoning of 10 Downing Street, breaking the windows of the prime minister's residence. Twenty-seven were arrested and locked away in Holloway Prison. The suffragettes' next stunt was to enter Parliament itself, with twenty-four being arrested.

All in all, of the two thousand members of Pankhurst's WSPU in 1914, fully one thousand of them had been arrested during the fight for suffrage. Salted away in the British prison system, the WSPU members declared themselves political prisoners, demanding conditions better than that of common criminals. They got no such relief, and widespread hunger strikes began in 1909. Instead of giving in, the guards were ordered by the government to force-feed them. It was for these trials and tribulations that the name *suffragette* was earned.

The Great War, as World War I was known, changed everything. The WSPU made a deal with the British government to suspend all actions and protests for the duration and furthermore to assist in the war effort. In return, the government released the prisoners and turned over two thousand pounds to help make a demonstration supporting the war. Thirty thousand women turned out for that march, instantly changing all the dynamics of the Suffrage Movement by carrying banners that read: "We demand the right to serve," "Let none be Kaiser's cat's paw," and "For men must fight and women must work." Women could now prove their worthiness of the vote by working hard to help win the war. As a result of their hard work replacing men in the factories, female citizens of the United Kingdom were soon awarded their prized goal in 1917—even though it was suffrage for only those over thirty who were householders, who owned property with an annual rent above five pounds, or who were graduates of British universities.

During the same years, the suffrage movement in the United States also grew militant. Larger and larger suffrage parades naturally caused the formation of a conservative countergroup, the National Association Opposed to Woman Suffrage, in 1911. Activist Alice Paul led five thousand suffragettes in a parade in Washington, D.C., on the same day as Woodrow Wilson's inauguration in 1912. She was trying to get his attention.

State after state in the United States meanwhile granted women suffrage, while members of the National Woman's Party picketed the White House. Paul and ninety-six others were imprisoned for obstructing traffic. After the arrest and rough treatment, the suffragettes started a hunger strike and, as in England, were force-fed by the authorities. In 1916, Jeannette Rankin—a Montana Republican—was sent to the U.S. House of Representatives as the first female member of Congress. After the Great War, as in England, the tide turned. A constitutional amendment was finally passed in 1920, giving women the right to vote in all U.S. elections.

It was a civil rights movement that had lasted a total of almost seventy-five years. And it was finally over—at least the part to win the vote. As the Organic Shift continued to develop, the female liberation struggle went on to economic and cultural freedoms.

Yet the new thinking did not stop there. The new definition of equality meant freedom for *all*.

WALT WHITMAN
AND GAY LIBERATION

The American bohemian poet Walt Whitman (1819–1892), living and writing in New York in the middle of the century, is considered a national treasure. Whitman, known for *Leaves of Grass* and many other poems, shook up American society with his freethinking worldview—which included writing about the reality of sex, money, and power. He eventually had an enormous influence on American life. Every high school student is now taught Whitman—not just for his poetry but also to teach what America is really about. Whitman revealed much about the country and its people that was true, yet entirely new.

Whitman had, however, a secret life that has, for the most part, been kept from the public. Known for his verse concerning his love of women, what is not known is that he erased *he* and wrote in *she* in order to sell those poems—for Walt Whitman was a rough-hewn gay man whose lover was Peter Doyle, an illiterate worker who could neither read nor write.[8]

Gay rights were part of the broader definition of freedom under the new worldview, which simply proclaimed full civil rights for blacks, women, and now homosexuals. Edward Carpenter, a gay rights advocate in England, wrote Whitman in 1876, thanking him for everything he had done toward the ultimate goal of gay liberation:

> The distance remains immeasurable. . . . As soon as I remember what the end is—however great the distance—I do not doubt. Dear friend, you have so infused yourself that it is daily more and more possible for men to walk hand in hand over the whole earth. As you have given your life, so will others after you—freely, with amplest reward transcending all suffering—for the end that you have dreamed.[9]

Although the struggle for gay rights would last more than a century, Whitman could see it becoming real in the distant future. The

Fig. 5.3. Walt Whitman and his lover, Peter Doyle

organic worldview always accepted the full equality of all people, no matter their sex, color, or sexual orientation, and nothing worried the old paradigm more than the freedom of these oppressed minorities. It saw within these liberation movements the seeds of destruction for the old bigoted culture and its eventual loss of power, for the old paradigm lived on the divisions and scapegoats that prevented the nonrich from unifying.

THE EMERGENCE OF SOCIALISM

Early unions, or guilds, made up of craftsmen like cabinet makers, cobblers, and carpenters first emerged in the American Colonies during the

1700s. Carpenters were at the center of the Boston Tea Party in 1773, while the Continental Congress met at Carpenters Hall in Philadelphia in 1776. The first strikes were from the New York printers in 1794, the cabinet makers in 1796, then Philadelphia carpenters, and so on. In the early 1800s, these primitive organizations tried to improve workers' conditions and succeeded in reducing the workday from twelve hours to ten. The concept of forming international union organizations was championed by the feminist Flora Tristan, and then by Karl Marx and Friedrich Engels. Unions were seen as the way to build the political and economic power of the working class, the way to lift the workers' standard of living.

Marx and Engels created "scientific" socialism in 1848—as opposed to the earlier utopian socialism of the French Revolution and the utopian commune movement. This eventually devolved in the twentieth century to a very mechanistic communism, tyrannies that treated people even more like machines than their capitalist counterparts did. It should be noted that by the 1880s, Marx himself had given up the notion of bloody revolution and had become an active Social Democrat, believing in elections as the way to change the world. Although it often preached violent revolution, nineteenth-century socialism was more altruistic than its twentieth-century Communist descendant. The ideals of the utopian socialists and the concept of the commune appealed to the bohemian young men, and many leaned in that direction—or toward total anarchism. Socialism and the organic worldview were thus often—but not always—interrelated during the nineteenth and early twentieth centuries.

THE RISE OF
THE ROBBER BARONS

Between the Civil War and 1900, the Industrial Revolution swept America, creating the first multimillionaire crony capitalists, the robber barons of the nineteenth century. Infamous market manipulators like Jay Gould repeatedly caused stocks to crash, so good companies could be bought for pennies on the dollar. The difference in wealth between the top 1 percent and the poor and middle class grew dramatically in these years, reaching a peak in 1900, when at least 45 percent of the wealth was owned by the top 1 percent of households. The stranglehold the robber barons had on the courts and the government of the United States at the time is now legend—an iron grip that proved extremely detrimental to ordinary workers and their families.

Abraham Lincoln himself worried that corporations had grabbed too much power during the turmoil of the Civil War and would use their money to deceive the people—until all the wealth had been "aggregated in a few hands." He was not far off.

> It has indeed been a trying hour for the Republic; but I see in the near future a crisis approaching that unnerves me and causes me to tremble for the safety of my country. As a result of the war, corporations have been enthroned and an era of corruption in high places will follow, and the money power of the country will endeavor to prolong its reign by working on the prejudices of the people until all wealth is aggregated in a few hands, and the Republic is destroyed. I feel at this moment more anxiety for the safety of my country than ever before, even in the midst of war.[10]

In the post–Civil War period, most of the robber barons made their fortunes on the railroads or in financing the railroads. Leland Stanford, Cornelius Vanderbilt, James Hill, and Collis Huntington were some of the first railroad moguls. Their influence on the government garnered them great tracts of land as a reward for laying track across the country, making them all richer still. Collis Huntington wrote in 1877 about how a robber baron got things done in Washington: "If you have to pay money [to a politician] to have the right thing done, is is only just and fair to do it. . . . If a [politician] has the power to do great evil and won't do right unless he is bribed to do it, I think . . . it is a man's duty to go up and bribe."[11]

THE LABOR UNION MOVEMENT

As we saw earlier, labor had organized unions and strikes here and there among certain occupations. Now the unions fulfilled the dreams of the 1830s and 1840s and forged international associations of unions, great brotherhoods and sisterhoods of unions. The National Labor Union (NLU) in 1866 was the first, a federation of labor unions. It managed to pass legislation mandating the eight-hour day for federal workers, a step toward an eight-hour day for all workers. The Panic of 1873, however, put the NLU out of business.

This depression impoverished the craftsmen and workers as well as the farmers. By the winter of 1873 to 1874, tens of thousands were literally starving to death, yet the bad times were only just beginning. By 1877, 27 percent of the population was unemployed, while those who did have work had suffered a 45 percent pay cut. Starvation and suicide were rampant.

Meanwhile, most workers were upset at the results of the U.S. presidential election of 1876, in which the Republican Rutherford B. Hayes lost the popular vote but won when twenty electors—from Florida of all places—were disputed by the Republicans and not allowed to vote. This threw the election into the House of Representatives, where the Republicans, in perhaps the most despicable deal in the history of American politics, agreed with southern Democrats to repeal Reconstruction. For their agreement to a Republican fix on the members of an electoral commission, Hayes would finally withdraw the occupying northern troops—who had made the South safe for freed slaves. The deal, the Compromise of 1877, was struck in secret.

The federal troops were withdrawn north by Hayes, leaving African Americans, who had risen to many public offices, including Hiram Revels and Blanche K. Bruce to the U.S. Senate seat in Mississippi, at the mercy of their former owners. Lynch mobs and the Klan were soon out in force, and civil rights were suppressed in the South until the 1950s and 1960s. African Americans, in fact, refer to the Compromise of 1877 as the Great Betrayal.

Another secret deal was made by the Republicans with Thomas Scott, president of the Pennsylvania Railroad, the nation's largest. In return for his influence on some congressmen, Scott was promised a federal bailout of his failing Texas and Pacific Railroad. In June 1877, Scott's Pennsylvania Railroad cut the pay of the rail workers by 10 percent, after having already made a previous 10 percent cut. The next month, the company announced that the number of all trains eastbound from Pittsburgh would be doubled—without any increase in work crews.

This proved to be the final straw. Furious workers threw switches, blocking the movement of all railcars. On July 13, another railroad, the Baltimore and Ohio, lowered by 10 percent the pay of those earning more than a dollar a day and cut the workweek to only two or three days. Fireman and brakemen struck three days later. Replacements were brought in, so the strikers assembled near Baltimore and blocked the rails. When the governor called out the militia, a gunfight and riot broke out, leaving eleven dead. Half the regiment deserted in the chaos, with the rest holing up inside Camden Station.

Also trapped inside were a number of B and O officials, the board of police commissioners, Mayor Latrobe of Baltimore, and Governor Carroll of Maryland. Outside Camden Station were fifteen thousand angry—and drunk—strikers, unemployed and workers of all types. One striker told a reporter from Philadelphia: "The working people everywhere are with us. They know what it is to bring up a family on ninety cents a day, to live on beans and corn meal week in and week out, to run in debt at the stores until you cannot get trusted any longer, to see the wife breaking down under privation and distress, and the children growing sharp and fierce like wolves day after day because they don't get enough to eat."[12]

After the rioters went home to go to sleep, the government was able to restore order the next day, but the rail strike spread all across the country. Similar battles between strikers, workers, and troops were repeated in West Virginia, Pittsburgh, St. Louis, Chicago, and elsewhere.

In Pittsburgh, local businessmen, upset at railroad freight rates, sided with the strikers. When the local police and militia also joined the strike movement, federal troops were called out. Greeted by a hearty stoning, the troops opened fire—killing twenty men, women, and children and wounding twenty-nine. The ground was littered with dead. Enraged, the entire town gathered, arming themselves by breaking into a gun factory. Twenty thousand people—five thousand of them armed—surrounded the roundhouse to which the army had retreated. After an all-night gunfight, the troops fought their way out of town. The mob merrily burned the railroad office building, the station, the yard roundhouse, and dozens of railcars. The *New York World* warned that Pittsburgh was "in the hands of men dominated by the devilish spirit of Communism."

In Chicago, where it had become a general strike, General Sheridan's men, recently recalled from the South, shot many workers; while in St. Louis the arrival of military force broke the strike without any violence. Backed by the U.S. Army, the strike was finished across the country by early August. "We were shot back to work," complained one striker. In all, out of one hundred thousand out on strike, more than one hundred died and more than one thousand were imprisoned. The United States was never quite the same, having come so close to total revolution.

The robber barons, determined to not lose control like that again, had the government start the National Guard we know today. While we think the National Guard is there to help in case of emergencies, its primary mission is actually to protect the wealth of the superrich. Huge armories in several large cities were built, complete with gun holes to fight off potential hordes of urban poor and unemployed. Meanwhile, the labor movement, seeing it could not fight the army, decided to get political and change the government. The Knights of Labor formed, and by the mid-1880s, they boasted a membership of 750,000.

Calling a general strike to shorten the workday from twelve to eight hours, 340,000 participated. Yet in trying to break up a labor meeting in Haymarket Square in Chicago, a bomb went off, killing a

Fig 5.4. In the latter part of the nineteenth century, economic inequality between the haves and have-nots manifested in numerous riots and served as the catalyst for the formation of the National Guard.

policeman. The Chicago police fired into the crowd, killing one and wounding more. Blaming the incident on the Knights, four organizers were hung, which eventually put the national union out of business.

The hangings of the union organizers did wonders for the morale of the robber barons, who were now emboldened to do as they pleased. The barons were soon being backed up by the federal army itself. After an 1892 battle at the Carnegie Homestead Mill, a robber baron now merely had to hold firm with company thugs and state militias until the federal troops arrived.

In 1893, the economy took another plunge, due to yet another panic arranged by the robber barons. It was the third depression in twenty years and made a bad situation much worse. Railroads cut and reduced in every area. The climax finally came in 1894, when Jennie Curtis, president of American Railway Union (ARU) Local 269, the "girls' union," and a worker of Pullman cars, spoke at a convention of the American Railway Union:

> We struck at Mr. Pullman because we were without hope. We joined the American Railway Union because it gave us a glimmer of hope. Twenty-thousand souls, men, women, and little ones, have their eyes turned toward this convention today; straining eagerly through dark despondency for a glimmer of the heaven-sent message which you alone can give us on this earth.
>
> Pullman, both the man and the town, is an ulcer on the body politic. He owns the houses, the schoolhouse, and the churches of God in the town he gave his once humble name.
>
> And, thus, the merry war—the dance of skeletons bathed in human tears—goes on; and it will go on, brothers, forever unless you, the American Railway Union, stop it; end it; crush it out.
>
> And so I say, come along with us, for decent conditions everywhere![13]

Of the rail workers, 125,000 agreed not to work Pullman cars, and 135,000 more union workers went out on general strike in the midst of America's worst depression ever. Battles between strikers and troops erupted in twenty-six states, and thousands were imprisoned for years, including Eugene Debs, the organizer of the ARU. The original Pullman strikers themselves were starved into submission, the back of the union movement broken.

Although the torch was passed in 1886 to the American Federation of Labor (AFL), which, of course, survives to this day as the AFL-CIO, the labor movement had basically been defeated by violence and would not have a real impact for decades. Never again would labor so directly challenge capital in the United States. Using court injunctions, thousands of thugs, and the full moral and military backing of the government, the robber barons were able to slow and even beat back the advances of the labor union movement during the rest of the nineteenth century. The unions themselves went international, agitating across national borders for change and a better life—yet progress in the work-place was slow indeed.

THE POPULISTS

As the laissez-faire economics of the nineteenth century gave free rein to the robber barons of the day, disaster came quickly—compounded by the outrageous prices the railroad magnates charged farmers to haul their grain and travel on the trains. Federal railroad land grants had deeded to the robber barons huge tracts of land on either side of the track, but this did not stop the rail companies from draining the farmers' only chance of profit.

The Panic of 1873 had deflated agricultural prices, and the farmers had been hurting for well over a decade. Desperation followed as economic ruin and depression ravaged the U.S. economy. Believing that increasing the amount of currency available would help the economy and raise prices, farmers started the Greenback Party and actually sent fourteen congressmen to Washington. Farmers wanted more greenbacks, more paper dollars in circulation; since the Civil War the U.S. government had withdrawn greenbacks from circulation, and their value in gold was only *half* their face value. Despite early success, the Greenback Party soon collapsed when new movements took its place.

One could not say that many farmers had philosophically dissected the struggle between the mechanistic and organic paradigms. Yet some farmers were radical socialist immigrants from Europe and their input—plus common economic sense—revealed a new progressive path. Farmers' alliances formed in the North and South, and the black farmers, not allowed to join the Southern Farmers' Alliance because of their skin color, formed the Colored Farmers' Alliance, led by the white minister Richard Humphrey. Both black and white, however, soon united politically—shocking the two major parties.

On election day in November 1890, the new People's Independent Party, the Populists, won clear control of the Nebraska senate, with eighteen seats—compared to the combined total of fifteen for the

Republicans and the Democrats. They also won complete control of the house, by fifty-four Populists to the combined Republican/Democrat total of forty-six seats. Yet the Democrat took the governor's race by a mere 1,144 votes over the People's Independent candidate. The Republican came in third. William Jennings Bryan, on the Populist and Democrat ticket, was meanwhile elected to the state legislature, along with Populist Omer Kem.

In the first state government of its kind, the Populists and Democratic Governor Boyd quickly passed a set of laws specified in the People's Party platform. These included:

- Free public education for every Nebraskan child
- Free textbooks
- Public fund deposit law
- The eight-hour day as a legal day's work, except farm labor

The main Populist goal, however—the regulation of the railroad freight rates—was vetoed by the Democratic governor. That proved to be the undoing of the uneasy alliance between Democrat and Populist. At the same time all this was going on, a wave of populism was sweeping through farming communities in the North, South, and West, and more state governments were soon taken over by the rural radicals. The preamble for the 1892 Populist platform for the presidential election read:

The conditions which surround us best justify our cooperation; we meet in the midst of a nation brought to the verge of moral, political and material ruin. Corruption dominates the ballot-box, the Legislatures, the Congress, and touches even the ermine of the bench. . . . The newspapers are largely subsidized or muzzled, public opinion silenced, business prostrated, homes covered with mortgages, labor impoverished, and the land concentrating in the hands of capitalists. The urban workmen are denied the right to organize for self-protection, imported pauperized labor beats down their labor, a hireling standing army, unrecognized by our laws, is established to shoot them down, and they are rapidly disintegrating to European conditions. The fruits of the toil of millions are boldly stolen to build up colossal fortunes for a few, unprecedented in the history of mankind, and the possessors of these, in turn, despise the Republic and endanger liberty.[14]

And that's just the beginning of the preamble! The Populist Party, also called the People's Party, nominated James B. Weaver for president, who took a million votes and twenty-two electoral college votes in the 1892 election. Those million votes got noticed, injecting the goals of populism firmly into the political debate. In the very next presidential election cycle, a young orator, the editor of a Democratic newspaper, captured the spotlight with hot Populist rhetoric, pledging to throw the Republicans out.

This was William Jennings Bryan, one of the most exciting speakers in all American history. At the 1894 Nebraska convention, Bryan and Silas Holcomb brought together the Populists and Democrats, with Bryan on the ticket for U.S. Senate and for every other major post a Populist—including Holcomb for governor. The Republicans were looking at a losing proposition if they stuck to the usual issues, so they exploited religious sentiment and took up a new issue: Prohibition. Due to local splits and the clever Prohibition campaign, the Republicans ended up controlling the Nebraska legislature for much of the decade. The last Populist Nebraska governor was William Poynter in 1898.

Bryan, meanwhile, went on in 1896 to ride the Populist wave—all the way to the Democratic presidential nomination. Bryan's nomination united the Democrat and Populist parties, effectively bringing to an end the revolt against the two-party system. Unfortunately, the Democratic Party of that day was so racist and southern-white oriented that black Populist voters were immediately offended, and populism as a force strong enough to win elections was finished. Bryan lost to McKinley.

Despite its demise, Populist ideas thereafter helped guide the Democratic Party. Meanwhile, insurgent Republicans formed a progressive wing, which was soon running the country in the form of Teddy Roosevelt. Although black and white farmers and workers had briefly united in this dramatic third party attempt, racism and economic issues kept them mostly divided for the rest of the century, and the more rural white communities returned to conservative voting patterns. Nevertheless, populist sentiment remains close to the surface in most farming regions to this day, waiting to be tapped by some future political movement.

FARMERS UNIONS AND THE LABOR WARS

Out on the farm, things actually got better after 1910, as the weather and the economy improved. Many organizations like the Farmers' Union and the California Fruit Growers Exchange started to give farmers a unified voice in the new century. In the 1920s, they made strong lobbies in Congress. At the same time, the farmers' cooperative movement spread from 1920 to 1932, pooling financial resources together to avoid the middleman. The populist feeling was still there for most, but as life returned to normal, the old conservatism reestablished itself over time. Drought and depression, however, would soon return.

Union organizers and the labor movement had a much harder time. In fact, the whole period from the 1870s through the 1930s was called the Labor Wars. The fighting was at its bloodiest in the early years of the twentieth century, as state and federal governments deployed armed might to put down strikes and demands for such radical things as a child labor ban and the eight-hour day. Just a small portion of the toll was horrific. Fourteen miners dead, twenty-two injured by scab herders in Pana, Illinois, in 1902; six union miners dead, fifteen taken prisoner by Colorado militia, and seventy-nine strikers deported to Kansas in 1904. In 1911, there was the notorious Triangle Shirtwaist Fire in New York, killing 147—mostly women and children. Bosses had locked sweatshop doors "to stop the interruption of work." Despite the owners being indicted for manslaughter, there was little reform of conditions, although shortly thereafter child labor was finally banned.

The Labor Wars then heated up in 1914 with the Ludlow Massacre in Colorado. Nineteen died, including twelve children, when Rockefeller guards and state militia machine-gunned a union tent. State militia also crushed a strike that same year in Butte, Montana. Industrial Workers of the World (IWW) leader Joe Hill was arrested and executed for murder in a controversial 1915 trial, while IWW organizer Frank Little was lynched in Butte in 1917. United Mine Workers (UMW) organizers Ginger Goodwin and Fannie Sellin were killed in 1918 and 1919. Also in 1919, there was the Great Steel Strike of over 350,000 steelworkers, as well as a general strike in Seattle, which shut down the city for six days. All this, and the Communist Revolution in Russia, provoked a powerful conservative counterreaction in the form of the Red Scare and the Palmer Raids, which deported hundreds of radicals and organizers back to Europe and Russia, while imprisoning key labor leaders for years.

The early 1920s witnessed even larger fights as several thousand miners took up arms in West Virginia—although most gave up or stole away in the face of federal troops and fourteen air force bombers, armed with tear gas and fragmentation bombs. Unrest continued until 1922, when the miners finally saw how unions could not stand up to the army. UMW membership in the district went from fifty thousand to six hundred by 1932.[15] As in the nineteenth century, the workers had been shot back to work.

The powers that be were clearly determined to stop unions at any cost. They finally succeeded when they put the movement on the defensive by using bombings and murders supposedly committed by organizers. This tactic was used again and again. Industry hired thugs while troops continued to kill strikers and organizers.

- 1922 Thirty-six killed at a coal-mine strike in Herrin, Illinois.
- 1924 An IWW hall in San Pedro, California, was raided and then razed to the ground.

- 1926 Textile workers battled police in Passaic, New Jersey.
- 1927 Organizers Sacco and Vanzetti executed.
- 1927 Picketing miners massacred in Columbine, Colorado.

The list goes on.

Directly confronting the powers that be took on other forms. The new thinking challenged the old worldview and power structure in a very different way. Protest and action against the nationalist/corporate grab in the late nineteenth century by European powers and the United States led to something new in the world, an organized and oftentimes government-supported movement to bring about world peace. For a while, Immanuel Kant's dream looked as though it might become a reality. Unfortunately, it did not turn out that way.

THE PEACE MOVEMENT
OF THE NINETEENTH CENTURY

During the last half of the nineteenth century, an influential and international peace movement helped the world avoid conflict, laying the foundation for today's peace organizations and demonstrations. Pacifism began to gather energy in the 1890s as a response to the land grabs of the Colonial era, when Western European nations divided up Europe and Asia among themselves. William Cremer of England and Frederic Passy of France led the formation of the Inter-Parliamentary Union of 1892, a multigovernment organization devoted to ending war. The establishment of the Nobel Peace Prize soon did even more to promote pacifism.

Then, in 1899, Queen Wilhemina of the Netherlands hosted the First Hague Peace Conference, attended by one hundred official representatives from twenty-six countries. They met to establish what we know today as international law. The most important result of the peace conference was to agree on a process to handle international disputes through a global Permanent Court of Arbitration. Disarmament, international humanitarian law, and the laws of war were also on the table, and other agreements soon followed.

Andrew Carnegie, the steel magnate, was so impressed with the First Hague Conference that he donated $1 million for the construction of a Palace of Peace, which is currently the home of the Permanent Court of Arbitration. The Netherlands provided the land, and each of the twenty-five other countries provided something—fine wood from here, stone from there, and so on. As as result, the construction of the palace was a truly cooperative effort. U.S. secretaries of state soon spent much time arguing arbitrations in front of the new court, which later evolved into the League of Nations and then the United Nations.

In 1898, the peace movement was confronted with a great challenge: the Spanish-American War, which the pacifists charged was a war of imperialism camouflaged as a war of liberation. The pacifists and new-paradigm thinkers of America united in an astonishing antiwar movement, a precursor to the anti–Vietnam War movement of the 1960s. Nineteenth-century nationalism and colonialism was now for the first time openly challenged by the new worldview—and it gathered much attention.

THE ANTI-IMPERIALIST LEAGUE OF MARK TWAIN AND WILLIAM JAMES

Remember the Maine! To hell with Spain!
RALLYING CRY FOR SPANISH-AMERICAN WAR

When President McKinley and the infamous yellow journalism provoked a war with Spain to grab away her remaining colonies of Cuba and the Philippines, a nascent peace movement quickly realized that the Republicans had a hidden war agenda: to acquire Spain's colonies in the name of freedom and liberation. As a result, something very unusual began—a strong and public anti-imperialism movement, which became the first strong antiwar movement and garnered much press at the time.

Following the death of over two hundred sailors in the controversial sinking of the USS *Maine* in Havana Harbor on Feburary 5, 1898, the United States used this event as a convenient excuse to declare war on Spain in April 1898. Although the sinking of the *Maine* was likely due to an accidental explosion, the yellow press accused the Spanish of the slaughter. On May 1, 1898, Commodore George Dewey led seven navy cruisers into Manila Bay in the Philippine Islands, where they succeeded in nearly destroying the entire Spanish naval force. A few months later, some notable New Englanders began organizing a letter-writing campaign criticizing the war as imperialist as the European power grabs. There were no demonstrations in this antiwar movement, just letters and opinion pieces. But there were lots of letters and lots of opinion pieces—pro and con. The movement called itself the Anti-Imperialist League.

> We deny that the obligation of all citizens to support their Government in times of grave National peril applies to the present situation. If an Administration may with impunity ignore the issues upon which it was chosen, deliberately create a condition of war anywhere on the face of the globe, debauch the civil service for spoils to promote the adventure, organize a truth-suppressing censorship and demand of all citizens a suspension of judgment and their unanimous support while it chooses to continue the fighting, representative government itself is imperiled.[16]

Finally showing his true colors, McKinley soon denied Filipinos their promised independence. The Philippine rebels turned against the Americans in February 1899—and the real fight was joined. In the end, five hundred thousand Filipino civilians were killed in a clear act of American genocide. Independence was buried deep in the mass graves of those half-million dead freedom fighters and civilians. The antiwar movement was aghast. Notables in the Anti-Imperialist League soon included celebrities like Mark Twain, William James, and Andrew Carnegie, who were horrified at the accounts of the fighting. These reports resembled Vietnam in many ways.

I am not afraid, and am always ready to do my duty, but I would like some one to tell me what we are fighting for.

ARTHUR H. VICKERS,
SERGEANT IN THE FIRST
NEBRASKA REGIMENT

Talk about war being "hell," this war beats the hottest estimate ever made of that locality. Caloocan was supposed to contain seventeen thousand inhabitants. The Twentieth Kansas swept through it, and now Caloocan contains not one living native. Of the buildings, the battered walls of the great church and dismal prison alone remain. The village of Maypaja, where our first fight occurred on the night of the fourth, had five thousand people on that day, —now not one stone remains upon top of another. You can only faintly imagine this terrible scene of desolation. War is worse than hell.

CAPTAIN ELLIOTT,
KANSAS REGIMENT

Fueled by such reports, the pacifist movement—combined with a strong attitude among Americans that the United States should never follow the Europeans in their fights for colonies and subjects—made the Anti-Imperialist League a vocal force. Nevertheless, it was derided as "mugwumpery" by the McKinley administration, who finally "pacified" the islands in 1901.

William James wrote:

We are now openly engaged in crushing out the sacredest thing in this great human world—the attempt of a people long enslaved to attain to the possession of itself, to organize its laws and government. . . . Why, then, do we go on? First, the war fever; and then the pride which always refuses to back down. . . . [Our] national destiny must be "big" at any cost. . . . We are to be missionaries of civilization! . . . The individual lives are nothing. . . . Could there be a more damning indictment of the whole bloated term "modern civilization" than this amounts to?[17]

Carl Shurz, editor of *Harper's Weekly,* declared that the United States cannot "play the king over subject populations without creating in itself ways of thinking and habits of action most dangerous to its own vitality." The media, by not holding the president accountable, allowed the Republicans to "purposely and systematically . . . keep the American people in ignorance of the true state of things at the seat of war, and by all sorts of deceitful tricks to deprive them of the knowledge required for the formulation of a correct judgment."[18]

Mark Twain had been away from America for ten years, and when he returned, he was besieged by the press wanting to know his views on imperialism, for they had heard he had turned anti-imperialist. In the *New York Herald* article of October 15, 1900, headlined "Mark Twain Home, An Anti-Imperialist," his change of heart was explained.

I left these shores, at Vancouver, a red-hot imperialist. I wanted the American eagle to go screaming into the Pacific. It seemed tiresome and tame for it to content itself with the Rockies. Why not spread its wings over the Philippines, I asked myself? And I thought it would be a real good thing to do.

I said to myself, here are a people who have suffered for three centuries. We can make them as free as ourselves, give them a government and country of their own, put a miniature of the American Constitution afloat in the Pacific, start a brand new republic to take its place among the free nations of the world. It seemed to me a great task to which we had addressed ourselves.

But I have thought some more, since then, and I have read carefully the treaty of Paris, and I have seen that we do not intend to free, but to subjugate the people of the Philippines. We have gone there to conquer, not to redeem.

We have also pledged the power of this country to maintain and protect the abominable system established in the Philippines by the Friars.

It should, it seems to me, be our pleasure and duty to make those people free, and let them deal with their own domestic questions in their own way. And so I am an anti-imperialist. I am opposed to having the eagle put its talons on any other land.[19]

Twain also wrote: "I thought we should act as their protector—not try to get them under our heel. . . . But now—why, we have got into a mess, a quagmire from which each fresh step renders the difficulty of extrication immensely greater."[20]

To solve the whole Philippines conflict, the pacifist Andrew Carnegie offered to buy the entire Philippines for $20 million and give them their independence. McKinley did not take him up on it.

There are many fascinating similarities in the political climate of then and the peace movement of today—despite being divided by more than a century of tremendous change. In both eras, the president's unbridled expansionism was basically unchallenged by the political establishment due to a traumatic event: the USS *Maine* in one case, 9/11 in the other. Instead of the drumbeat to war of twenty-first-century television and newspapers owned by Rupert Murdoch, there was the maniacal yellow press of William Randolph Hearst, which exploited and sensationalized the sinking of the *Maine* to no end—all for Republican and colonial gain.

Many of the Anti-Imperialist League arguments against expansionism of 1898 hold true today. In 2005, however, the United States does have the power—with a draft—to control Central and Southwest Asia, seen as the current prize thanks to its oil and gas reserves. In 2003, Donald Rumsfeld referred reporters to the Project for a New American Century (PNAC) when they asked what the Republican agenda was. (The PNAC plan is the neoconservative's scheme for America to dominate Central Asian oil and so the world.) Both the Philippines and Iraq were touted as wars of liberation, but in reality they were aggressive invasions to gain precious resources and regional power. Theodore Roosevelt expressed this attitude in a speech he gave on September 7, 1900, on giving independence to the Philippines: "The policy of expansion is America's historic policy. We have annexed the Philippines exactly as we have annexed Hawaii, New Mexico, and Alaska. They are now part of the American territory and we have no more right to give them up than we have the right to restore Hawaii to the Kanaka Queen or to abandon Alaska to the Esquimaux."[21]

In the end, the need of industry for raw materials and the Republican transformation of America into a colonial land grabber rather than a peacemaker dashed all hope of avoiding war by the early twentieth century. Only this time, it was a war like no other before it. It was the Great War. Millions upon millions would die, scarring the collective psyche for decades to come.

WORLD WAR I:
THE LOST GENERATION AND
THE METAPHYSICS OF DESPAIR

Despite the Franco-Prussian War of 1870, Europe lived in relative peace for nearly a century. Progress, the conquest of science and reason over nature and want, was taken for granted. It was thought that diplomacy and the new Permanent Court of Arbitration would head off disputes before they could become major conflicts, that new energy sources would allow industry and agriculture to provide for all.

But the sunny optimism of the nineteenth century was suddenly confronted with the Great War, World War I, and the ugly reality of nations and corporations making a final grab for the remaining colonies and resources of Earth. Worse, this was the first truly mechanized war, with poison gas, machine guns, airplanes, tanks, the bombing of civilian areas, and insane infantry charges into machine-gun nests. The result was over 15 million dead. World trade plummeted, not reaching its pre–War World I levels again *until 1980,* thanks to the Depression and then the true devastation of World War II. Tariffs and protectionism also kept world trade levels low from the 1950s up until the 1980s.

Remember the shock of September 11, 2001, when 3,000 died? Now imagine the trauma of 15 million being killed—and then another 8.5 million dead in the Russian Revolution after that. Beyond the trauma of such a great loss of life, World War I and the Russian Revolution transformed the world in many ways. During the war, the work of women was especially noticed, and suffrage came soon after. In the United States, returning soldiers found little work, while union organizing was now declared communistic and unpatriotic, and the oppressive Palmer Raids of 1919 imprisoned, deported, or intimidated thousands of labor organizers and members, as well as some bohemians. Meanwhile, history's best shot at a solar-powered world went to the grave with Frank Shuman, as oil-powered machinery became essential to every nation's security.

Perhaps the most significant result of World War I was its effects on people's outlook. Optimism and a belief in progress were replaced by a metaphysics of despair, a depressing new view of the future by mechanist and radical alike. Hope was replaced by pessimism and a new belief that war and technological catastrophe were now inevitable, that science and technology would be the undoing of humanity rather than its savior. The cynical conclusion of all this was to give up on true reform and progress, live one's life for its fleeting pleasures, and await the final end along with the rest of a whimpering human race. The uncaring Jazz Age was the result—yet the music, dancing, and libertine ways of the Roaring Twenties also served to break down the traditions of the old worldview.

Since the mechanistic worldview had no real philosophy of life, the metaphysics of despair became the belief of most scientists and technologists, despite their lip service to progress in technology. This great pessimism only deepened with the onslaught of World War II, atomic weapons, and the Cold War. There was no thought of transcending the paradigm, of deep change, of switching from coal, nuclear, and oil energy. As there was no thought of looking at the world differently, there could be no plan for change and no real hope. Despite all the lip service to "progress" mouthed by politicians, the metaphysics of despair is basically the intellectual foundation of the mechanistic dilemma we live in today. Always dressed as realism, this pessimistic philosophy is designed to inactivate resistance to the mechanistic corporatist agenda. For if change is not possible, why even try?

OIL-POWERED CORPORATISM VERSUS ENVIRONMENTALISM

In 1908, the Swedish chemist Svante Arrhenius calculated that the burning of oil and coal was rapidly increasing carbon dioxide in the atmosphere and would eventually create a planet-wide warming of the atmosphere. He figured that doubling the CO_2 would increase average global temperatures by 5 to 6 degrees Centigrade. Yet instead of being worried, Arrhenius looked forward to the day. From his book *World in the Making*: "[W]e may hope to enjoy ages with more equable and better climates, especially as regards the colder regions of the earth, ages when the earth will bring forth much more abundant crops than at present, for the benefit of rapidly propagating mankind."[22] Of course, Arrhenius could not imagine the climate disruption models of today.

For the new oil-powered corporatism—that had just taken over the economy by becoming the primary energy source—the predicted global warming was just more good news. World War I, the mechanization of the army, and the incredible increase in gasoline-powered cars had made the oilman king. Oil was soon discovered in many different parts of the world, especially in the Mideast. Just through sheer economics, cheap oil and pipeline infrastructures ended the solar engine era by the teens and then undermined solar hot water heating by the end of the 1930s. During the 1920s, John D. Rockefeller Jr. and the other energy producers became so rich they were able to influence the government and the Congress to the point where they were unstoppable. During the Depression, they bought up *even more* of the industry and resources of the world. Despite all of Teddy Roosevelt's attempts to reduce the power of monopolies and trusts, the industrialists now garnered all the influence they needed to deepen their hold on the world.

A long list of ecodisasters caused by fossil-fuel dependence or the chemical industry meanwhile started in the early 1900s. Each time, there was a counterresponse from an environmental movement, and each time that response was smothered in a sea of lobbying, lawyers, and newsprint by the corporations. For example, in 1909, in Glasgow, Scotland, one thousand people died of air pollution from coal, leading to the coining of the word *smog* in 1911. Horrible oil pollution and sewage in the oceans, meanwhile, led to the creation of the National Coast Anti-Pollution League in 1922, which forced some cleanup but basically let the oil companies off the hook. The *Newark Evening News* noted at the time that "[a]t last the forces working for the ending of the oil nuisance . . . have concentrated on a program and have created a national organization to carry it out."

These early antipollution groups were mainly led by women.

The addition of lead to gasoline as an antiknock additive—despite the proven alternative of ethanol from corn—was also a huge political fight during the teens and 1920s. The oil industry, PR men, and the Republican Party covered up known health problems and stalled and lobbied for lead additives until their efforts were successful.[23] That loss caused terrible health problems over the next half a century, especially for children, as thirty million tons of unnecessary lead were spewed into the atmosphere. Big businesses won nearly all their fights with environmental organizations in the 1920s.

The Russian revolutionaries, too, were mechanists when it came to the environment, leading to the horrific degradation of air, water, and soil in that country. In 1918, Leon Trotsky said, "The proper goal of communism is the *domination of nature by technology* and the domination of technology by planning, so that raw materials of nature will yield to mankind all that it needs and more besides"[24] (italics added).

Despite their propaganda-powered reach for the absolute control of society, the mechanists were challenged by a small creative minority of organic scientists and technologists. Although they were heavily outnumbered and outgunned, as we have seen, the sheer genius of the organic thinkers allowed them to make one profound discovery after another—discoveries that still have the potential to one day end the mechanistic dilemma.

Unfortunately, the mechanist/corporatist side had their own geniuses.

THE MECHANIST SECRET WEAPON: THE PR MIND CONTROL OF EDWARD BERNAYS

Those who manipulate the unseen mechanism of society constitute an invisible government which is the true ruling power of our country. We are governed, our minds molded, our tastes formed, our ideas suggested largely by men we have never heard of. . . . In almost every act of our lives whether in the sphere of politics or business in our social conduct or our ethical thinking, we are dominated by the relatively small number of persons who understand the mental processes and social patterns of the masses. It is they who pull the wires that control the public mind.

EDWARD BERNAYS,
PROPAGANDA, 1928

Why is present-day corporatism so successful? How have corporations managed to pull off lies like smoking tobacco is good for you, nuclear power is safe and too cheap to meter, or the jury is still out on global warming? Since the 1920s, the mechanists and their crony capitalist patrons have had a secret weapon: a sophisticated method of public relations based on the science of psychology. Backed by the corporate media, this insidious form of PR is literally a type of mass mind control.

The man who helped the corporations in their quest for ultimate gain and profit was Edward Bernays, the nephew of Sigmund Freud, known as the "father of spin." Bernays used psychology to build image and to help corporations position their products, even if the product was not good for people—or might even kill them. Uncle Sigmund had taught Bernays one thing: emotions have a profound affect on the subconscious. So Bernays cleverly used psychological images to sell his clients' wares, tapping deep-seated emotions to bend the perceived reality about the product. He wrote the book on propaganda, literally. Bernays's book *Propaganda* had an impact within all large companies and governments, including the Nazis, and they were all soon using his methods. The politicians were not far behind in learning how to spin campaigns for themselves and their corporate contributors, creating an intricate web of spin and mechanistic big lies and little lies—a web of deceit that was impenetrable to the casual observer. The public was only told what the spin machine wanted them to know.

The result is the world you see today.

After advising the Republicans and big oil on how to win the fight to suppress ethanol and put lead into gasoline, Bernays scored another coup during the 1929 Easter Parade in New York City when he arranged for the "Torches of Liberty" to march. The Torches were cigarette-smoking suffragettes who proclaimed that cigarettes were part of women's liberation—that women have the right to smoke like men. Ever since the Torches of Liberty, women have seen smoking as a sign of their independence. Score one for emotion, zero for reason.

Bernays stated that emotion could be used to triumph over fact and that a subtle lie repeated over and over will eventually be believed by the majority of the listening, viewing, and reading public. His books *Crystallizing Public Opinion, Propaganda,* and *The Engineering of Consent* provided the foundation for public relations, giving corporations and governments a tremendous new power over the mind of the general public.[25] Since advertising worked by influencing subconscious emotions, people didn't even know they were being subjected to PR mind control. Advertisers could now manipulate a person into thinking almost whatever they wanted about a product, such as convincing someone he or she absolutely had to buy it. In his book *Propaganda,* Bernays likened the public to a herd that needed to be led: "Because man is by nature gregarious he feels himself to be member of a herd . . . " It is this herdlike quality, this pliable group mind that makes the population susceptible to leadership and makes it possible to "control the masses without their knowing it."[26]

One of the greatest triumphs of Bernays was forging the decades-long marketing alliance between the American Medical Association (AMA) and the tobacco industry. Despite many opinions to the contrary, the AMA agreed to state in ads that cigarettes were actually beneficial to health. As mentioned earlier, smoking was recommended as a treatment for nerves. Even doctors believed it, and a huge number took up cigarettes themselves. This particular deception was successful for a half a century, until the surgeon general's report on smoking came out in 1965. Still another health campaign resulted in all of America *eating bacon for breakfast.* Bernays has been literally responsible for the deaths of tens of millions over the decades.

Bernays realized that by using "independent" third-party associations, corporate propaganda could be fed through a seemingly objective source, which then issued press releases saying the product was safe. The safety was determined, however, through falsified studies by the phony institutes and foundations that Bernays had his clients set up. As communication technology delivered more and more power into the hands of fewer and fewer media owners, who typically sat on the boards of major corporations themselves, the circle of deception was complete. The security of modern corporatism and mechanistic institutions was for the time assured—no matter what outrage greed and mechanistic technology produces. It would be decades before the organic worldview could even begin to compete with the corporatist media and tell the whole truth about the world. Abraham Lincoln had worried that corporations would be "enthroned" and would exploit "the prejudices of the people until all wealth is aggregated in a few hands and the Republic is destroyed." It seemed the republic was well on its way to that terrible future.[27]

CORPORATISM, WORLD WAR II, AND THE COLD WAR

Fascism should more properly be called corporatism because it is the merger of state and corporate power.

BENITO MUSSOLINI

Part of understanding the current state of the world is to know the history of corporatism, as full corporatism is the future that the regressive mechanist conservatives have long been implementing. Mussolini's fascim in Italy was a corportate fascism, which allowed, as is happening today, certain corporate owners to become fundamentally partners with the rulers at the top of the power pyramid. Ordinary voters then become second-class citizens and lose the real democratic power that they enjoy in a capitalist state. This is the danger we face today from the nearly unbridled power of multinationals to corrupt and influence the political process all over the world. It has degenerated to the point in the United States where lobbyists literally write the legislation concerning their industry.

Corporatism is not capitalism, as capitalism merely regulates the market, while in corporatism the select corporate elite dictate government policy and shape the market for their own greedy purposes. Under corporatism, free markets are controlled and manipulated by the few, with little or no thought for the public good. In fascist Italy, the corporations actually had their own government legislative body, the Camera dei Fasci e delle Corporazioni.[28] Today, corporatism in America and around the world is subtler, pulling the strings through lobbying and campaign contributions, which in the United States makes up two-thirds of the money going to politicians.

In the 1920s and 1930s, the media and the political elite stood by as Mussolini and then Hitler overthrew democracy in Italy and Germany and began rearming to conquer the world. A U.S./British industrialist support network even financed that rearmament, funded and directed in part by certain millionaires in those countries and in the United

States and Britain. Included were many well-documented contributions from Rockefeller and Ford.[29]

Although Ford was an agrarian and researched chemurgy, politically he was a troglodyte and known to be extremely anti-Semitic, having funded the notorious *Elders of Zion* book. Ford Motor alone helped Hitler build one-third of the war machine that became the *blitzkrieg*. That was how a war-ravaged Germany was able to spend billions of dollars on rearming itself. It was paid for through *American and British banks*. The industrialists and bankers also funded the German American Bund (an American Nazi organization) and a fascist movement in England, which almost succeeded in taking over the United Kingdom. In the United States, the silence of the corporate media on fascist horrors, combined with a strong pacifist movement, natural Republican isolationism, and the powerful America First Committee, which opposed entry of the United States into World War II, prevented President Franklin Delano Roosevelt from joining in the conflict—*even after Hitler took over Europe.*

The war wrought many, many changes. Along with the democratization of former enemies Germany and Japan, the United States and the Soviet Union put the civilized world back on its feet through the United Nations and created a new, slightly less nationalistic world. Kant's dream of a world government had suddenly become a reality.

Women, who had once again participated heavily in the war effort, won still more respect during the war but were then expected to promptly return to home and kitchen to raise children. Nevertheless, there were now more women in the workforce. And the Depression was finally ended by the war budgets. The United States has since become addicted to weaponry and bomb manufacturing as a way to supposedly "grow" the economy in perpetuity. This led to the creation of the military-industrial complex, but what was intended to be a strong defense system has been corrupted, turning it into the ravenous corporate war-profiteering machine of today.

More than five thousand top Nazis were allowed to escape all over the world, with seven hundred scientists immigrating to the United States and going to work for the defense industry.[30] In return for his life and a steady job, General Reinhard Gehlen, in charge of Hitler's spy network in Russia, arranged a deal with chief American spy Dulles to turn over all his considerable intelligence and the entire spy network. In this way, the new CIA acquired an instant intelligence system in place throughout the Soviet Union. By making this deal with the devil, the right-wing industrialists who had supported Hitler secretly won the day after all.

Despite all the changes brought about by World War II and the defeat of the pure fascism of the Axis powers, the world was still fundamentally trapped in the mechanistic dilemma, in the mind-set of outmoded nationalist politics and dangerous fossil-fuel technology. Only now there were weapons of mass destruction, which soon included monster 10-megaton hydrogen bombs. It was reported the Soviets had actually designed and wanted to test a gargantuan 100-megaton device in the atmosphere. The physicist Andrei Sakharov objected to Khrushchev on environmental grounds, and for many reasons, the shot never took place.

ENVIRONMENTAL DISASTERS INCREASE IN SIZE AND SEVERITY

During the 1930 to 1954 period, environmental disasters and the resulting social response were nothing less than legendary. In 1934, the famous dust bowl began in Oklahoma and the Midwest, mostly the result of poor farming practices. It was clear that an environmental change had affected the global economy. Franklin Roosevelt responded with successful programs that taught farmers methods to prevent land and water erosion. His secretary of agriculture, Henry Wallace, literally saved American farming by applying science and conservation, which then was taught to developing nations after World War II. Wallace's international agricultural programs transformed the lives of the billions of people to come.

The issue of environmental change was also on the mind of an assistant professor of geography at the University of Wisconsin, Glenn Thomas Trewartha, who coined the term *greenhouse effect* in his 1937 book *An Introduction to Weather and Climate*. Like Arrhenius before him, he thought the greenhouse effect would *benefit* humanity.

During temperature inversions, however, thousands died directly at this time from air pollution. Twenty died and thousands more were affected by the 1948 Donora, Pennsylvania, smog—with six hundred dead in London that very same year. Then there came the great 1952 killer smog of London. Four thousand died from pollution so thick that all transportation except the subway was shut down. Although, according to the journal *Environmental Health Perspectives*, twelve thousand may have been the real death toll in the 1952 killer smog.[31] Visibility was so low that driving cars and buses became impossible. An additional one thousand Londoners died in 1956 from another inversion.

At the same time, the Minamata disaster in Japan saw sixty-seven die horrible deaths from mercury poisoning thanks to a Chisso Corporation chemical plant, with more than three thousand disabled over the years. Mercury has an insidious effect on the body, even in small amounts, and is the suspected cause of multiple sclerosis and other diseases. This element, long a favorite of alchemists and traditional doctors alike, is still used today in numerous industrial and medical processes. The mechanists love mercury and routinely ignore its dangers. They even add mercury to tooth fillings. Although holistic dentists deplore mercury's continued use, amalgams with mercury are still declared safe by the American Dental Association (ADA).

Dr. Andrew Landerman, DDS, a Santa Rosa dentist representing the American Academy of Biological Dentistry, had the following to say about mercury amalgam fillings: "The amount of Mercury in each dental filling is colossal by medical standards. Mercury amalgam is dangerous before it goes into the mouth, and it is a hazardous material when it comes out. Each filling has 750,000 micrograms of Mercury. A person with four fillings has three grams of Mercury in his or her mouth, enough to shut down a lake, a school, or a business."[32]

The ADA is another one of Bernays's "independent" groups, which, like the AMA, is available for a price—in this case grants from the amalgam industry.[33] Mercury was also added to vaccines, in the preservative

Thimerosal. Thimerosal is in vaccines given to babies and young children. These vaccines have been linked (but disputed) to the startling increase of autism in children. Due to public concerns, Thimerosal was reduced or eliminated and mercury-free vaccines were developed starting in 1999. Reports in 2002 reported a significant decrease in new autism cases. However a new preservative containing aluminum was substituted and is still in vaccines. Aluminum is a known neurotoxin, which has also been linked to autism (and Alzheimers and other disorders). This and the fact that mercury is still in flu shots given to babies and pregnant women in their first trimester would account for the more recent reports that autism rates have continued to rise. These reports have mistakenly assumed that this means that mercury is not linked to the increase of autism.[34] At the same time, the AMA in the United States no longer had any competition.

Meanwhile, the chemical industry was creating a multitude of toxic and unnecessary fertilizers, pesticides, and herbicides, resulting in a massive pollution runoff. The solar power and electric car companies from earlier in the century had been put out of business by cheap oil. Thanks to the silence of the media and the highly effective PR techniques of Bernays and his disciples, the casual observer had little idea of the horrific problems that were being created, or of the solutions that had been lost.

The mechanistic dilemma deepened in the 1940s and 1950s when the decision was made to make energy from nuclear power, never stopping to ask if that was really a good idea. Technologists were so enamored of nuclear reactors that they actually designed and built a nuclear-powered jet: the NB-36H nuclear test aircraft. Apparently, the regional contamination that would occur if the aircraft should crash was never seriously entertained. Flown with nuclear jets, it had a 12-ton lead-shielded crew compartment in the nose cone and cameras so the crew could check on the reactor and other parts of the plane. The NB-36H made forty-seven test flights between 1855 and 1957. Kennedy ended the Aircraft Nuclear Propulsion program in 1961, writing "Nearly 15 years and about $1 billion have been devoted to the attempted development of a nuclear-powered aircraft; but the possibility of achieving a militarily useful aircraft in the foreseeable future is still very remote."[35]

The nuclear plane is most instructive in revealing the illusion of technique and the main flaw of the mechanistic paradigm. Yes, science can technologically accomplish something like a nuclear jet, but the question is never asked, Should the project be done at all?

The size of environmental disasters soared with the 1950s nuclear atmospheric tests, which we now know radiated vast tracts of the world, especially Nevada, Utah, Colorado, Wyoming, and Montana, as well the South Pacific and the eastern Soviet Union. Radiation fell on the grass, the grass was eaten by cows, and soon radioactive cesium 137 and strontium 90 could be found in milk all over the Northern Hemisphere. Around 250 bombs were set off in the years of the Mutually Assured Destruction (MAD) arms race. The atmospheric atomic tests also wreaked havoc on the soldiers who were radiated for exposure experiments, as well as the inhabitants of nearby Utah. An inexplicably large number of soldiers, over fifty thousand in all, were contaminated. Government and defense industry scientists, being mechanists obsessed with both technology and national security, were willing to put aside all principles and continue the radiation experiments throughout the 1950s, even with the enormous H bombs. In a secret session held on November 10, 1958, to discuss Los Angeles's hazardous levels of radiation, Lauriston S. Taylor, head of the Atomic Radiation Physics Division of the National Bureau of Standards, was reported saying, "If you ever let these numbers get out to the public, you have had it."[36]

Add to this the horror stories connected with uranium mining and the World War II and Cold War dumping of poison gas and other toxics in the ocean. Eleven nuclear bombs were even lost during these years, falling out of jet bombers—*one of which is still sitting right off the coast of North Carolina.*[37]

Threatened by environmental groups and the science of the Organic Shift, the mechanistic establishment in the twentieth century used Bernays's methods to belittle and ignore positive organic solutions. As

Thomas Kuhn showed in his paradigm shift theory, the old scientists maintain power by ignoring data from the new theorists in favor of raw ideology and personal attack. Although that was to be expected from any old paradigm, the mechanists were empowered by the new diabolical powers of modern public relations. A host of phony institutes, paid by corporate masters to create studies supporting the industry, were particularly effective in covering up many contradictions and lies. At the same time, the underlying metaphysics of despair following World War I and II conveniently reinforced the dismissal of any organic solution—like solar power—as a hopeless pipe dream. For a time, this slowed the progress of solutions like renewable energy, organic agriculture, and alternative medicine.

EINSTEIN:
THE NEED TO CHANGE MAN'S
MODE OF THINKING

In 1924, Albert Einstein held up a glass of water in Manhattan and declared that it held enough power to blow up the entire city. Following the dropping of the bomb on Japan, Einstein warned, "The unleashed power of the atom has changed everything, save our modes of thinking, and thus we drift toward unparalleled catastrophe."[38] To save ourselves, said Einstein, we need a new type of thinking.

Most know that Einstein was a genius in physics, his theory of relativity once again showing the power of Kant's realization that the observer is part of the observation. Yet Einstein was much more than that. He became an avowed pacifist and was quite active in using his celebrity to push his concept of a different world. He went right to the core of the problem, challenging the military mentality and the limited ways it thinks: "[C]ould not our situation be compared to one of a menacing epidemic? People are unable to view this situation in its true light, for their eyes are blinded by passion. General fear and anxiety create hatred and aggressiveness. The adaptation to warlike aims and activities has corrupted the mentality of man; as a result, intelligent, objective and humane thinking has hardly any effect and is even suspected and persecuted as unpatriotic."[39]

The nuclear leveling of Hiroshima and Nagasaki were in fact so horrific that a planetary self-preservation instinct was triggered, as the world created the United Nations—an attempt for nonnationalistic politics. Nuclear weapons were a frightening new reality, helping to inspire a deeper desire for peace and the abolition of total war. The mechanistic/ nationalist worldview, which had worked fairly well for such a long time, now seemed doomed, made obsolete simultaneously by the new physics and nuclear weapons. A whole new paradigm in world politics was needed, as the mechanistic structure helplessly stumbled through the Cold War, approaching mutual assured destruction. Nuclear holocaust came closest in October 1962, during the Cuban missile crisis.

Given the stakes, the need for a new peace movement and a new mode of thinking was unprecedented. As we shall see, in the later 1950s and 1960s, and again at the turn of the millennium, this new mode of thinking finally did arrive.

POLITICS IS NOT DEEP ENOUGH

Feminists, abolitionists, Populists, union organizers, suffragettes, pacifists, environmentalists—all were developing different aspects of the new paradigm's redefinition of freedom. Yet all the political, economic, and liberation movements of the nineteenth and early twentieth centuries do not tell the whole story, for the change of the Organic Shift went deeper still. The new philosophy and politics of civil rights and environmental science had a profound affect on a small part of the population. Whole new lifestyles were created. An entirely different culture evolved. Although they were not sure what to call this phenomenon in the 1800s, today we see this social evolution as the *counterculture*—a culture that sees mechanist/nationalist civilization as only an incomplete and deeply flawed predecessor to a more evolved holistic environmentally aware and spiritually enlightened global culture.

In the end, politics and all the social movements did not constitute a change that went deep enough. Deep change had to come from within; it had to transform the whole person. Many people underwent this transformation as the Organic Shift developed from the early 1800s to the 1950s. As previously discussed and as Kant determined, it was essentially the change from *homo phenomenon* to *homo noumenon*. Rather than dealing solely with the external world, *homo noumenons* also explore the realm of inner space; they contemplate and mystically connect with the whole of nature. The counterculture was therefore not only a broader worldview, it was also an entirely new way to live.

The members of this new lifestyle practiced what they preached. They were aspiring *homo noumenons,* seeing the world in all its layers, from the unknowable noumena to the world of illusory phenomena. Rather than the inside-out thinking of *homo phenomenon*—from the internal self first and then to the outside world—the counterculture types usually thought from the outside-in, possessing a spiritual respect for the unknowable and nature that always guided them in their rambunctious personal lives. The Organic Shift was the very opposite of the endless materialism and the anthropocentric exploitation of nature that drove the mechanistic worldview. The revolution in philosophy, science, and politics thus went hand in hand with a revolution in lifestyles and also in art. In fact, for the first time, art itself became revolutionary, reflecting the counterculture worldview of the new thinking.

There are many things involved in changing humanity's mode of thinking. There are many questions to be answered, many new words and sets of vocabularies that need creation. Looking at the human situation from the outside in, from noumenal consciousness, casts aside the old anthropocentric view of the universe, raising a whole different list of important questions. Who are we really? What is humanity's true potential in a universe in which it is not the center? What would the future be like if enough of the world shifts to the new paradigm? These are the questions that drove the new counterculture in its quest for a different kind of knowledge, a journey to seek a deeper wisdom.

William Blake

6

THE ORGANIC COUNTERCULTURES AND THE NEW ARTIST

The tree which moves some to tears of joy is in the Eyes of others only a Green thing that stands in the way. Some See Nature all Ridicule & Deformity . . . & Some Scarce see Nature at all. But to the Eyes of the man of imagination, Nature is imagination itself.

WILLIAM BLAKE,
LETTER TO REVEREND JOHN TRUSLER, 1799

THE ASPIRING *HOMO NOUMENON*

In the previous paradigm shift, from the medieval to the mechanistic, a new culture and lifestyle quickly developed, one quite different from the one that prevailed during the Middle Ages. The Renaissance humanist was born, in a rebirth of science and learning. Rather than stuffy medieval church scholars, the Renaissance humanists were incredible polymaths, versed in the knowledge and techniques of the

Romans and Greeks, as well as being masters of artistic leaps such as the use of perspective and "scientific" painting. Leonardo da Vinci comes to mind. Like the different lifestyle of the Renaissance and Copernican Revolution humanists, a new counterculture was a major part of the Organic Shift. Along with the new science and philosophy, this vibrant movement thought only with the organic worldview, finding transcendental truth by simply communing with nature.

As this new sense of being came about, civilization and mechanistic science were critiqued from a broader perspective. The new way to think immediately led many to an intimate connection with nature and a sense of wonder about the unknowable, mystical whole. It was the very beginning of modern organic living.

Kant's view of the two types of people—*homo phenomenon* and *homo noumenon*—is perhaps the best description of the difference between the old culture and the new counterculture. Beyond science, beyond all measurement, beyond Kant's limits of reason, were the very real intangibles of feeling, the beauty of nature, and the love of true peace and freedom. This inner sense of being began to express itself among the wave of new thinkers, along with shocking politics such as feminism, pacifism, civil disobedience, and the abolition of all slavery.

It was to these idealist goals and feelings that the organic counterculture attached itself—from the 1780s right up until today. Following the breakthroughs of Kant and Goethe, early aspiring radicals—the organic pioneers—rebelled against the machine by mystically finding themselves in the contemplation of nature, peace, freedom, and the unknowable whole. Although no one may ever be able to become a full *homo noumenon,* at the least people now directly or indirectly understood the broader awareness and a great many aspired to it.

In previous worldviews, nature was seen as something evil. Forests, for example, were seen as dark and foreboding. In the organic and Romantic worldview, nature was now good and beautiful, able to renew the spirit and creativity itself. Many attitudes were turned on their heads. The early Romantics, especially the English poet and artist

William Blake (1757–1827), fundamentally transformed the role of the artist in society. Artists had traditionally glorified man and God, but in Blake and the Romantics we now see the revolutionary artist, one who uses vision to create art that wakes up the "slumbering masses," so that the innate goodness of humanity can reform the world.[1] The Romantics would be followed by the bohemians in the 1830s and 1840s, a much larger counterculture that attracted many to Paris and soon other major Western cities. Bohemian counterculture was in turn succeeded by the beats of the late 1940s and 1950s. It was all very, very different from the sedate, mainstream culture of the old worldview. The continuum of the status quo was increasingly challenged.

WILLIAM BLAKE

In London of the late 1700s, we saw how Mary Wollenstonecraft was part of a small circle of radical thinkers and artists, sort of a proto-counterculture. In the circle were writers like Richard Price, who in 1758 wrote that, rather than the fear of hell, reason and one's own conscience should guide moral choices. There were also artists like Charles Fuseli, whose painting *The Nightmare* was an exploration of the inner mind and even female sexuality rather than a depiction of the outer world. With a few exceptions, art up until that time had strictly glorified God, the royalty, the rich, and the nobility.

Fuseli was the very first to truly break that mold—and the sound of that particular mind-set breaking could be heard all over England and the continent. *The Nightmare* was painted in 1781, the very same year Kant published his *Critique of Pure Reason*. This artistic exploration of the inner mind went hand in hand with a new politics, which supported the American and the French revolutions, as well as radical spiritualism of different types such as gnosticism, which sees the Christian God Jehovah as a false deity and worships the Christ consciousness within. A straight historical line can clearly be drawn from this radical circle through the many countercultures and on to the changes of the 1960s and beyond.

One member of the London radical circle, William Blake, a mystic poet and engraver, did more than explore the inner mind—he lived in it. After his brother Robert died, Blake had constant visions of the "other side," declaring he communicated with Robert, who had taken a piece of his heart with him to heaven. Through his mystical visions, Blake was the first to give an organic definition of the modern artist, one that continues to this very day. He did this by envisioning his own mythology for a now-industrialized world.

To Blake, humanity is an unconscious giant—put to sleep thousands of years ago by the invention of the wheel and the resulting spell of money and materialism. Only the arts can heal and awaken this "slumberous mass." Once the sleeping giant is awake, he will sweep away the materialist thoughts that have taken over the world and start a new golden age. Blake especially railed against the pollution of the early London factories. Smog was now a great problem, even though it was only the late 1700s. The famous fog of London was not natural but caused by smog mixing with humid air, a new phenomenon in Blake's time. He prayed that London would one day see the sun again—for the black smoke of the factories and chimneys would surely be banned by an awakened humanity. Blake sang out for that "glad day" to come quickly. "Awake! awake O sleeper of the land of shadows, wake! expand! I am in you and you in me! mutual in love divine."[2]

Artists in previous worldviews had supported the institutions of the world. Yet in the organic worldview, the mission of the artist was to expose and transform them. This new role of art, to awaken the sleeping giant of humanity, became the guiding principle of progressive artists everywhere. The spirit of Blake's mythology became an underlying foundation for all the subsequent countercultures, from the Romantics to the hippies of the 1960s.

Like Goethe, William Blake was also one of the first to explore what we would now call psychology and Jungian archetypes. In *The Four Zoas,* Blake paints a mind ruled by four aspects or Zoas (Tharmas,

Urizen, Luvah, and Urthona or Los). These aspects lie within every human mind and, if realized, can awaken one to powerful currents of creativity and knowledge. The Zoas can thus be seen as a guide to self-understanding, to unlocking one's full divine potential. In Blake's mythology, Albion, or primeval man, was put to sleep by Urizen (your reason), with the invention of the wheel. The materialism that followed has kept humanity entranced ever since.

The Industrial Revolution was Urizen's mechanization of the wheel, the gears of the factories being wheels upon wheels. Urizen, in a fit of mindless greed, was poisoning Vala, who represents nature, through the nightmarish pollution of the Industrial Age. Humanity's only hope of reviving Albion lies in Los, the Zoa of creativity and spirit. Los has spent all this time trying to reawaken his beloved friend Albion, yet the spell of materialism still leaves the amoral but very powerful Urizen in charge.

And so the collective consciousness sleeps. Yet individuals, if they can understand their mind in this new way, can awaken themselves—a necessary first step to awakening the world. "Four Mighty Ones are in every Man," said Blake, naming Tharmas as the hands, Urizen as reason, Luvah the heart, and Los the spirit. Truly living life to the fullest requires waking up the three sleeping aspects of the mind—Tharmas, Luvah, and Los—and letting them loose to try and awaken the sleeping giant Albion.

The Zoas are thus not only a spiritual history, they also represent the spiritual state of each person, showing a journey to "perfect unity": hands, mind, heart, and spirit. Most of all, let Los loose, for only art and the Muses can awaken the sleeper. The four Zoas were, in their way, the ancestor to the human potential movement of much later times.

Mary Wollstonecraft knew Blake and was very close to his friend Fuseli, the painter of *The Nightmare*. In 1788, Blake and Wollstonecraft collaborated on a progressive schoolbook for children, *Original Stories from Real Life,* with Blake providing six engravings.

Containing tales to explain the world to children, this was the most popular of all Wollstonecraft's books and was used by many women to help homeschool their daughters. It advocated liberal education for all—a truly radical idea at the time. By 1800, *Original Stories from Real Life* had gone through five printings.

In his own time, however, Blake was repeatedly rejected and ended up living in abject poverty—kept alive only by sympathetic liberal patrons. He was almost certain to be forgotten by all when, in his last decade, he was discovered by a group of early Romantics: the Ancients. The Ancients, as they liked to call themselves, were the first English Romantics to return to beards and long hair, so common among later countercultures. John Linell, a landscape painter, was the leader, a young artist whose goal in life was to change landscapes by bringing vision into the picture. Like other early Romantics, the Ancients loved nature, seeing it as pure and good—something spiritual to inspire creative vision.

Most of all, the Romantics discovered the power of nature, feeling, and love in their art and poetry. In the early 1800s, idealism was the leading philosophy and Romanticism the up-and-coming literary and artistic movement. It was at this moment that John Linell discovered the elderly William and Katherine Blake living in their one-room hovel in a London slum. For the Ancients, finding Blake was an unexpected divine revelation. Calling him "the interpreter" for his meaningful visions and his hovel "the house of the interpreter," the Ancients sat at Blake's feet for his last decade on Earth.[3] This was a welcome final recognition for the man who first forged the path artists have followed ever since.

Blake taught the Ancients that the most important thing to an artist is imagination. The outer world can inspire only so much. Turn to your inner imagination and envision your art, said the interpreter—very much an English version of what Goethe was writing in German about science and poetry. Linell became Blake's friend and student for nine years, until the day of Blake's death. Every time he saw the old visionary, the young Romantic appreciated him more and more. Linell went on to become one of the most successful artists of his time. Despite the appreciation of the Ancients, Blake's greatest influence came later, in the twentieth century, when his enormous contributions were more fully recognized.

THE FIRST ORGANIC COUNTERCULTURE: THE ROMANTICS

After being confined in the prison of the mechanistic Enlightenment, the liberation of thought in the Romantic era was breathtaking for those involved. Liberated from the soulless laws of nature, from materialism, and from the inanimate model of the machine, the Romantics began exploring a true philosophy of life, developing the new organic worldview. It was positively exhilarating. Percy Bysshe Shelley expressed the ecstasy of the new freedom to think in *Prometheus Unbound.*

> *The joy, the triumph, the delight, the madness*
> *The boundless, overflowing, bursting gladness*
> *The vaporous exultation not to be confined!*
> *Ha! Ha! The animation of delight*
> *Which wraps me, like an atmosphere of light,*
> *And bears me as a cloud is borne by its own wind.*[4]

Where the mechanist *philosophes* were cold, mechanical, logical, and unfeeling, the Romantics were human, organic, intuitive, and passionate. Instead of seeing nature as something to be conquered, communion with nature was sought, as it nurtured warmth in the heart, causing a regeneration of the spirit. Like Blake and Goethe, they believed in the power of imagination. One could take in creation and then set loose imagination to create one's own universe. An unbridled artistic freedom eventually became possible—all part of the new organic thinking, the new way of being. Organic living, as an aspiring *homo noumenon,* was now a reality in the form of this small but vocal community of Romantic thinkers and followers.

Feeling, imagination, and intuition trump reason and logic, declared the Romantics. There are clear limitations to mechanistic science. Machines cannot evolve, love, feel, or reproduce, so how can mechanism explain life—which does all of those things? An organic naturalism developed, one that recognized the influence of nature and the environment upon life.

Although there were many different kinds of Romantics, from the astounding English painter John Turner to the German medievalists and the American Hudson River school of art, almost all were opposed to the mechanistic science and technology of the day, as well as the capitalism that accompanied it. Romantics saw the world as an organic, living thing—like modern ecologists do. So not only did the Romantic era transform art and culture, it also turned science from the study of planets to the contemplation of plants and Nature—and this led to Goethe, Darwin, evolution, and then the discovery of ecology itself.

Being such lovers of nature, the Romantics were, of course, horrified at the Industrial Revolution and the massive pollution that the first coal-powered factories produced. Believing in the ideals of the French Revolution, they especially despised the economic egoism of the newly rich factory owners, who thought their money made them better than others. Romantic spirituality prevented them from even thinking along the lines of the old mechanistic paradigm.

In America, this new spirituality became visible to all in the philosophy of a well-known movement that started in New England.

THE TRANSCENDENTALISTS

How does one transcend the phenomenal world and gain the broader awareness of *homo noumenon*? What exactly is involved? Ralph Waldo Emerson (1803–1882) brought the nineteenth-century exploration of these questions to the next level. He taught how to escape from civilization, how to replenish oneself through communion with nature. Out of Kant's and Goethe's philosophy, out of the Romantic belief in spiritual regeneration, vision, and the transcendence, and out of a fascination with Buddhism, Emerson created the transcendentalism of New England. It was an organic movement consciously intended to reform the world. He urged everyone to resist the cultural status quo, to search beyond appearances for the real truth.

Fig. 6.1. Ralph Waldo Emerson said to transform the world of man, we must commune with the "oversoul."

The transcendentalists had a deep interest in Eastern thought, displaying a reverence for the statues of Buddha they acquired from sailors and merchants who had been to India and China. This fascination for wisdom from the East, however, was only part of their spirituality. Emerson blazed his own path, calling for individual transformation by communing with nature, with the oversoul. The oversoul is the unity of all nature, the object of what religions and philosophy have always talked about. He explained that "nature is transcendental."[5]

Emerson believed that people needed to ascend to the transcendental plane through periodic communion with nature, where they would then be recharged and be able to go back and transform the world of man. In the 1830s and 1840s, transcendentalists flocked to places like Niagara Falls so that they could experience its transformational power. They even called these natural wonders the transformers. Many spread out into the world, all advocating the new thinking.

How does one know what is the real truth? The answer was through a "liberation by vision," proclaimed Emerson. Personal intuition—and a vision of what could be—are far more important than the fossilized ethics of the status quo. Like Blake and the Romantics, there was a strong belief in the creative power of vision. Vision allowed you to find yourself—and so your life.

The best way to experience the oversoul—and become one with the whole—was to let oneself get carried away by the astonishing beauty of nature. Then one could receive the vision, the spark of momentary intuition that all require. This was the primary goal of the transcendentalist—to acquire moments of visionary enlightenment through contact with the oversoul.[6]

HENRY DAVID THOREAU: NATURE, THE SIMPLE LIFE, AND CIVIL DISOBEDIENCE

Out of transcendentalism came a towering figure in the Organic Shift: Henry David Thoreau. Acknowledged for inspiring the conservation movement, Thoreau must also be recognized for giving us much of what the counterculture of today takes for granted: a fervent, spiritual environmentalism, the concept of civil disobedience—so important to the next century—and the philosophy of voluntary simplicity.

In opposition to the typical materialism of the time, Thoreau fought vigorously with his pen against the status quo. "Any fool can make a rule, and any fool will mind it," he wrote. Fighting the blight of industrialism and the frantic manner of modern urban life, he pleaded, "Our life is frittered away by detail . . . Simplify, simplify."[7] He practiced what he preached, living for a while in a bare shack at Walden Pond on a meager amount of money, proving that one could live in nature with a simple life—and without a visible means of support. Henry David Thoreau had envisioned a whole new lifestyle. "[I]f one advances confidently in the direction of his dreams to live the life which he has imag-

*Fig. 6.2. Henry David Thoreau gave us
the concepts of voluntary simplicity
and civil disobedience.*

ined, he will meet with a success unexpected in common hours. . . . In proportion as he simplifies his life, the laws of the universe will appear less complex."[8]

With Walden, Thoreau rejected the materialism of civilization, proposing what we know today as voluntary simplicity. Try to live life with little impact on the world of nature, said Thoreau, an ecological concept diametrically opposed to the rapacious exploitation of the mechanistic technologists.

To the later transcendentalists and the bohemians that followed, the plea for an ascetic simplicity was in concert with the trend toward Easternization. Voluntary simplicity was added to the counterculture mix, a concept so important to the future as our planet grapples with the triple threat of environmental degradation, overpopulation, and a shortage of resources. We must wake up from the entire anthropocentric series of worldviews if we are to survive. Henry David Thoreau saw it coming.

Thoreau also invented another crucial aspect of the organic

worldview. When asked to pay a war tax for the invasion of Mexico in 1846, his pacifism led him to jail instead—and his famous essay on civil disobedience. This new method—of changing the world in a nonviolent way—later guided the women's suffrage movement and the civil rights movement, as well as Gandhi and the anti–Vietnam War movement: "If [the injustice] is of such a nature that it requires you to be the agent of injustice to another, then, I say, break the law."[9] Henry David Thoreau made an enormous contribution to the Organic Shift, creating peace and human rights warriors who fight their battles without shedding a drop of human blood, as well as inspiring voluntary simplicity and a new level of communion with nature. The direct influence of Thoreau's writings continue to this day.

THE ARRIVAL OF THE BOHEMIANS

They were the young Romantic artists and writers who gravitated to Paris, who loudly dedicated themselves to their art—or perhaps just the spirit of their art. Their numbers grew, and a fairly large and very radical artistic scene soon appeared, centered around pleasure and against order of any kind—the very opposite of the rigid, dull lives of the mechanistic bourgeoisie.

The bohemians had arrived. A publication of the time had this to say about the bohemians: "[T]he term 'Bohemian' has come to be very commonly accepted in our day as the description of a certain kind of literary gipsy, no matter in what language he speaks, or what city he inhabits. . . . A bohemian is simply an artist or littérateur who, consciously or unconsciously, secedes from conventionality in life and in art."[10]

In the Copernican Revolution of the sixteenth and seventeenth century, the new secular scene in the coffeehouses and inns had provided a fertile place for intellectual debate and development. The bohemian café society of the nineteenth century played the same role in the Organic Shift. New paradigms require an atmosphere in which new ideas can be expressed and nurtured with others of the same worldview. The Parisian world of bohemia filled with that rarefied air.

Romanticism had opened up a new love of nature, and in that love, young artists and writers explored their own human nature. Like all Romantics, their new secular religion was art, in which they worshiped imagination and devoted themselves to expanding their capacity to create the new and the visionary. Seeing the middle-class "Code of Behavior" as a prison of the mind, as an obstacle to true artistic freedom, the bohemians resolved to go one step beyond the Romantics and find an entirely different way to live.

Inspired by the original meaning of bohemian, a gypsy vagabond type styled on a fifteenth-century rogue-poet named François Villon, the first true counterculture bohemians began to gather in the Parisian cafés and restaurants of the 1830s and 1840s. It was a mutual, ego-driven love fest and intoxicated debating society—all at the same time.

The world of the bourgeoisie was held in such disdain that bohemians would typically not carry a watch, as time and regular schedules are things to be avoided. And just as time should not control you, neither should money, so they tried to have as little to do with that as well. Any extra cash that did come their way was usually spent on art supplies and intoxicants. Young artists and writers often roomed together, all creating their works in different parts of the apartment by day and carousing with the crowd until the early hours of the morning. Then it was up at 7 or 8 a.m. for another day of writing and art. The writer Arsene Houssaye described the scene: "One was writing by the fireplace, the other sitting in a hammock; Theo, always caressing his cats, wrote his chapters lying on his belly; Gerard, always elusive, came and went with the vague unrest of someone who is looking for something without finding it. Beauvoir appeared now and then with his burning rhymes."[11]

Although they are all supposed to be at work on their creative projects, Houssaye said that, more often than not, they would put down the work and joyfully interact with one another for the rest of the day. When night came, it was time again for wine, women, song, and maybe some hashish or opium brought back from the East by sailors. Everything was shared communally—even the female companions. For accompanying the men of bohemia were the bohemian women. The flamboyant women of mid-nineteenth-century Paris were nicknamed *la lionne*. These "lionesses" drew their inspiration from Romantic poetry and the novels of George Sand, eschewing typically feminine behavior and pursuing masculine activities.

"[*La lionne*] knew how to ride in Arab fashion on horseback, to tipple down burning punch and iced champagne, to manipulate the riding whip, to draw the sword, to fire the pistol, to smoke a cigar without having vapors, to pull an oar in case of necessity; this was the *enfant terrible* of fashion, alert, dashing, intrepid, never losing her stirrups."[12] The lionesses modeled themselves after George Sand and the heroines of her novels. Born Amandine Aurore Dupin in 1804, under the pen name George Sand she became a well-known bohemian novelist. Like Mary Wollstonecraft before her, the later works of Sand advocated full equality for women. Yet unlike the Romantic women who preceded her, who dressed in elegant medieval gowns, George Sand often appeared with a short haircut and in men's clothing. This was something that could be tolerated only in Paris.

Most of the time, however, the women of bohemia were typically art models and very feminine companions of the men, although they could often be much like *la lionne*—boisterous, exuberant, and sexual, breaking all the rules for a proper lady. Although they would retire behind a curtain for a little privacy, in many ways the women of bohemia were certainly on the way to what we today would consider liberated. Many of them, the *grisettes,* came from working-class backgrounds and toiled as seamstresses. For them, the bohemian scene in the cafés was a way to climb the social ladder and congregate with men from all classes. Often a young bohemian would return in his later years to his middle-class or upper-class life and take his *grisette* with him.

Although few were artists or writers themselves, these women nevertheless played important roles in the lives of the people they inspired. They praised their paintings and were touched by their poems—giving support for their attempts at creativity. Like the men, the bohemian women totally rejected the bourgeois code of behavior and lived large.

Ordinary Parisians were naturally aghast at the goings-on, which were noisily visible in a few key neighborhoods, especially the Latin Quarter where the students lived. Although the 1830s and the 1840s no doubt comprised the golden age of the bohemians, this was only the beginning of the bohemian movement, for Parisians would have to put up with the bohemians and their descendants for quite a bit longer—thanks to one Henry Murger.

Henry Murger, the son of a concierge, was born in 1822 and ended up in the bohemian scene with the likes of the painter Courbet and the poet Baudelaire. In 1845, Murger—also known as the first bohemian—wrote a series of stories in a Paris magazine detailing the bohemian scene and its shocking rejection of bourgeoisie manners. His 1849 play *La Vie de Boheme* (The Bohemian Life) was enormously popular and became the basis of Puccini's opera *La Boheme* in 1895. A book called *Scenes de la Vie de Boheme* soon followed, and this became the bible for artists in the later nineteenth and early twentieth centuries.

The play and books idealized the Latin Quarter and the bohemian scene, which had been underground and unknown, making it irresistible to many young people. The neighborhood was now famous throughout the Western world. One at least had to go to the Latin Quarter and see what the fuss was all about. Anyone bored with their life knew they could head to Paris and join in on the scintillating scene. And so they did. Amateur and true artist alike flocked into the Latin Quarter, whether they had any talent or not.

All this became part of the nineteenth century in Western Europe and then New York, with the bohemian segment of the population exploring all the ramifications of an expanded worldview. As the direct forebears of the current counterculture, they understood the world through much the same organic analogy that progressives use today. And like their counterculture ancestors, many progressives in their younger days had a very similar bohemian-type lifestyle.

EXPANSION OF
THE BOHEMIAN REALM

In the second half of the nineteenth century, bohemia and the Latin Quarter became even more well known, and the influx of amateurs, artists, and the merely curious grew dramatically. Bohemian thinking and attitudes in the last half of the century spread beyond art and literature to science and then politics. It also spread geographically, from the heart of Paris to London, New York, and other cities in the West. Bohemia was now truly an international land of the mind, a vibrant, artistic state of being, as expressed by Walt Whitman:

> This is what you shall do: Love the earth and sun and the animals, despise riches, give alms to everyone that asks, stand up for the stupid and crazy, devote your income and labor to others, hate tyrants, argue not concerning God, have patience and indulgence toward the people, take off your hat to nothing known or unknown or to any man or number of men . . . re-examine all you have been told at school or church or in any book, dismiss whatever insults your own soul, and your very flesh shall be a great poem.[13]

For many, entering bohemia meant a short life and eventual financial ruin. A new pattern emerged, in which a young artist or writer would enter bohemia and make a determined attempt to achieve mastery of his craft. But if his art did not blossom in his early years, he would return to the bourgeoisie and live the life of time, money, and the code of behavior. Many London artists and writers were known bohemians, including Oscar Wilde, Aubrey Beardsley, and the poet Ernest Dowson. Other bohemian scenes in the later nineteenth century could be found in Berlin, Vienna, and Barcelona.

IMPRESSIONISM
AND THE AVANT-GARDE

In Paris, among the heady artistic brew of Romantics, bohemians, realists, and revolutionaries of various stripes, a new sort of visionary emerged in the latter part of the nineteenth century. Rather than painting what the ordinary eye sees, the impressionists revealed the reflection of the inner eye of imagination and feeling. On April 15, 1874, Claude Monet organized a show of the new art, an exhibit that earned the new form its name, taken from Monet's painting *Impression Sunrise.* Mary Cassatt, an American woman impressionist living in Paris, also exhibited her special work alongside the men, her theme being the life of women. By itself, her exhibit demanded more attention for women artists, and they slowly began to receive it.

Rather than trying to exactly imitate what the eye saw, the Impressionists felt the emotion of the light and color in a scene and then attempted to reflect that feeling. They made the intangible visible. In some impressionist paintings, it looks as though the artist has captured the vibration of the atoms themselves. The nonartist could look at these works and realize that the world is mere appearance—that underneath the real world is another realm of vibration and raw emotion.

Through the art of geniuses like Georges Seurat and Monet, impressionism became a tremendous influence on art. Many young artists started out as impressionists but then went on to create the new forms of pointillism, postimpressionism, cubism, expressionism, and modernism. Thanks to popular exhibits, modern art became an entry point for many Parisians into the nascent organic worldview. By going beyond the phenomenal world, modern art makes one realize there are other unknown dimensions to reality—by its very existence, it forces contemplation of a larger whole, of the unknowable.

As early pioneers like William Blake had hoped, visionary art was becoming a force for enlightenment, for waking up the slumbering middle class. The Organic Shift was gathering momentum in many different areas at once, although it was never viewed at the time as any one single change, or united as any one movement. It was instead many movements—all heading toward a grand synthesis in a far-off future.

BOHEMIAN AMERICA

Bohemia next landed in America. Pfaffs opened in New York in 1855, a saloon visited frequently by Walt Whitman and other early bohemians. By 1859, the crowd at Pfaffs had nurtured the birth of America's first counterculture newspaper: the *New York Saturday Press,* printing all sorts of radical articles on freedom, politics, sex, art styles, and so on.

Some of New York's first bohemians, including the Gothic writer and poet Edgar Allan Poe, were members of the literary Algonquin Club in the 1870s, which met at the Algonquin Hotel in Manhattan. In 1894, the bestseller *Trilby* further romanticized the freedom of the bohemian life, and the scene in New York grew rapidly. Similar to *La Vie de Boheme, Trilby* concerned a love triangle among English art students in Paris and clearly piqued the interest of a new generation.

In the 1890s, bohemian New York grew into larger circles, centered around restaurants and cafés like Schwabs, Maria's, and Mould's Café. There, in between plates of bratwurst or spaghetti, there would be loud conversations about the state of the world and everything in it, especially the coming revolution. By the mid-1890s, another severe depression had led to what Henry James described as two classes: the satisfied class and the swindled. As we have seen, many held the state of the world in the 1890s in deep disdain—the bohemians were not alone. Yet the bohemians not only preached change, they lived it.

The art critic James Gibbons Huneker wrote, "In New York as in Paris, the café is the poor man's club. It is also a rendezvous for newspaper men, musicians, artists, Bohemians generally. . . . It is the best stamping ground for men of talent. Ideas circulate. Brain tilts with brain. Eccentricity must show cause or be jostled."[14] These men—and unchaperoned ladies—spent much of their lives in these places, often residing in the nearby neighborhood. Many left their former roles and made the attempt to live as true bohemians. Others, as in Paris, were "weekend bohemians." In any case, a small scene began to thrive.

Some were also involved in progressive or radical politics. European nations, such as Germany in the 1880s, had been swept by social reforms and the progressives brought back new ideas such as worker housing, municipally owned streetcars, and municipal playgrounds. They began to agitate for such things in New York City.

As in Europe, the New York bohemians also expanded their awareness by interacting with all the diverse types and cultures within the city, from the workers to the Wall Street gentlemen, from the Yiddish theater to all the different forms of art. Although American bohemia grew rapidly and spread to Boston, Chicago, New Orleans, and San Francisco, the other bohemian enclaves couldn't begin to match the New York scene in its level of intensity and diversity. Also to its credit, New York had the Hudson River and the nearby Catskills for country escapes, which many bohemians took advantage of.

THE NEW WOMAN

During the 1890s, America was astonished by a new type of person in the cities: the new woman. Living beyond the traditional female roles, women became doctors and lawyers, taught college, and worked within the ministry. And they also became illustrators and artists of many different kinds. There were suddenly tens of thousands of young women who could paint, sculpt, compose music, and—most important—write. The ideal of educating women was beginning to bear fruit.

Romance novels with new woman characters and morals filled the private lives of young females of the time. While most enjoyed them as an escape, many were inspired by these books. Henry James can be said to have started this trend in English and American fiction, followed by Arnold Bennett, George Gissing, Mary Chavelita Dunne Bright (writing under the pen name George Egerton), Edith Wharton,

Kate Chopin, Theodore Dreiser, and Willa Cather. The modern hero-ine in these stories was always on a quest to experience the full poten-tial of life and culture. She was always looking to develop herself as a person—to make a real difference in the world.

This quest to live outside the traditional female role became ingrained in these new women—and it naturally led many of them to the Lower East Side and the New York bohemian scene. After find-ing the circuit, one young female artist, Mary Heaton Vorse, wrote in her memoirs, "I am part of the avant garde! I have overstepped the bounds!"[15] The new mind-set was certainly different from what she had ever experienced in quiet Amherst, Massachusetts. The new woman would have a profound influence on the twentieth century to come.

A NEW SPIRITUALITY

The end of the nineteenth century also saw a different spirituality emerge in America and Europe. Back in the seventeenth century, Baruch Spinoza had said that true spirituality was measured not in the amount or kind of rituals that are performed but by how much better one left the world through the life one lived. Spinoza's notion evolved over the centuries and combined with transcendentalism and Eastern beliefs in the late nineteenth and early twentieth centuries to make a new spirituality. This was actually a range of beliefs and philosophies, from traditional Hinduism and Buddhism to mysti-cal Christianity and philosophy to the séances of the spiritualists and many other groups. Often those of the organic worldview might believe in several at once.

In 1875, Madame Blavatsky formed the Theosophical Society, declaring that the paranormal could bridge the physical world with the realm of metaphysics. Her controversial séances, whether real or not, spurred great interest in all things spiritual and were an early attempt to bring science to the study of psychic phenomena. From 1879 to 1884, Blavatsky traveled to India to seek enlightenment. She continued to lead what came to be known as the spiritualists for some time.

In 1878, a book by Edwin Arnold sold nearly a million copies. It was called *The Light of Asia, or The Great Renunciation, Being the Life and Teachings of Gautama, Prince of India and Founder of Buddhism*. This introduction to Buddhism fascinated many in the West. Nine years later, the first Buddhist magazine appeared in Santa Cruz, California: *The Buddhist Ray*. A year later, in 1888, Van Gogh painted himself as a Japanese Buddhist monk.

In 1893, a Parliament of World Religions was held, and the first Hindu guru from the East arrived in Chicago and other cities, causing a sensation among many bohemians, spiritualists, new women, and intellectuals. Swami Vivekananda was said to have electrified the audiences at the parliament with his talks. He soon had many disciples in both the United States and in England, where he traveled next. After returning in triumph to India and establishing an influential mission in Calcutta, Vivekananda returned to the West in 1899 and continued to expand interest in Eastern thought and beliefs.

Also speaking at the Parliament of Religions in 1893 was the Japanese Buddhist master Soyen Shaku. Shaku was accompanied by a young translator named D. T. Suzuki, who would go on to write a popular introduction to Zen. In the 1890s, Buddhist missions, youth associations, temples, and schools were started in San Francisco by Japanese immigrants. All of this Eastern thought, based primarily on negative logic, on understanding the world by what you cannot know, fit in well with the contemporary neo-Kantian movement in academia and philosophy, which saw the world in terms of phenomena and noumena, the knowable and the unknowable. The tendrils of change were coming around from the other side of the planet and reinforcing the transformation already underway. Eastern concepts found ready ground in the West and in the nascent Organic Shift of the 1880s and 1890s.

THE NEW THOUGHT MOVEMENT

Science and religion, East and West, also came together in the American new thought movement. Started by a clockmaker with the unlikely name of Phineas Parkhurst Quimby (1802–1866), new thought is seen as a direct ancestor of the American New Age movement of the twentieth century. For many in the young Organic Shift, new thought gave them a center that they otherwise could not have had. In 1902, William James in *The Varieties of Religious Experience* praised new thought to no end, calling it America's "only decidedly original contribution to the systematic philosophy of life."[16]

Using the power of positive thinking, Quimby was supposedly able to heal many people. He felt that human potential was so enormous, being the manifestation of all creation, that by maintaining positive thoughts you were already on the way to solving all your problems. By dint of this way of thinking, we could invoke the power of God or the life force—whatever you call it—to reach your goal of health, wealth, and love.

During the 1880s, new thought was recognized as having many similarities to Ralph Waldo Emerson's transcendentalism, giving it an even stronger intellectual foundation. Although Quimby was not fond of traditional religions, one of his disciples, Mary Baker Eddy, "borrowed" his ideas and in 1879 began the Church of Christ, Scientist—which continues to this day in the form of churches, reading rooms, and the *Christian Science Monitor* newspaper.

At the Parliament of World Religions in 1893, Swami Vivekananda profoundly affected the new thought movement. Many in new thought took up the Hindu concept of reincarnation, which becomes a main theme in later New Age spirituality. Hinduism was considered similar to early Christianity and much admired. Many of the new thought leaders and writers were new women—and many were also African Americans. As we shall see, this movement in the next century became very widespread, especially through the Christian Science sect and the writings of several famous authors and ministers.

JOHN MUIR AND THE SIERRA CLUB

Most people are on the world, not in it. . . . [They] have
no conscious sympathy or relationship to anything about
them—undiffused, separate, and rigidly alone like marbles
of polished stone, touching but separate.

JOHN MUIR, "EXPLORATIONS IN THE
GREAT TUOLUMNE CAÑON,"
OVERLAND MONTHLY, AUGUST 1873

The year was 1863. A twenty-five-year-old John Muir, transformed by the writings of Emerson and Thoreau and suffering from an obsession with botany, decided to leave his studies at the University of Wisconsin and take a botanical foot tour of Wisconsin, Iowa, and Illinois. Later on, in 1867, Muir walked a thousand miles to Florida, studying the botany and geology of the Southeast all along the way. The next year, he ended up in California and traveled immediately to Yosemite Valley, which had been ceded by President Lincoln to California as a state park. There he spent years exploring, hiking, and working as a guide.

Going beyond Thoreau, Muir wrote brilliantly and with great insight on many things, particularly how the environment is a fragile thing and must be protected from the pursuit of money. He also explained how the world is an interconnected whole: "When we try to pick out anything by itself, we find it hitched to everything else in the Universe."[17] Muir felt that one could plug into this whole, that nature could recharge one's very soul: "Climb the mountains and get their good tidings. Nature's peace will flow into you as sunshine flows into trees. The winds will blow their own freshness into you, and the storms their energy, while cares will drop off like autumn leaves."[18]

In 1871, by chance, Ralph Waldo Emerson actually met John Muir, a joining of two great change agents in the midst of the transforming presence of nature, as Emerson would say. The thirty-three-year-old Muir was beside himself—his hero had just arrived in the Yosemite Valley! Muir described the encounter:

When he came into the Valley I heard the hotel people saying with solemn emphasis, "Emerson is here." I was excited as I had never been excited before, and my heart throbbed as if an angel direct from heaven had alighted on the Sierran rocks. . . . I saw him every day while he remained in the valley, and on leaving I was invited to accompany him as far as the Mariposa Grove of Big Trees. I said, "I'll go, Mr. Emerson, if you will promise to camp with me in the Grove. I'll build a glorious campfire, and the great brown boles of the giant Sequoias will be most impressively lighted up, and the night will be glorious." At this he became enthusiastic like a boy, his sweet perennial smile became still deeper and sweeter, and he said, "Yes, yes, we will camp out, camp out"; and so next day we left Yosemite and rode twenty five miles through the Sierra forests, the noblest on the face of the earth, and he kept me talking all the time, but said little himself. The colossal silver firs, Douglas spruce, Libocedrus and sugar pine, the kings and priests of the conifers of the earth, filled him with awe and delight. When we stopped to eat luncheon he called on different members of the party to tell stories or recite poems, etc., and spoke, as he reclined on the carpet of pine needles, of his student days at Harvard.[19]

In the following years, Muir solo-climbed both Mt. Shasta and Mt. Whitney, which led to his two-volume *Picturesque California*. He next wrote articles on protecting Yosemite and, in 1894, came out with *The Mountains of California*. In 1892, John Muir founded the Sierra Club, advocating wilderness preservation over "wise use," which, he warned, was the way to clear-cutting and the degradation of forests. He took it upon himself to preserve Yosemite and other wild places by starting a conservation organization that was truly activist. The Sierra Club immediately advocated the establishment of Yosemite as a federally managed park and later pushed for National Parks at Mt. Rainier, the Grand Canyon, and in the vast Sierra Forest. All efforts were eventually successful.

Building on the foundation started by Emerson and Thoreau,

and aided in the great mission by the other great child-soul of nature, William Burroughs, John Muir made an invaluable contribution to the Organic Shift. He taught us how to preserve nature's greatest wonders, while they are still there to save. Yet Muir taught us so much more, especially how we could develop ourselves as people and as a culture if we just try to absorb and understand the nature that is all around us— rather than always trying to make a dollar.

There is much that the world of today can learn from Muir's writings—and of course the Sierra Club is still very active in conservation. Most of all, his writings reflect what a keenly developed organic worldview sees, a philosophy honed by the teachings of nature—and an inner contemplation inspired by an unmatched degree of isolation from the world. Many environmentalists of today see Muir as a role model, his singular devotion to nature and reform being their guiding principle. The world owes much to this man who walked every trail and climbed every mountain. As Emerson realized after meeting him: "Muir is more wonderful than Thoreau."[20]

ROOSEVELT AND CONSERVATION

Teddy Roosevelt later visited John Muir in Yosemite and learned from him some of the subtler points of conservationism. Although he was an incorrigible imperialist Republican in foreign policy, Roosevelt's accession to the White House had started a powerful conservation movement in government. By 1909, President Roosevelt had set aside forty-two million acres of national parks, including Yosemite and the Grand Canyon—and much more in wildlife refuges. In 1916, President Wilson signed the Organic Act, starting the National Park Service.

In his progressive crusade, Teddy Roosevelt also got assistance from two particular authors. Ida Tarbell's exposé on John D. Rockefeller in *The History of Standard Oil* in 1904 helped Roosevelt in his quest to establish antitrust law. During 1906, Upton Sinclair's *The Jungle* told of the lethal and unhealthy working conditions in the meat-packing industry and how pollution was casting a pall on the land. This book aided Roosevelt's push for the establishment of the Food and Drug Administsration (FDA) and regulation of industry in other areas. When the possibility of a coal strike in 1902 meant those in the Northeast would freeze, Roosevelt called a meeting of the robber barons and the coal miners—the latter winning the right to organize and new safety regulations as a result.

The beginning of the twentieth century did indeed have a progressive start.

THE NEW PARADIGM BURNS BRIGHT

By the turn of the twentieth century, the future looked bright. Pacifism, the neo-Kantian movement, the new environmental concern, the science of ecology, the realization of the evolution of consciousness, the wave theory, the new woman, the labor and populist movements, progressivism, early environmental protection, the success of the solar engine, the new thought movement, the rapidly expanding bohemian scenes, art nouveau, psychology, and cultural anthropology—all these late-nineteenth-century trends and discoveries soon transformed people's understanding of what the true potential of humanity could really be.

As the twentieth century began in 1901, the organic worldview had grown beyond the early pioneers and the many movements of the later nineteenth century. All these movements reached a brief apogee before the horror of World War I shook society with its traumatic convulsion of millions dead. Before that happened, however, there was a fascinating period before cheap oil and World War I.

If you were in Florida or California in the early years of the twentieth century, you could wake up, take a nice hot bath with your day-and-night solar water heater, and then drive your electric car to work. You might read books on Zen or William James's *Varieties of Religious Experience* or buy art nouveau art and furniture. Or you could attend a lecture by John Muir or a pacifist—or hear an Indian guru speak.

Meanwhile, the progressive movement had landed in the White

House when Teddy Roosevelt took office in 1901. Although primarily a mechanist nationalist intent on building an American empire and definitely (as previously stated) an imperialist Republican when it came to foreign policy, Teddy Roosevelt was a progressive Republican when it came to matters of the U.S. economy, and he quickly moved to rein in the robber barons with antitrust laws, improve the food and drug supply with the FDA, and conserve the environment through an expanded national parks system—all the while pushing for reform, reform, reform. Progressivism was not only in the air, it was also ensconsed in the halls of power. If you lived in New York before World War I, Greenwich Village was a thriving bohemian neighborhood, where you could meet significant change agents like Mabel Dodge and John Reed. Although most of the world was old paradigm, the inhabitants of several bohemias in large cities around the world clearly were not. The new thinking simultaneously expanded to many places beyond these artistic ghettos.

AMERICAN BOHEMIA
AND THE ARMORY SHOW

One place bohemia spread to was the sleepy village of Woodstock in the New York Catskills, which the philanthropist Ralph Whitehead made into an art colony in 1902. Here was a refuge from city life and the obligations of civilization. Artists were invited to come and create, oftentimes art being accepted as payment for rent. Soon writers, theater people, and whole entourages were moving up to Woodstock, which did indeed become a small center for the arts and music. Later a group of maverick artists led by Hervey White broke off from Whitehead and formed the Maverick Colony at Woodstock, which gave artists total freedom and also welcomed musicians, writers, and dancers.

Provincetown, Massachusetts, on Cape Cod was another place the creative minority would escape to. Mary Vorse lived there and provided the wharf building where the Provincetown Players put on Eugene O'Neill's *Bound East for Cardiff* in 1916. This play was acclaimed as the beginning of original American drama, complete with the innovations of dramatic lighting, working-class characters, and biting social commentary. Provincetown would continue to nurture the arts and the counterculture in many different ways.

Woodstock and Provincetown were escapes, however, as the real scene was in New York City. The bohemians began to concentrate on the Greenwich Village area, with scores of tea rooms scattered throughout storefronts in the old neighborhood. Many colorful characters became known to one and all, such as Ralph Bourne, a radical writer; Papa Strunk, a landlord who sometimes accepted art for rent; and Sonia the Cigarette Girl. Sinclair Bobby Edwards, a musician-singer, had a hit in the Village Follies called "Why Be an Industrial Slave When You Can Be Crazy?" John Reed published *The Masses,* a socialist magazine, and Max Eastman *The New Republic,* while the talented entertained all with shows like Sinclair Lewis's *Hobohemia* at the Greenwich Village Theater. The tearooms started serving food and were soon more crowded than ever. A large community found itself—and partied throughout the first two decades of the new century.

Percolating above the froth of Greenwich Village was the determined group of artists and exhibit organizers who first brought modern art to America. At the time, the National Academy of Design was the art establishment; it laid down the law on what was allowed in exhibits and what was not. Modern art, such as cubism and futurism, was definitely out. No less a figure than Teddy Roosevelt had declared modern art "fake"—as phony as P. T. Barnum's sideshows.[21]

By 1911, the new artists were so frustrated they started their own organization: the American Association of Painters and Sculptors. Realism and noble subject matter were no longer restrictions, and the only requirements were that you create art with vision and an unfettered imagination. This small group of radicals was about to have a tremendous impact on New York and America—for they decided to hold an international exhibit in the New York Armory.

MABEL DODGE,
BOHEMIA'S QUEEN TAKES MANHATTAN

The main organizer of the Armory Show was one Mabel Dodge, heir to the Dodge fortune, who could best be described as a full-time change agent for the Organic Shift. Mabel Dodge seemed a veritable force of nature to the journalists and others who met her, with all quickly entranced by her ability to make you instantly feel like the most important writer or thinker in the world. Within a few months of her decision to shake things up, Dodge helped send both the counterculture and the established culture into new realms of thinking.

After her arrival on the scene in 1912, she quickly took on not only the post of vice president of the radical Artists Association, she was funding the planned exhibit with her fortune and marketing the show to all New York. Before the Armory Show opened, however, Dodge established a salon in her elegant apartment on Ninth Street and Fifth Avenue, declaring, "Let the town pour in!" She wanted to know everybody, and she wanted to challenge the old order. "Many rules are being broken—what a wonderful word—'broken'!" wrote Dodge. "Nearly every thinking person nowadays is in revolt against something, because the craving of the individual is for further consciousness, and because consciousness is expanding and is bursting through the molds that have held it up to now. . . . It seems as though everywhere, in that year of 1913, barriers went down and people reached each other who had never been in touch before; there were all sorts of new ways to communicate, as well as new communications. . . . Men began to talk and write about the fourth dimension, interchangeability of the senses, telepathy, and many other occult phenomena."[22]

Starting in January 1913, journalists, writers, artists, radicals, poets, and the genteel crowd in their elegant gowns and suits all gathered together in a scintillating series of evenings. There was the "Dangerous Characters Evening," for instance, with Big Bill Haywood extolling the virtues of the IWW (the Industrial Workers of the World or the radical Wobblies), followed by Emma Goldman and her views on anarchism and English Walling on socialism. Two hundred guests—half in evening dress—listened and then argued late into the night over a sumptuous buffet. It was the original radical chic.

One month after she set New York on its ear with her salon, she became the greatest publicist modern art has ever seen, with her relentless campaign to bring one and all to the Armory Show. A tour de force, the exhibit displayed styles in impressionism from Cézanne to Marie Cassatt and on to cubism and many others. The effect of seeing forty years of astonishing modernism all at once was breathtaking. Women were especially taken by it all and thereafter were big buyers of modern art, especially in the big cities. Although many may not have fully understood the new worldview, modern art made you think—and that often led to new directions.

Dodge met Gertrude Stein in Florence, and Stein had written a "cubist word poem" about her. Dodge took the poem and made it part of the exhibit, with an explanation about Stein and the new literary style. The display of Stein's cubist poem attracted much attention, immediately making Dodge and Stein famous. Dodge became so well known she was one of the first real celebrities of the new century, and Stein grew quite jealous.

The Armory Show also displayed the works of many other women artists besides Cassatt, elevating their station to new levels of acceptance. Yet the establishment critics were, as a whole, opposed to the Armory Show, seeing it as a circus like the Wild West Shows of the day. *Nude Descending a Staircase*—the most controversial painting—was compared by the *New York Times* to "an explosion in a shingle factory." The *Times* warned, "It should be borne in mind, that this movement is surely a part of the general movement, discernible to all the world, to disrupt, degrade, if not destroy, not only art but literature and society too . . . the Cubists and the Futurists are cousins to anarchists in politics."[23] This was all very true. In the early twentieth century, modern art and radical politics united, and they clearly did intend to destroy art, literature, and society—but only that of the old worldview. One of the kinder critics sent a telegram to a party celebrating the success of the Armory Show: "It was a good show. But don't do it again."[24] Yet it was too late, and America was never quite the same.

For the next three years, Mabel Dodge's salon served as an attractor point for the gathering forces of the Organic Shift. There was the Sexual Antagonism Evening, which talked openly about sex. Dodge herself helped made popular the acceptance of the female orgasm—and even named her dog Climax. She also aided Margaret Sanger in her crusade for birth control and sexual liberation. Other evenings introduced writers such as Gertrude Stein and Jack Reed, while the New Psychology Evenings spread the ideas of Freud's "talking cure," the new miracle of psychoanalysis. Jung, too, was discussed. Jack Reed, known as the poet laureate of Greenwich Village for his poem "A Day in Bohemia," one day returned from Europe with Mabel Dodge, the queen of bohemia, making them the most illustrious hip couple of all. They were the talk of New York. Reed, however, soon left to write about Pancho Villa—breaking Dodge's heart.

THE FAMOUS 1914 PEYOTE CEREMONY IN GREENWICH VILLAGE

One night in the spring of 1914, a certain Raymond Harrington told Dodge of a unusual medicine he discovered while conducting ethnological research among Native American cultures in Oklahoma. When she heard that eating these peyote buttons allows you to go beyond everyday awareness, she declared it the next "in" thing—and invited several select friends to participate in a real peyote ceremony.

On the night ordained, the group took not one but two peyote buttons—except Dodge, who ate the first one but palmed the second one. Everyone had a different experience. "Everything seemed ridiculous to me," she wrote in her memoir, "utterly ridiculous and immeasurably far away from me. . . . Several little foolish human beings sat staring at a mock fire and made silly little gestures. Above them I leaned, filled with an unlimited contempt for the facile enthrallments of humanity, weak and petty in its activities, bound so easily by a dried herb, bound by its notions of everything—anarchy, poetry, systems, sex, and society."[25]

Hutch Hapgood meanwhile felt "God drunk" at first, and then "[i]t didn't seem strange to me when Raymond left his seat and ascended through the air to the ceiling." Hapgood ended up imagining himself to be an Egyptian mummy "making a complete review of my whole life, applying to it an intense criticism, which amazed me for its complete unworthiness."[26]

Meanwhile, Hapgood's nephew Raymond Harrington—last reported on the ceiling—said later that he had gone on "a long voyage to wander for months in a tropical valley full of huge birds and animals of hitherto unknown colors." Terry, the anarchist alcoholic, was the last to speak. "I have seen the Universe!" he declared with the "most illumined smile," "Man! It is wonderful!" Some others had bad trips, including Dodge, who regretted the escapade in the end, yet noted years later that "every one of the others who had been at the apartment that night talked about it for years . . . undoubtedly that legend has encircled the world."

THE TAOS COUNTERCULTURE
OF THE TWENTIES

With the onset of World War I and anti-Communist hysteria in the form of the 1919 Palmer Raids and the Red Scare, many left the village and migrated east to Paris, where "the lost generation" of writers and artists took up residence. For years, Gertrude Stein joyfully entertained them and networked them in her Parisian apartment. Mabel Dodge, too, decided to move on—but west, to the new Taos art scene in the New Mexico desert. She became a major patron there, making it a cultural center, and there she also met her fourth husband—whose face she had seen in a dream. Upon meeting her, this tall, mystical Pueblo Indian told her that he had sung a song for her to come to see him and that he too had seen her before—which Dodge took as clairvoyance. Quite swept off her feet by all this, Dodge soon dumped husband number three and married "Tony" in 1923. "Why Bohemia's Queen Married an Indian Chief" screamed a headline in the *Pittsburgh Post,* with a full article explaining the shocking development.

Dodge, of course, transformed little Taos into Village West and was soon entertaining a list of luminaries including Willa Cather, D. H. Lawrence, Thornton Wilder, Georgia O'Keefe, Ansel Adams, Edna Ferber, Robinson Jeffers, Leopold Stokowski, John Marin, Jean Stafford, and Thomas Wolfe. Tony, meanwhile, loved taking the glitterati out for the Taos tour in Dodge's Cadillac. Georgia O'Keefe went on to become one of the most famous women artists in history and put Taos and Native American themes permanently on the art and culture map.

For all of the events and accomplishments listed above, Dodge, Stein, and the other leading members of the bohemian counterculture deserve a special place in the history of the Organic Shift.

Fig. 6.3. Mabel Dodge and her husband
Tony in Taos, New Mexico, in 1920

THE INSCRUTABLE WISDOM
OF THE EAST

During the first years of the new century, Eastern religions and concepts based on negative logic continued their influence on the organic creative minority in the West. Swami Vivekananda was still the biggest influence, but in 1907 D. T. Suzuki published the *Outlines of Mahayana Buddhism,* which introduced Zen. Readers in 1914, meanwhile, learned about yoga in *The Yoga-System of Patanjali* translated by James Haughton Woods. In 1920, Paramahansa Yogananda arrived in Boston, spreading Hinduism further into America. Another book by Suzuki, *Essays in Zen Buddhism,* continued to make Zen thought more accessible in 1927, and in that same year, Oxford published *The Tibetan Book of the Dead,* revealing the exotic ways of Tibetan Buddhism.

Western writers of the time compared Eastern thought with that of the West. William James's *Varieties of Religious Experience* was published in 1901, comparing all the major beliefs in the world for their similarities, their differences, and their wisdom. Caroline Rhys Davids in 1914 showed how Buddhism and Western psychology are compatible in *Buddhist Psychology,* while Rudolf Otto in 1917 explored mysticism from a Kantian perspective. In 1926, Otto's *Mysticism East and West* looked at similarities between mystical practices around the world. All this interest in Eastern philosophy literally dissolved the old worldview away. The negative logic of it was so appealing, and it all seemed to support what the new physics was discovering about the world. We shall see that the influence of Eastern thought grew exponentially when the new paradigm entered the mainstream in the mid-1950s and 1960s.

One new thinker, from India, offered not only a new philosophy of living together but also showed the world how it could achieve peace without bloodshed. The world required massive change, and he decided to devote the rest of his life to make it come about. This one man summed up the teachings of the Organic Shift, plus the wisdom of the East—and ended up inspiring the world.

GANDHI AND THE GRAND UNITY

Earth provides enough to satisfy every man's need, but not enough for every man's greed.

MAHATMA GANDHI

Mahatma Gandhi, a great change agent of the new paradigm, was one who thought globally and acted locally on an unprecedented scale. Influenced by Thoreau's essay on civil disobedience and the nonviolent principle of *ahimsa,* Gandhi started his activism for civil rights while working as a lawyer in South Africa. Ironically, it was Prime Minister Smuts who had him deported to India. There Gandhi preached the voluntary simplicity of Thoreau and the emancipation of women, declaring that Indians must become self-sufficient and suggesting that the spinning of cloth was a good occupation to learn and practice. Without voluntary simplicity, explained the holy man, the resources of the world will become used up so quickly by the exploding population that war and famine were sure to result.

Following World War II, Gandhi's nonviolent campaign against British rule in India succeeded as no civil disobedience ever had before. Gandhi not only liberated the subcontinent, he elevated the idea of pacifism and world peace to new levels. Most important, he predicted that, as the world develops, communities of peaceful and enlightened beings would form unities, as he called them. He envisioned these unities growing larger and larger, eventually joining together and working in concert until they created a great planetary awakening, a coming together of all peoples in a grand unity: "The circle of unities will ever grow in circumference until at last it encompasses the whole world." A new era of peace and wisdom would then be born. Gandhi's assassination in 1948 was truly a tragedy of global proportions.

Thanks to his example, however, and the establishment of the United Nations, the end of war became a firm hope of all new-paradigm thinkers, a goal they worked for in many different ways. In the early 1950s, antiwar sentiment grew again following the devastation of

Fig. 6.4. Mahatma Gandhi,
great change agent of the world

World War II, and then Korea and the Cold War. Tiny liberal and left-wing protests against the Korean War were held in front of the White House, receiving great attention in the media. By the mid-1950s, civil disobedience, learned from Thoreau, the suffragettes, and Gandhi, was poised to have a huge impact in the United States in the form of the civil rights movement and the nuclear test ban movement.

Despite all the deep changes in science, technology, and in society itself, there was yet another set of questions that still needed to be answered. The subatomic world was now far better understood, but what about the physics of human energy: How could they be harnessed and channeled to build a better world? Gandhi spoke of the unities becoming a grand unity, with the world ascending to a new level of awareness and cooperation, suggesting that through organizing and nonviolence this wonderful end would be achieved. Yet how can this global awakening be carried out in a modern world, ruled by mechanistic science, big government, and the profit motive of big business? The answer to this question is crucial.

Fortunately, another great thinker, a Jesuit priest who was also a paleontologist, confronted this question directly. Perhaps no one thought more about the future than this man—who receives his own chapter in order to cover all of his many contributions to the Organic Shift that would ulimately bring about a global awakening. Truly, he was one of the deepest thinkers of the entire Organic Shift, almost a reincarnation of Kant or Goethe for the 1940s and 1950s. What did this spiritual scientist say about what is to come and how to get there? The answer is at once exhilarating and entirely plausible and practical.

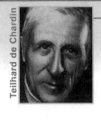
Teilhard de Chardin

7

THE FUTURE ACCORDING TO TEILHARD DE CHARDIN

We are not human beings having a spiritual experience.
We are spiritual beings having a human experience.

Pierre Teilhard de Chardin

PIERRE TEILHARD DE CHARDIN: THE FUSION OF SCIENCE AND SPIRITUALITY

One of the great geniuses of the twentieth century was the Jesuit priest and paleontologist Pierre Teilhard de Chardin (1881–1955). In *The Aquarian Conspiracy,* Marilyn Ferguson in 1980 surveyed transformational thinkers, asking who their biggest influence was. Teilhard de Chardin came out number one. In a century that produced Einstein, Gandhi, Heisenberg, Smuts, and so many others, Teilhard de Chardin still stands out. Why?

The answer lies in the importance and the quality of the discoveries that Teilhard made. In short, his critical advances made it possible to establish global purpose and will—a collective purpose that would remove the impulse to war. By worshiping evolution, Teilhard de

220

Chardin gave us a scientific yet spiritual basis upon which a new world could be created. In fact, spirituality in his science became something that can actually be measured and is visible in evolution and in the real world in many different ways. He literally laid down a new scientific foundation for the workable future, transforming the social sciences. The world would finally be changed, he predicted, by the spontaneous combustion of "homo progressivus"—by progressive thinkers coming together through human interaction and technology on a new "planetized" level, a level in which the population begins to think globally.

In recent years, Teilhard de Chardin has come under criticism from deep ecologists as being anthropocentric and for having spoken favorably of megatechnologies as possible solutions. Yet we must not ignore his undeniable contributions. There is validity to the anthropocentric argument, especially as far as Teilhard's earlier writings go. In an interesting quote from *The Future of Man,* however, the later Teilhard emphasized that humanity has no special place in the universe other than being the first species to be able to reflect upon their own awareness. "[I]f a given Phylum X . . . preceding the anthropoids, had succeeded in passing the barrier separating reflective consciousness from direct consciousness, *Man would never have come into existence. Instead of him, Phylum X would have woven and constituted the Noosphere*"[1] (italics added).

The center that holds in Teilhard's universe is thus not a divine humanity, but rather complexity and consciousness. If a different branch of evolution had been the one to reach intelligence first, said Teilhard in 1950, *Homo sapiens* would have never even evolved! Phylum X would have rapidly formed civilization and the collective consciousness, the noosphere, precluding the rise of any other sentient species. This is not an anthropocentric universe; it is a complexity-centered worldview.

How can you have an anthropocentric universe in which humanity may not even exist? Teilhard de Chardin is simply too important to dismiss as being anthropocentric—especially when he was not anything of the kind in his later years. He established the foundation for much useful thinking about the future, from proving the existence of Bergson's *élan vital* to identifying many parts of reality previously hidden from us.

REVERSE ENTROPY

Teilhard's greatest discovery, in fact the greatest discovery of the twentieth century, leads to all the rest of his theories. In contrast to the entropy of the physical universe, Teilhard explained that life was driven to evolve ever more complex and conscious forms through "reverse entropy." Rather than falling apart through the second law of thermodynamics (entropy), the realm of life is always synthesizing itself and leaping ahead—to new levels of being and form. Teilhard never actually used the term *reverse entropy*; he only postulated that the reverse of entropy must be true in the evolution of life and consciousness and called it "the cosmic movement in the reverse sense" of entropy.

Making scientific what Bergson had mystically called the élan vital, reverse entropy, which can be measured, is a far better description of the driving force behind evolution. Reverse entropy was the most important advance of the entire Organic Shift—as the concept led directly to the reinvention of our entire universe. Teilhard explains:

> In the downward direction of entropy, we find that matter becomes diffused and energy is neutralized. This is something we have long known. But why should we not take into account *the cosmic movement in the reverse sense,* toward the higher forms of synthesis. . . . Beneath our eyes, extending from the electron to Man by way of the proteins, viruses, bacteria, protozoa and metazoa, a long chain of composites is forming and unfolding, eventually attaining an astronomical degree of complexity and arrangement.[2] (italics added)

Reverse entropy, which is the same as Buckminster Fuller's syntropy, is a powerful discovery indeed. With one theory, all of life, evolution, consciousness, and cultural development was explained. The second law of thermodynamics, entropy, thus reveals the workings of the physical realm, while reverse entropy is the force underlying life and all living things and systems. With the concept of reverse entropy in hand,

Teilhard could understand the past and evolution anew and could thus also describe a whole new future for humanity.

Reverse entropy may even exceed the importance of Einstein's formula $E = MC^2$, for the formula merely gave us a new understanding of physics for power sources, nuclear weapons, and research. Without reverse entropy and Teilhard's other discoveries, we would not have the essential scientific foundation for the rapid world change that is required for planetary survival itself. Without reverse entropy, the new paradigm would not have the intellectual basis to successfully challenge the mechanists. With Teilhard's discovery, it does. Teilhard uncovered one of nature's primary open secrets, the cause of the very unfolding of evolution that Goethe himself had first talked about.

THE LAW OF COMPLEXITY AND CONSCIOUSNESS

Through reverse entropy, life is continually synthesizing itself, attaining greater and greater levels of complexity. Consciousness itself is a result of reverse entropy. More and more complex life-forms have increasing levels of consciousness, giving them new mental abilities, which Teilhard called the law of complexity and consciousness. Thanks to the integrated interactions of about ten billion neurons in the human brain, our species possesses *reflective awareness,* the ability to think about our own awareness.

Teilhard saw the evolution of consciousness as having three major thresholds. The first threshold of consciousness was animal thought, from the earliest organisms up to the most complex animals: see food, eat it. The second threshold of consciousness was a significant synthesis and came when human beings reflected on their own awareness: I am. In doing this, a special circuit closed and humans became sentient, able to think in an organized way and create a culture and to even think about the future.

The third threshold we have not yet reached, but it comes in our near future when we finally realize *we are.* This third threshold of consciousness, Teilhard predicted, is the inevitable result of reverse entropy, which keeps bringing society and our individual minds together through technology. The planet is acquiring what he saw as a vast electronic neural network, much like a brain but on a global scale.

THE NOOSPHERE AND
THE 1947 PREDICTION OF THE INTERNET

Teilhard called the collective mind the noosphere. We stand on the *geosphere,* Earth itself, we live in the *biosphere,* the outer layer of life, but we think, said Teilhard, in a collective *noosphere.* It was a collective effort, for example, that invented the airplane. The Wright Brothers merely advanced a series of basic inventions made by others before them in the field of gliders, engines, and model airplanes. The noosphere invented the airplane—and in the same way created our entire culture.

Today, the noosphere grows close to awakening and the third threshold of consciousness. Ever since the invention of the telegraph, to the interconnections of today's worldwide web, there has been a closing of the electronic circuit of communication, bringing the world closer and closer together. Today, the power of the nascent blogosphere as a new political, social, and cultural phenomenon is accepted as real. Teilhard basically predicted it. The evolution of the collective consciousness, which Teilhard called noogenesis, is now about to close the special last circuit leading to full global awakening.

Let us take a rapid glance at the structure and functioning of what might be called the "cerebroid" organ of the Noosphere. . . . I am thinking, of course, in the first place of the extraordinary network of radio and television communications which . . . already link us all in a sort of "etherized" universal consciousness. But I am also thinking of the insidious growth of those astonishing electronic computers which . . . enhance the essential (and too little noted) factor of "speed of thought". . . . As in the case of all organisms preceding it, but on an immense scale, humanity is in the process of "cerebralis-ing" itself.[3]

What we are experiencing today is clearly an acceleration in the "speed of thought," in which the two largest megatrends, the paradigm shift and the Internet/communications revolution, are combining to take thought to qualitative new levels. In short, the closing of the final circuit in this process would allow new properties of collective think-ing and collective action to develop. What is impossible in our current day would then become routine. It will all be based on a new global ability to reflect, which Teilhard spelled with an *x, reflexion,* signify-ing collective reflection as opposed to the reflection of individuals upon their own awareness. If what Teilhard is predicting here actually occurs, attaining the third threshold of consciousness would easily be the most significant event in human history.

What happens if the noosphere does awaken from its dream of his-tory? Where would we all be headed then? Is it utopia? No, not utopia, the perfect place. But clearly a better place, a workable future—if enough people become active in trying to create that better future. Teilhard's future would be an end run around the mechanistic dilemma, and all the useable information and technology of the old worldview would then be absorbed into the much larger whole of an organic global soci-ety. It is this awakening of the noosphere/blogosphere that is so critical to transcending the dilemma and finally overcoming the old mechanis-tic worldview.

THE COMPRESSION OF
HUMANITY AND INFORMATION

What does all this collectivity mean in terms of the individual? Something quite wonderful, actually. Teilhard understood that because the human population explosion was taking place on the small sphere of Earth, humanity was in effect being compressed, especially since the entire world was now settled. He predicted this compression of the exploding population on the closed sphere of the globe would quickly lead to an unprecedented rate of change for both individuals and collective humanity.

None can deny that the rate of change has accelerated dramatically in the last few decades, but Teilhard said it will continue to speed up until a tremendous shift or awakening occurs. The largest part of this transformation comes about because "human molecules" change "state" as they are compressed together, just as gas being forced inside a cylinder changes from gas to frozen liquid—the molecules simultaneously form droplets on the inside of the cylinder, which then join with other droplets and stream downward to form a pool at the bottom of the cylinder. Like the transformation of gas to liquid, the new state immediately leads to new properties; only these are primarily mental properties rather than physical characteristics. When this transformation arrives, what had been impossible suddenly becomes routine.

Individuals will seek and find more of what makes them human—and they will, according to Teilhard, be able to develop their full creative potential in a greatly expanded nurturing environment. In the new state of humanity, Teilhard predicted there would be a way for each like-minded person on Earth to connect with each other in a mutual support network. Teilhard explained how the new collectivity would actually develop the individual: "[T]he medium will be established in which a basic affinity may be born and grow, springing from one seed of thought to the next, canalizing in a single direction the swarm of individual trajectories. In the old Time and Space a universal attraction of souls was inconceivable. The existence of such a power becomes possible, even inevitable, in the curvature of a world capable of noogenesis."[4]

It is thus compression of both the population and information that drives the powerful coming together of many individuals into an unprecedented collective purpose, a collective mind ready to act. If the compression of gas analogy holds true, as the human race enters its final level of compression, we will gather our minds together first in associations, clubs, and blogs, and then go on to join these entities in a vast worldwide alliance. This will not submerge the individual, believed

Teilhard, but rather will bring out the true personality of each person. Compression will "superpersonalize" each one of us through a "new kind of love." Teilhard is here getting a sense of the human potential movement to come, a jump to a new level of personal development. This new state of nurturing will allow us to explore more deeply the potential of what it means to be human and create—and it leads to a totally different future.

> True union, the union of heart and spirit, does not enslave, nor does it neutralize the individuals which it brings together. It *superpersonalizes* them. Let us try to picture the phenomenon on a terrestrial scale. Imagine men awakening at last, under the influence of the ever-tightening planetary embrace, to a sense of universal solidarity based on their profound community, evolutionary in its nature and purpose. The nightmares of brutalization and mechanization which are conjured up to terrify us and prevent our advance are at once dispelled. It is not harshness or hatred but a new kind of love, not yet experienced by man, which we must learn to look for as it is borne to us on the rising tide of planetization.[5]

The power of the exploding human population on the finite geography of the globe, and the resulting effect of compression on society, is essential to understanding Teilhard's predictions about the future. It is what energizes the whole transformation of the future, rapidly bringing forth new inventions, new thoughts, and new solutions, making the impossible of today a reality tomorrow. It means that the usual view of the future presented in movies, the techno/totalitarian nightmare, will actually never come about on a global scale. Compression forces the final battle between good and evil, between love and hate, and between peace and war. What the Hollywood script writers don't realize—and what they leave out—is that a coming transformation of consciousness, caused by the compression of humanity, will *transcend* the war, greed, and stupidity of the twentieth century. "From the first beginnings of History . . . this principle of the compressive generation of consciousness has been ceaselessly at work in the human mass. . . . What is the automatic reaction of human society to this process of compression? Experience supplies the answer—*it organizes itself.*"6

It is a profoundly instructive and mysterious phenomenon. The human mass is spiritually warmed and illuminated by the iron grip of planetary compression; and the warming, whereby the rays of individual interaction expand, induces a further increase, in a kind of recoil, of the compression which was its cause . . . and so on, in a chain-reaction of increasing rapidity.7

Each increase in pressure leads to further opening up between people and nations; each increase makes the development of new social structures and new technologies a reality. This process accelerates until it reaches a critical mass, Teilhard said. Then we will be subject to "forces of attraction between men which are as powerful in their own way as nuclear energy."

When the dual process of the compression of information and the completion of the neural net is complete, explained Teilhard, something new under the sun will emerge: an awakened collective consciousness, an unprecedented transformation of individual minds into a collective will. This global mind will find a united purpose and be able to take collective action, transcending the old war reflex by uniting to restore our ravaged Earth.

The Global Awakening and the enormous leap to a collective future outlined by Teilhard would certainly be just that.

And if Teilhard is right, the waking point leading to this great transformation lies in our near future. In the twenty-first century, Teilhard predicted we will reach the third threshold of consciousness, the point at which the noosphere actually becomes conscious of itself and reflects on its own being, achieving the state of *reflexion*. As stated, there will then be a new global purpose, a new ability to act in a collective manner—somewhat like Gandhi's convergence of the unities. Teilhard envisioned a tremendous raising of consciousness, an era of peace and wisdom. It would be a new enlightenment based on a new Copernican Revolution. He even compared it to the paradigm shift of the seventeenth century, seeing that scientific revolution would precede social transformation. We would enter a "new space of complexity," meaning a broader understanding of the universe brought about by the new paradigm. This new level of scientific awareness would be as significant a change as the Copernican Revolution was, and it would happen within two or three generations after 1950:

> What is happening in the world today is as though, four hundred years later and at a higher turn of the spiral, we found ourselves back in the intellectual position of the contemporaries of Galileo. . . . Following the moment when a few men began to see the world through the eyes of Copernicus all men came to see it in the same fashion. . . . So it requires no great gift of prophecy to affirm that, within two or three generations, the notion of the psychic in-folding of the earth upon itself [the noosphere], in the bosom of some new "space of complexity," will be as generally accepted and utilized by our successors as the idea of the earth's mechanical movement round the sun, in the bosom of the firmament, is accepted by ourselves.[8]

For his multiple contributions, Pierre Teilhard de Chardin deserves a special place in the history of the Organic Shift, as the impact of his philosophy on the decades to come was tremendous. Recent history has already unfolded according to many of his predictions, the Internet and the 1960s being just two examples. Although he wrote many essays that have been made into many books, his thoughts on the noosphere and its awakening can best be found in the last two-thirds of *The Future of Man*. It remains to be seen whether his main prediction of planetary awakening will now unfold. If the world can but understand what this one Jesuit paleontologist-philosopher was trying to tell us, that we are about to come together in an unexpected global catharsis, then a bright new future can be seen.

There are, however, many other factors needed to create a workable organic society and thoroughly transcend the mechanistic dilemma. As long as American culture continues to follow the old consumer lifestyle and the rest of the world emulates the conspicuous consumption and oil addiction of the United States—one person in his SUV driving to the mall—we will never succeed. We need a new culture, an organic one, based on voluntary simplicity and low-impact technology and lifestyles—a culture that heals the world rather than destroys it. We need to change, each one of us, deep down inside ourselves, and open up to the wonders of nature and each other. There is much to learn, much to transform, on the inside and in the world itself.

175 YEARS OF PREPARING THE GROUND FOR DEEP CHANGE

From Kant and Goethe to Blake, Wollstonecraft, and Hahnemann; from Emerson, Thoreau, Stanton, and Haeckel to Dilthey, James, and Bergson; and finally to Smuts, Heisenberg, Gandhi, and Teilhard—the big picture thinkers of the Organic Shift prepared the scientific and intellectual ground for a new enlightenment. At the same time, the political ground had also been made ready for great change by the many movements for women's and minorities' rights as well as the union and populist movements. Consecutive countercultures also blazed a path for others to follow, from the Romantics to the bohemians and the beats.

In response, both older paradigms fought the Organic Shift. The medieval worldview found a more sophisticated voice in the form of the young Earth geologists of the eighteenth century who defended the concept of a recent creation, the Princeton Theology (the intellectual foundation of today's fundamentalism), and then the development of scientific creationism in the 1920s and thereafter.

All were a counterreaction against secular change and especially the theory of evolution, which directly contradicted the Bible. At the same time, the mechanistic paradigm, in the form of the AMA and other similar organizations, as well as most of society's institutions, reacted by blocking and ignoring organic change in many different areas, from medicine to energy, culture, and personal freedom, as we have seen. Nevertheless, by the mid-1950s, the new paradigm became strong enough to begin to challenge some of the foundational assumptions of the old mind-set, such as war, poverty, pollution, limited civil rights for minorities, and women. The organic worldview was now poised to enter the mainstream, to blossom in the first phase of the social tansformation pattern, the early enlightenment.

Teilhard had said the compression of information and of the population would start an awakening process that would end in a tangible collective consciousness. Certainly in the decade to come, the new ability of television to daily broadcast the horrors and images of war into the living rooms of everyone did just that. Teilhard predicted a time would come when people would begin to break through the barriers that have always repulsed them and learn to love all other members of the human race as equals, as partners in building a peaceful world. This completely different thinking promised to thoroughly shock the unsuspecting and unprepared nationalist mechanistic world. A time of turmoil nothing short of mind-boggling was about to begin, as one change followed another in rapid-fire fashion—a mighty clash of culture and counterculture.

We know it as the 1960s.

Rachel Carson

8

THE EARLY ENLIGHTENMENT

1950–1975

We shall overcome
We shall live in peace
We shall overcome someday!
I know that deep in my heart,
I do believe!
That we shall overcome someday!

SONG OF THE CIVIL RIGHTS MOVEMENT

THE NEW ENLIGHTENMENT

By the middle of the 1950s, a rare situation had developed. An enormously significant scientific and cultural paradigm lay in the offing—yet the mainstream, especially in the United States, understood little or nothing about it. What happened to change the world so dramatically in the 1960s? How could the sedate 1950s be followed by such a dynamic decade of social transformation? Could it be that we then entered into the enlightenment pattern, the start of the four stages of social transformation?

232

It certainly appears so. In particular, the 1960s and 1970s correspond to the early enlightenment phase, the 1980s and early 1990s to the conservative backlash phase, and the years after about 1990 to the intensive phase trends. The main trends and changes of those times clearly followed the century-long pattern. As happened in the lead-up to the Enlightenment, the three essential questions at the center of every paradigm were asked in the 1960s: who are we, how did we get here, and where are we going? A scientific, civil rights, and cultural revolution suddenly entered the mainstream, prompting a great clash between old and new. In the 1700s, the humanist counterculture was *really* dangerous to the old worldview, for without sin and guilt, the theocracy of medieval times could never stand. Both the king and the church were stripped of their power over the secular world. Two hundred and fifty years later, the mechanistic worldview saw the series of organic countercultures with much the same kind of fear, as the old beliefs were made irrelevant.

THE SPIRAL OF HISTORY

History repeats itself in broad patterns, rather like a spiral. The general pattern is similar but each turn of the spiral brings with it multiple levels of complexity that are new. Corporations and an imperial presidency have replaced the aristocracy and kings of old, but the same enlightenment pattern dynamics are at work. In the Copernican Revolution, or the mechanistic shift, the change in worldviews culminated in the Enlightenment. Recognizing the four phases of the social transformation pattern helps us understand exactly how the mechanistic worldview was able to overthrow the power structure and the institutions of the church and the monarchy. It was the early enlightenment that started this great transition, a time when shocking new ideas burst into the mainstream. Today, the same historical pattern can decode the basic trends of change affecting the world.

The 1960s have often been ridiculed by the old worldview, yet it must be remembered this ridicule is part of the struggle, part of the war between the worldviews. In truth, the 1960s are the beginning of not only the modern attitude toward civil rights for all but also toward women's liberation, gay liberation, Native American rights, disability rights, animal rights, and even prisoner rights. In many ways, the 1960s and 1970s shocked society as much as the early enlightenment ideas did in the first decades of the 1700s. Environmental awareness and a movement for world peace also went mainstream in the 1960s, an enormous transformation that will likely dominate the twenty-first century. The 1960s, in short, were very different from the decade before—they were a crucible for change.

Many, many parallels are obvious between the two early enlightenment phases, yet the 1960s had layers of complexity that did not exist in the early 1700s. Nevertheless, the basic social transformation pattern was there. Before the 1960s, the mainstream had been able to mostly insulate itself from the new ideas of the organic worldview. That was no longer possible. With new means of communication, such as paperback books, television news, documentaries, and a rapidly growing music industry, progressives were suddenly able to put their views and agendas before the public. As Teilhard had predicted, the compression of information was indeed transformational.

The impact of the new thinking on the social and political realms was tremendous. Yet despite all the turmoil, shocking new concepts, and movements of all kinds, change was limited wherever possible by the power of the old paradigm and the "establishment." Many of the up-and-coming generation now thought with new models, however, and many dedicated themselves to social transformation as their life's work. The Organic Shift became a mass movement in the 1960s, affecting the social and political realms on a global level, not just in Europe and the Americas.

All the previous history made by bohemians, environmental groups, civil rights groups, pacifists, feminists, the new thought movement, radical art and music, organic scientists and ecologists, Eastern spirituality, transcendentalism, voluntary simplicity, populism, and various forms of socialism and anarchism now surfaced, attracting the young of the late 1950s and the 1960s. What had been underground and ignored by the corporate-owned media for decades was suddenly being televised and printed in cheap paperback books. Struck by a profound "affluenza," youth all over the industrialized world hungered for more than a regular job, a house in the suburbs, and the latest model car. Raised from birth to "duck and cover" in case of nuclear attack, many wondered if they would reach adulthood before being vaporized in a Soviet attack. "Is the suburban commute and possible nuclear annihilation all there is to life?" they asked themselves. Looking for a new outlook, they turned in droves to one or more schools of thought based on the organic worldview.

Once begun, the process of social transformation by the new paradigm could not be stopped. The old establishment had lost its relevance. Despite its success in initially holding off the organic worldview, by the 1960s the eventual demise of the mechanistic worldview was visible on the distant horizon.

Although it may sound strange, the 1960s actually began in the 1950s—with a series of dramatic and influential seed events. Change was slow relative to the rapid pace of the actual 1960s, yet from 1950 on, the new thinking piece by piece entered the mainstream. The calm "straight" world of the 1950s was punctured five times by the new worldview—the civil rights and antinuclear movements, the emergence of rock 'n' roll, James Dean, and the beats and the "Beatniks."

THE CIVIL RIGHTS AND ANTINUCLEAR MOVEMENTS

May 17, 1954, was the day the Supreme Court decided against segregated schools in *Brown v. Board of Education,* the most significant Court decision of the century for civil rights. Whereas the cry for "liberty, equality, and fraternity" was the rallying point of the early enlightenment in the eighteenth century, in the organic early phase, a similar push for greater equality and freedom started with this Court decision. *Brown v. Board of Education* set in motion the movement to end segregation in all its forms—not just for the schools.

Racial tensions flared as ultraconservative whites felt their control slipping away. In 1955, a fourteen-year-old African American boy, Emmett Till, was murdered in Mississippi after whistling at a white woman. The killing brought intense media coverage, resulting in a new determination by southern blacks to rid the United States of discrimination. On December 1, 1955, Rosa Parks was arrested for refusing to move to the back of that famous bus in Montgomery, Alabama. Four days later, the Montgomery Improvement Association elected a preacher as their new president, making him the official spokesman of the Montgomery bus boycott. His name was Martin Luther King Jr.

King, an organizing genius and the greatest speaker of modern times, began a nonstop effort for civil rights, winning one battle after another. On November 13, 1956, bus segregation was ruled unconstitutional by the U.S. Supreme Court. In 1957, King traveled 780,000 miles and made 208 speeches, including one before 15,000 people in Washington, D.C. Early in that year, he was also elected head of the new Southern Christian Leadership Council. In September, President Eisenhower intervened in Little Rock, Arkansas, using federal troops and the National Guard to enforce integration at Central High School, setting a precedent for the rest of the nation. The next year, King's book, *Strides toward Freedom,* was published, and Congress passed the first civil rights act since Reconstruction. King next met President Eisenhower and many others in Washington. During 1959, King traveled to India to study the nonviolent philosophy and methods of Gandhi. This study thoroughly infused his thinking with Gandhi's views, and he was now poised to fully bring Gandhi's approach to the streets of America in the tumultuous decade to follow. King said of Gandhi: "Gandhi was inevitable. If humanity is to progress, Gandhi is inescapable. He lived, thought, and acted, inspired by the vision of humanity evolving toward a world of peace and harmony. We may ignore Gandhi at our own risk."

The civil rights movement, empowered by the powerful vision and oratory of King, was thus changing America long before the 1960s officially began. At the same time that the modern civil rights movement was gathering momentum, the antinuclear testing and world peace movement emerged, along with early protests against nuclear power.

To make up for the horror of the nuclear bomb and to improve the image of nuclear research, the mechanists in the 1950s decided to crack the atom and make electric power, no matter what the cost. In 1957 Congress even passed the Price-Anderson Nuclear Industries Indemnity Act (commonly called the Price-Anderson Act) to limit corporate liability in case of a meltdown, which might otherwise lead to lawsuits in the hundreds of billions of dollars. This peacetime use of the atom was positioned as part of the package for entering the future—it was to be the age of the atom. The small antinuclear movement of the time thought it a terrible mistake. Making electricity with the atom, which was supposed to have been "too cheap to meter," was instead an extremely difficult task requiring miles of complex plumbing, pumps, and equipment. The United States narrowly escaped a full meltdown in 1955, when a breeder reactor near Idaho Falls, Idaho, began to melt because of operator error. History recorded the first nuclear power plant protest in Detroit at the experimental Fermi plant in 1956, following a partial meltdown. These early protestors were correct in their concerns, of course: the very next year one of the worst ecodisasters in history took place in the Ural Mountains when a radioactive explosion irradiated a whole region, which, unfortunately, was populated by 250,000 people.

Meanwhile, a small but vocal nuclear test ban movement fought back against the onslaught of bombs. In 1957, the Committee for Non-Violent Action (CNVA) was formed by members of the War Resisters League and the Catholic Workers Movement. Linus Pauling, Homer Jack, and Norman Cousins in that same year started National Committee for a Sane Nuclear Policy, or SANE, an organization that began a noisy and effective campaign for an atmospheric test moratorium and then a complete test ban.

In 1958, Albert Bigelow, a CNVA member, sailed his boat *Golden Rule* near a South Pacific island scheduled for H-bomb tests. After Bigelow's arrest, Earle Reynolds took his *Phoenix* into the same testing site. These acts helped draw attention to the cause, and the very next year, Eisenhower implemented a moratorium. Both mechanistic superpowers then resumed nuclear "shots" after the moratorium, and testing reached an all-time high just before the full ban commencement date in 1964. This higher radioactivity was all duly recorded in U.S. Public Health Service studies. The Cold War so threatened either side that the influence of the military basically spun out of control, and U.S. progressives have been trying to rein it in ever since.

President Dwight Eisenhower himself worried about the power of the military and the corporations allied together, issuing this warning in his farewell address: "In the councils of Government, we must guard against the acquisition of unwarranted influence, whether sought or unsought, by the Military Industrial Complex. The potential for the disastrous rise of misplaced power exists, and will persist. We must never let the weight of this combination endanger our liberties or democratic processes."[1]

Young people became involved when the Student Peace Union, a nationwide student organization founded in 1959, joining a swelling peace movement at a demonstration outside the Polaris nuclear submarine station in New London, Connecticut. The climax of the SANE campaign came at a rally at Madison Square Garden in 1960 that drew twenty thousand people. This new activism, inspired by the model of the civil rights movement, would, in a few years, turn from nuclear weapons to war itself when the United States, seduced by its own military might, got sucked into the Vietnam War.

During 1960, Al Haber and Tom Hayden formed what became the center of the New Left in America, the Student for a Democratic Society, or the SDS. The New Left first appeared in Britain following Khrushchev's secret speech exposing Stalin and the invasion of Hungary in 1956. Focusing on peace, global justice, and the nuclear disarmament issue, many in the New Left in Britain formed the International

Socialist Tendency. In the United States, the SDS was similarly focused on peace and disarmament but went beyond those issues to advocate a participatory democracy, using nonviolent civil disobedience to oppose "the establishment."[2] Those who rejected the authority of the prevailing power structure were the "antiestablishment." Universities, in particular, were seen as the organizing centers to create the new democracy.

ROCK 'N' ROLL AND JAMES DEAN

The seed events of the 1960s also included music, as it did in the beginning of the eighteenth century with the power of Handel and Johann Sebastian Bach's compositions and larger orchestras. Culture is always an integral part of a paradigm shift, and in the case of the late 1950s and the 1960s, music was again transforming the population while gathering the change community together.

It was not, however, classical music.

This time, the entire baby boom generation was swept up in the pressure to change and be different. A veritable youth revolution was the result, with teenagers grabbing all the attention.

In 1955, Elvis Presley exploded on the national scene with rock 'n' roll and black dance moves. Music would never be the same. One reporter tried to describe the loose, sexy new style of dancing that Presley brought to the stage: "He kicks, slinks, shimmies and gyrates. At times he resembles a man who is desperately wrestling his way out of a particularly confining batch of saltwater taffy."[3] It was a complete opposite from the sedate performances of all earlier singers in the 1950s, and it especially affected female teenagers. Countering Presley and his lewd dance moves, Ed Sullivan would only show him from the waist up on his TV show, while churches, radio stations, and parents smashed thousands of his records all over the country.

On April 15, 1956—after Presley had unlocked the repressed sexuality of twelve thousand screaming female fans during two shows in San Antonio—a "near riot" ensued when an army of teens tried to get into his dressing room. Two days later, after a show in Corpus Christi, local officials banned concerts that featured this horrifying new music called rock 'n' roll. Despite the backlash against early rock, the music survived, going on to help ignite the change of the 1960s.

Although they didn't have Presley, the early 1700s did have Bach and Handel, and masses of people back then were also transformed by music. While the 1960s had the electric guitar and the big beat of the bass and drums, the early enlightenment phase of the early 1700s thrilled to the new instruments of the piano and violin in powerful full orchestras. Large dances were held, and people in both early enlightenments were literally swept away by the new music, giving them a real love for change and the new. The music also brought people together in new communities of mind and spirit, nurturing a new culture and sensibility. Sexuality also became freer in both periods, another parallel between these two early enlightenments.

At the same time Presley was rocking the world, James Dean blazed across the silver screen, becoming a role model for young men around the globe. In *Rebel without a Cause,* directed by the legendary Nicholas Ray, the young twenty-four-year-old actor portrayed a neurotic yet perceptive teen, one who saw through the hypocrisy and lies of suburban 1950s life. He was hip, cool, smart, and extremely sexy. James Dean inspired the young, including Presley and the later Beatles.

Dean had an incredible influence. His premature death only magnified his popularity; fans flocked to the new teen god in mourning. The premiere of *Giant,* his last film, was one of the larger openings of the time. Although Dean was hip and criticized American society, his few roles did not reflect any really deep thinking about the difference between worldviews and the social, personal revolution that was needed to change things. That task of the early enlightenment was left to the early beat poets.

THE BEAT GENERATION

Following the lost generation of the 1920s and 1930s, the next great wave of counterculture came in the late 1940s and early 1950s. The bohemian movement by the 1940s had long burned out and was no longer providing the drive behind the new thinking. So a new counterculture began—one that soon would burst into the mainstream of Western life. It was a hip, new movement: the beats. Very small at first, the beats grew rapidly, becoming the powerful catalyst that initiated the social transformation period to come.

The beats deliberately defied convention in the 1950s—that most conformist of decades—by always using the organic analogy to expose the many anomalies and contradictions of the mechanistic dilemma. Allen Ginsburg's poetry such as "Howl," Gary Snyder's poems on nature and ecology, abstract art and jazz, Jack Kerouac's books—all reflected the new organic consciousness on a more sophisticated hip level. Eastern philosophy was read extensively by the beats, with Snyder becoming a world-class expert on Zen.

Before 1955, it was all just starting—they were all still underground and meeting each other at the new City Lights bookstore in San Francisco. Yet these unknown writers and poets were the ones that would help ignite a new enlightenment by thrusting the organic worldview into the mainstream—transforming society and creating the struggle of the three worldviews that we live in today.

By the mid-1950s, the toxic pollution of agriculture and industry, the corrupt AMA still defending tobacco use, the horror of atmospheric nuclear testing and environmental destruction, the power of the military-industrial complex and crony capitalism, and the many other parts of the mechanist worldview—all began their historic reach for absolute power. At the same time, the organic worldview was about to enter the mainstream of public awareness—an irresistible force that would confront the seemingly immovable object of the old thinking and power structures, for by the mid-1950s, the long scientific and cultural revolution phase of the Organic Shift was now finally over. The full-blown social transformation pattern, a new enlightenment, was about to generate multiple mass movements, with wave after wave of significant change.

THE BEATS CREATE A SCENE

In 1955, on October 13, five young poets—Allen Ginsberg, Philip Lamantia, Michael McClure, Gary Snyder, and Philip Whalen—staged a reading at the Six Gallery, an artist's co-op in San Francisco. Poetry in America, and America itself, would change that night as the beats challenged the old worldview with well-written radical organic thought. Michael McClure talked about the Six Gallery:

> That night changed all of us. We were each filled with dissent or radicalism in our own ways. Allen was one type. Gary at the time was a Wobblie.
>
> I was an anarchist, to the extent that I understood anarchism. Philip Lamantia was a Surrealist Catholic. Philip Whalen was clearly a Zen master-to-be. People were standing up and saying things that the audience actually knew, but it was being said out loud. The audience was cheering for you, agreeing with you, leading you on. I had the sense that nothing like it had happened before, at least not in this country. You had the feeling that things weren't going to be the same.[4]

Gary Snyder said of the Six Gallery reading: "The reading at the Six Gallery jump-started a whole new round of poetry readings around the world. With the quality of the poetry, Allen's 'Howl' not being the least, it was a galvanizing event. People were listening, laughing, people were moved by it. There was a community feeling of discovery in the room, a feeling of, 'Gee, I didn't know there were that many of us.'"[5]

Ginsburg gave his first public reading of "Howl," which was completely different from traditional poetry. "Howl" was a near-primal scream, filled with sex and four-letter words, releasing a generation's pent-up frustration at the world. It broke down old mind-sets, expressing the deep alienation of the organic beats—an alienation that had never really been fully revealed.

Michael McClure again: "In all our memories no one had been

so outspoken in poetry before. We had gone beyond a point of no return—and we were ready for it, for a point of no return. None of us wanted to go back to the gray, chill, militaristic silence, to the intellectual void—to the land without poetry—to the spiritual drabness. We wanted to make it new and we wanted to invent it and the process of it. We wanted voice and we wanted vision."[6]

Jack Keruaoc was at the Six Gallery that night, and listening to Gary Snyder's poems about nature and the environment, he wrote his next book, *The Dharma Bums,* about going into the mountains with Snyder. In San Francisco, the poets congregated at the City Lights bookstore, where owner Lawrence Ferlinghetti might read a poem and publish it. Then, in 1956, City Lights published *Howl and Other Poems* by Allen Ginsburg—and was soon sued under the obscenity laws in San Francisco municipal court. Successfully defended by the ACLU, the Howl case set a legal precedent for artistic expression and free speech under the First Amendment in that the judge ruled that even the slightest social redeeming value of a work meant it was art and therefore protected by the Bill of Rights. This cleared the way for novels that were previously banned, such as *Tropic of Cancer.*

Meanwhile in New York, the beat scene arrived, expanding rapidly in Greenwich Village, attracting curiosity seekers and new poets and artists as the bohemian scene had done before it. A new counterculture landed East and West, and it thrived. Jazz and be-bop were the beats' favorite music, and there were soon more bars, clubs, coffeehouses, and strip joints in Greenwich Village than any other place in the world. The beats had their own places like the White Horse Tavern and the Gaslight. Other clubs had comedians and different types of music, like folk, which was soon to grow in popularity.

EASTERN PHILOSOPHY IN THE 1950S

In the late 1950s, there was a turn to Eastern philosophy and spirituality among a very small part of the population. *The Way of Zen* by Alan Watts, published in 1957, brought new awareness of the Japanese

Buddhist sect and its practice of meditation to intellectual America. Two years later, Maharishi Mahesh Yogi arrived in San Francisco, the leader of the Transcendental Meditation movement, which would become so influential during the summer of love. In that same year of 1959, Shunryu Suzuki moved to San Francisco to become a priest for the local Japanese Zen Buddhist community. He would go on to create the San Francisco Zen Center, a counterculture hub in the late 1960s and the 1970s.

These Eastern teachings focused on enlightenment in the Buddhist sense, a mystical breakthrough allowing one to understand everything in a new light. What that new light is, however, cannot be described in words—a paradox expressed by Zen *koans* that ask unanswerable questions like "What is the sound of one hand clapping?" Extensive meditation is prescribed for those who wish to experience enlightenment, years and years of silence being required. In contrast to the constant noise and chatter of civilization, the quest for this state of silence and enlightenment would become a major preoccupation for many in the 1960s.

Another change came with the learning of Eastern thought. Zen is decidedly antianthropocentric in its view of reality. The usual Western scientific and religious view that the world is here to be exploited by the human race is thoroughly rejected. Civilization is seen more as a disease than as an accomplishment, an attitude that can be traced all the way back to Lao-tzu in the sixth century BCE. These organic paradigms of the past were clearly reflected in the writings of Aldo Leopold and other environmentalists, who view nature as something to be respected, not destroyed for the sake of humanity. The organic worldview is centered on the primacy of the biosphere and the universe, rather than on human beings. The common sense of negative logic, introduced to the West so long ago by Immanuel Kant, finally, in the 1960s, began to answer the sophisticated questions of a mass movement of counterculture beats and hippies.

THE GREAT DIFFERENCE
BETWEEN THE PARADIGMS

Science and its values were thus radically transformed by the Organic Shift. As in the Enlightenment, when the science of the Copernican Revolution led to a vast change in the value system of the day, the Organic Shift brought in an entirely new set of values, one that was in many ways the opposite of the old worldview. All the seed events of the late 1950s grew quickly in the 1960s, many leading to massive protests and huge demonstrations. There was the well-known civil rights march of 1963, where King gave his "I Have a Dream" speech, and the free speech movement on campuses in 1964. Around the same time, opposition to the Vietnam conflict began in the form of teach-ins. These direct confrontations were caused by the new worldview's broader definition of freedom and democracy, much as mechanistic thinkers who declared that the world should be based on liberty and equality shocked society in the early 1700s.

But the old paradigm was not about to just give up its power just because a better idea had come along. As the organic worldview and its broader definition of peace and freedom entered the mainstream in the late 1950s and early 1960s, the true struggle between the worldviews was finally joined. This time, however, nearly the entire baby boom generation—the largest generation the world had ever seen—was caught up in the paradigm shift. Although the old worldview controlled all the institutions of society, some 80 million young people in the United States, and many more around the world, powered the incredible drive for change that was the 1960s. It was a spectacular clash of culture and counterculture, and it caught the attention of one and all.

THE EARLY 1960s

The 1960s began in the United States with the election of John F. Kennedy, who galvanized the United States and led it through the Cuban missile crisis, the closest the human race has ever come to all-out nuclear war. Meanwhile, the anti–nuclear war movement continued to grow as the missile crisis revealed to everyone the real possibility of total annihilation. After the test moratorium fell apart and the two super-powers set off more bombs than ever, SANE started an advertising campaign featuring a photo of the famous Dr. Spock, the baby author. The ads said "Dr. Spock Is Worried." The Spock ad appeared in the *New York Times* and was reprinted in seven hundred papers worldwide. SANE finally succeeded in 1963 when JFK negotiated the atmospheric test ban with Khrushchev.

Around the same time, in 1963, the civil rights movement grew white hot. NAACP field secretary Medgar Evers was murdered in Mississippi, Martin Luther King held the March on Washington, and four young girls were killed when a bomb exploded in a Birmingham church. Two more black youths were killed in the rioting that followed the Birmingham bombing. The old worldview was now forced to make a decision: to continue to fight civil rights for all or take a step toward a new future. Martin Luther King said it best in his "I Have a Dream" speech before 250,000 people:

> Let us not wallow in the valley of despair.
>
> I say to you today, my friends, that in spite of the difficulties and frustrations of the moment, I still have a dream. It is a dream deeply rooted in the American dream.
>
> I have a dream that one day this nation will rise up and live out the true meaning of its creed: "We hold these truths to be self-evident that all men are created equal."
>
> I have a dream that one day on the red hills of Georgia the sons of former slaves and the sons of former slave owners will be able to sit down together at a table of brotherhood.
>
> I have a dream that one day even the state of Mississippi, a des-ert state, sweltering with the heat of oppression, will be transformed into an oasis of freedom and justice. . . .
>
> And if America is to be a great nation, this must become true.[7]

Finally, in 1964, Congress passed the Civil Rights Act, which prohibited discrimination in many ways. Though racism and barriers still exist today, King had succeeded more than any black leader in history, except for perhaps Frederick Douglass. King did not just free African Americans, however, he freed America from an apartheid system that was poison to democracy.

The civil rights movement meanwhile inspired all the other minorities. Latinos were among the first to begin agitating for their rights. Cesar Chávez became the leader of the United Farm Workers Organizing Committee in California and started to unionize Latino pickers in 1962. Initially, most workers were too intimidated by their American bosses to take a proactive stance but by 1965 the UFW had a national grape boycott going, and the next year, union members made a 340-mile march from Delano, California, to Sacramento. *La Causa* (The Cause) united many in Latino communities around the United States for years to come.

GARY SNYDER, RACHEL CARSON, AND ECOLOGY

JFK's election and the 250,000 who came for the 1963 civil rights march on Washington confirmed the fact that great change was in the air. The new organic definition of freedom and equality for all—soon even for the birds and the trees—was a direct challenge opposing the old mechanist-corporatist-nationalist mind-set.

In San Francisco, City Lights published *The Journal for the Protection of All Beings* in 1961, a radical ecology magazine. The beat poet Gary Snyder wrote for it, pleading that the loss of the environment to greed and limited thinking meant that even Buddhists, who usually abhor politics and social activism, should now get involved. Snyder recommended the joining of Eastern enlightenment with Western movements: "The mercy of the West has been social revolution; the mercy of the East has been individual insight into the basic self/void. We need both." He urged Buddhists and everyone to realize that all of nature is sacred and must be protected. Yet people must first develop their minds through meditation and loss of ego so they can have moral clarity as they work to change the world. Here was the connection between spirituality and ecology, which would continue to grow in the coming decades. "Wisdom is intuitive knowledge of the mind of love and clarity that lies beneath one's ego-driven anxieties and aggressions. Meditation is going into the mind to see this for yourself—over and over again, until it becomes the mind you live in. Morality is bringing it back out in the way you live, through personal example and responsible action, ultimately toward the true community (*sangha*) of 'all beings.'"[8]

Combining with the older Sierra Club and the conservation movement, the new environmentalists explained that deeper thinking was required, that the whole anthropocentric worldview was to blame. Rachel Carson made a huge impact when she was interviewed for her 1962 book *Silent Spring*. Carson's ecological plea catalyzed a new awareness, one that eventually became the environmental protection movement.

As usual, the old worldview and corporate science had little defense against the ecological data and reasoning, so they simply attempted to win holding actions that avoided the overthrow of profitable businesses and policies. For example, in the case of DDT, although the chemical was finally banned in the United States in 1972, its use was allowed overseas—giving the chemical companies the freedom to exploit those markets (and continue degrading the biosphere).

The rise of environmental awareness toward the end of the 1960s was dramatic. The *Torrey Canyon* wreck off England in 1967 was especially lethal, killing twenty-five thousand birds and keeping rocks oily for years. This made the Europeans ecosensitive. Next it was America's turn.

On January 29, 1969, the ecological nightmare known as the Santa Barbara oil-well blowout was televised to a horrified nation. An offshore drilling accident on a Union Oil rig led to a rupture on the seabed, and an underground pool of oil rushed to the surface. Forty miles of beach were despoiled, and the toll on wildlife was enormous. The color footage of the attempt to rescue the oil-covered birds and seals was

Fig. 8.1. Oil slicked animals like this one were commonplace after the Torrey Canyon *wreck off the coast of England in 1967.*

so heart-breaking that a huge shift occurred, and America, too, became ecosensitive.

Every incoming tide brought in the corpses of thousands of dead seals and dolphins, killed when the oil clogged their small blowholes. Nearly four thousand birds were estimated to have died. The reputations of the oil companies were never the same. Within a few days after the blowout began, an antioil group formed, called Get Oil Out (GOO), demanding a boycott of oil companies that drilled offshore. Water quality laws in California were passed, and the following spring, Earth Day was born, with many pointing to the Santa Barbara blowout as the major reason. Fred Hartley, the president of Union Oil, being solidly of the mechanistic anthropocentric worldview, was meanwhile mystified at the uproar. After all, he figured, no human beings were killed. "I don't like to call it a disaster," he said. "I am amazed at the publicity for the loss of a few birds."[9]

Still yet another environmental disaster took place on June 22, 1969, when a *five-story-high fire* ignited on the polluted Cuyahoga River in Cleveland. In August, *Time* magazine ran an article on it, and the strange nature of a river catching on fire became a rallying cry for environmentalists. Within three years, the Clean Water Act passed, a decades-long cleanup of America's waterways began, and the Environmental Protection Agency was formed. It was the beginning of the present-day environmental movement, another shocking development to the old worldview in this early enlightenment phase.

On the issue of the environment, the difference between the new and old mind-sets grew more distinct as the 1960s developed—as it did in many other areas. Almost two centuries earlier, Goethe had pointed the way to a new way to do science, one that would have avoided the pollution and needless destruction of the environment. The world did not take to Goethe's scientific method then, but today we still can— and must. We need only ask nature to reveal its open secrets and wait intuitively for the answer, and the ecological solutions of the future will become obvious. Mechanistic science, in contrast, does not notice or care when chemical, radioactive, or fossil-fuel pollution occurs—it is in the pocket of the heartless and endlessly greedy crony capitalists.

WOMEN'S LIBERATION

Still another new-paradigm concept upset the status quo in the early 1960s: women's liberation. In 1960, the pill was introduced, allowing women to take a simple drug and not have to worry about pregnancy. Within a few years, the pill was in wide use, adding to the loosening of sexuality already underway, thanks to movies and rock 'n' roll. The new woman of the 1890s had indeed come a long way, baby—and the movement was about to take it one step further.

In the 1960s, the cause became present-day feminism—as expressed in Betty Friedan's *Feminine Mystique* in 1963. Friedan was the first to provide a moral and intellectual foundation for feminism, and she undertook seminal consciousness raising on the issue. The civil rights struggle taught that if the customs of society do not fit with the desire for equality on the part of the individual then the society must change, not the individual. Friedan applied this to feminism. Women's liberation and active struggle thus entered the mainstream, igniting change all over the planet.

We can see this as a parallel to the rise of the first women philosophical writers in the early 1700s, such as Lady Masham (1658–1708) of England, one of the first English women to publish philosophical writings. These early female advocates—who pushed for the education of women—were just as shocking to Europe as the feminists of the 1960s had been. They broke the continuity of the status quo, forcing a change to a new set of rules, namely the broader definition of freedom under the new paradigm.

THE EARLY ANTI–VIETNAM WAR MOVEMENT

After JFK's assassination shocked the world, the Vietnam War began to heat up. The folksinger Joan Baez and the antiwar priests Daniel and Phillip Berrigan held a small protest in Washington in 1964. Antiwar teach-ins were held the next year in Ann Arbor, at the University of Michigan, then in colleges around the United States. The media presented the war as a battle of democracy versus communism, but the teach-ins conveyed the message that Vietnam was really more a continuation of European-style imperialism. To the students, the United States had simply replaced the French in South Vietnam.

In 1964, the SDS joined the student coalition known as the Free Speech Movement and became leader of the campaign to allow political activity on the Berkeley campus. Soon the SDS was involved in the antiwar movement, and within a short time, the organization was leading the coalition opposing the war. In April 1965, twenty-five thousand attended the first of mass rallies against the war in Washington, D.C., landing themselves on the front page of the *New York Times*. The *Times* went on to note that the protestors really had no idea how to end the war. In November, the SDS cosponsored another demonstration that drew thirty thousand.

On October 21, 1967, one of the most prominent anti-war demonstrations took place, as some 100,000 protesters gathered at the

Lincoln Memorial; around 30,000 of them continued in a march on the Pentagon later that night. After a brutal confrontation with the soldiers and U.S. Marshals protecting the building, hundreds of demonstrators were arrested. One of them was the author Norman Mailer, who chronicled the events in his book *The Armies of the Night,* published the following year to widespread acclaim. Also in 1967, the anti-war movement got a big boost when the civil rights leader Martin Luther King Jr. went public with his opposition to the war on moral grounds, condemning the war's diversion of federal funds from domestic programs as well as the disproportionate number of African-American casualties in relation to the total number of soldiers killed in the war.[10]

The antiwar movement had begun in earnest.

DYLAN AND THE BEATLES

All this apocalyptic maneuvering created a new trend: protest music. Joan Baez, Judy Collins, Tom Paxton, Eric Andersen, Phil Ochs, and Tom Rush were some of the new songwriters, and like Woody Guthrie and Bob Dylan, they expressed themselves through a broadened awareness—and with a new progressive politics. Weekly hootenannies at several coffeehouses in Greenwich Village allowed many new singers to join in, and many different types of musicians, from jazz to blues to folk, jammed with one another at all the venues. The beats—being sophisticated jazz lovers—were put off by the folk music invasion. Greenwich Village finally became too hectic for many of the beat poets, and in the late 1950s and early 1960s they packed up and moved to the quieter realm of San Francisco.

Protest demonstrations and folksingers went hand in hand in the middle 1960s. Bob Dylan, in particular, prompted the baby boom generation to start thinking about a different kind of world, where the generals and warmongers were no longer in control. "Masters of War," "The Times They Are A-Changin,'" "The Eve of Destruction," and many other songs transformed millions of young people, and they began thinking more along the lines of the organic worldview than the old nationalist, mechanistic worldview. Then, in 1963 and 1964, while the public was still depressed about the death of JFK, along came the Beatles.

Beatles' songs ignited the new generation, like nothing before or since. When booked on the *The Ed Sullivan Show,* 50,000 requests for 728 tickets were received. Nielsen ratings went through the roof, with 45 percent of the United States—73 million Americans—tuning in. This was the largest nonsports broadcast in history. The legendary Carnegie Hall concert took place three days later. After the Beatles made appearances on three consecutive Ed Sullivan shows, the youth of the United States, and soon the world, were never the same again.

In *The Beatles Come to America,* author Martin Goldsmith agreed that the Beatles' influence on America is key to understanding that time, explaining: "In many ways, the Beatles' arrival set off a wave of changes."[11] By April 4, 1964, the top five singles on *Billboard* were all Beatles tunes. From music, hairstyles, and then drugs to their pleas for world peace and an end to the Vietnam War, the influence of the Beatles was unprecedented. John, Paul, George, and Ringo became change agents on a global scale.

The unique excitement they generated only began to dissipate when the group officially broke up in 1970. However, by then a lot of their charismatic energy and pacifist message had rubbed off on every other musical group to follow, as the social revolution had been fully united with rock music. The band became a primary driving force in the 1960s, riding atop an enormous wave of counterculture music calling for massive social change, calling for a change in the very way people thought.

It was a mass cultural revolution.

THE NEW PARADIGM GOES WITHIN: ESALEN AND MASLOW

Organic science continued to develop in the early 1960s—going deep within. While Russia and the United States explored outer space,

California's Esalen Institute, founded in 1962 as a center devoted to a new vision of human development and philosophy, became a main center for dissemination of new theories on consciousness. Charlotte Selver (1901–2003), who developed sensory awareness, taught at Esalen, with husband and colleague Charles Brooks, starting in 1963. The following year, Fritz Perl held seminars on Gestalt. Gestalt philosophy and methods brought holistic thinking into the field of psychotherapy.

Psychologist Abraham Maslow (1908–1970) soon after offered a new, humanized psychology, a revolution based on self-actualization, or self-development. During a 1965 brainstorming session with Michael Murphy, one of Esalen's cofounders, George Leonard, coined the term *human potential movement*. So the path blazed by transcendentalism and the new thought movement in the nineteenth century now bore some real fruit in the 1960s. During the 1970s, the human potential movement became its own industry, when human potential trainers became regular consultants to corporations. They were hired as part of the corporate team or put onboard in human resources departments or they became consultants to the corporate or advisory boards; in other words, they were now "on the inside."

THE ESCALATION IN VIETNAM AND THE DRAFT IGNITE THE COUNTERCULTURE

Vietnam had meanwhile simmered away, the war growing in size and ferocity. Since all young men were subject to the compulsory draft, the impact of the war was immediate and distributed among every community and family—except the rich and well-connected, who could almost always find a way out. In 1961, there were only 3,200 U.S. troops deployed. By 1963, that number had grown to 16,300—still considered a minor deployment. Then, in 1964, the falsified Tonkin Gulf incident gave Lyndon Johnson the "right" to basically declare war on the small country. By 1965, 184,300 troops were there, and hundreds of thousands more would soon be on the way.

This escalation, of course, ignited the student-folksinger-peace movement. In the same year that Phil Ochs wrote the song "I Ain't Marchin Anymore," the Viet Cong attacked Pleiku airbase and Johnson reprised with a heavy air attack on North Vietnam. The war soon became a sustained bombing campaign, fully supported by the media as a fight for democracy against communism. Following a major battle at Ia Drang Valley, General Westmoreland requested a huge escalation in troops; LBJ obliged.

Toward the end of the year, Norman Morrison immolated himself in front of the Pentagon in a suicide protest. Another antiwar protestor set himself on fire at the UN building in New York. The immolations came as a great shock to the American mainstream: this was a clash of two completely different value systems. The old nationalist mechanists were offended by the very idea that their thinking was imperialist rather than democratic and that peace was deemed to be a priority no matter what. This great difference in values between the organic and mechanistic worldviews became the most heated topic in the famed generation gap of the 1960s.

Vietnam and civil rights were central in this struggle between the worldviews. Even the organic progressives and the mechanistic liberals were split by the war. While the liberals still supported the war for the most part, the organic/progressive thinkers were dead set against the fighting. When the mechanists continued to believe that U.S. superiority in weapons would win the peace, the organics argued that this was false, that withdrawal was the only viable road to peace. Vietnam could not be "won" in the traditional sense of the Allies conquering Japan or Germany.

In the final analysis, the Vietnam War was a war of independence against the colonial powers, just as the teach-ins claimed. The progressives claimed that true peace and freedom—world peace—must be grown and nourished over decades through international development, education, and aid. It could not be forced on countries by America. It was all very similar to the views of Immanuel Kant from two hundred years earlier. The world had not changed much in this respect.

TURMOIL, PROGRESS, AND MURDER

When a bomb blew up four children in a church in Birmingham in 1963 and three civil rights workers were murdered in 1964, outrage grew over the situation in the South. These horrific events helped pass the Twenty-fourth Amendment forbidding the poll tax, which had kept black voter registration in single digits, and the Civil Rights Act, which forbade discriminatory laws in education, housing, hotels, and employment. There was thus much progress amid the turmoil.

Then, between July 18 and August 30 in that year, in the heat of the summer, large riots began in Harlem and spread to six other American cities. The Watts riot in Los Angeles started on August 11, leaving thirty-four dead, with 3,500 arrested. An estimated $225 million in property was lost to the flames. Malcolm X was assassinated in February of the next year, raising the emotional temperature in the ghetto.

At the same time, the Voting Rights Act passed in 1965. With a summer registration drive, black registration in Mississippi soared from 7 percent to 67 percent, forcing many politicians to give up their overt support of white supremacy groups. President Johnson, who signed the new laws, privately predicted that passage of the civil rights legislation would turn the South over to the Republicans for "a generation."

Riots erupted again in Detroit in the summer of 1967, the worst since the draft riots in New York during the Civil War. Forty-three people were killed and federal troops had to be called in to reinforce the guard. A total of forty riots broke out in other cities. Watts, in Los Angeles, and several other inner cities, including one in Detroit, were not rebuilt, and the suffering in the ghetto grew worse. Even today, many other inner cities have not been rebuilt, and as a result, they resemble war zones. Once the riots occurred, a crime wave typically followed in a massive, decades-long social breakdown, a "slow-motion riot," which has basically continued to this day.

THE SUMMER OF LOVE

While the black inner cities burned, young white people were also overthrowing the establishment, though in a less violent way. In the 1950s and early 1960s, the beats were the vanguard of the new, but they were a tiny minority, a network of cliques primarily in New York, San Francisco, Austin, and Seattle. In 1966, however, the Organic Shift became a mass movement when the hippies came of age. A majority of the largest generation in history, the baby boomers, went organic. The generation gap, a huge chasm between the two paradigms, now grew into a grand canyon of differences in values. While the old worldview defined spirituality through the traditional Western religions, the new organic worldview turned to the East for its spirituality or to a transcendentalist mysticism.

The first known use of the word *hippie,* meaning "young hipster," was in a San Francisco newspaper story by Michael Fellon on September 6, 1965. All through 1965 and 1966, the Haight-Ashbury district developed a unique subculture, a scene based on a series of drug-induced dance parties. Augustus Stanley had begun to mass-produce LSD the year before, and soon the drug was being used recreationally rather than for mystical exploration, as undertaken by the likes of Timothy Leary.

The first acid test—one of a series of parties held by Ken Kesey and his Merry Pranksters—was in November 1965 in San Francisco. This party, which centered around the use of LSD, was followed by many more in 1966. The parties grew progressively larger until they were being attended by thousands—dancing to psychedelic rock bands like the Warlocks, an early incarnation of the Grateful Dead. Far more than a dance, the acid tests were social, cultural, and political statements challenging the status quo. For the music-loving youth, it was all a great excuse for yet another party.

In 1966, disturbed by plummeting church attendance levels, the cover of *Time* magazine asked "Is God Dead?" In October 1966, before a new law made LSD illegal, a huge acid test was held in San Francisco, at which the entire crowd danced to the so-called psychedelic San

Francisco Sound—music performed by San Francisco rock groups—and drank punch laced with LSD. Some there reportedly believed they were God—never mind thinking you had merely *met* God.

Then it happened. On March 26, 1967, Easter Sunday, human be-ins were held in Los Angeles and New York. Something new was in the air—and it wasn't just the marijuana smoke. In San Francisco, a self-appointed Council on the Summer of Love held a press conference on April 5, declaring 1967 as the summer of love—inviting everyone to come to Haight-Ashbury for a summer-long "love-in." A May 15 announcement reinforced the new idea.

On June 1, the Beatles released their seminal, multitracked album *Sgt. Pepper's Lonely Hearts Club Band*—and the world changed once more. Appearing on the famous cover dressed in colorful and outlandish marching band costumes, the Beatles threw down the gauntlet to the old worldview and to the new one. We have discovered a new realm of inner space, they seemed to be saying, and it's time to follow us into it. Almost the entire generation of boomers happily did just that, and the young counterculture began to discover so much more about the world. The excitement about the summer of love conflated with the multitracked musical revelation of *Sgt. Pepper,* and all eyes turned to the Haight. *Sgt. Pepper* had an incredible impact on the local music scene in San Francisco, where the Jefferson Airplane had already played at the world's first be-in, and where the Family Dog and the Fillmore regularly held psychedelic concerts.

Simultaneously, the song "San Francisco" became a major hit in mid-June, which made San Francisco sound so intriguing that hundreds of thousands of young people decided to descend on the Haight-Ashbury district of the city—and the summer of love was on. They were no longer the more elderly beatniks; most of them were in their teens.

This was a unique moment in history, and Haight Street was now filled to the brim with the new counterculture. A lush alternative newspaper, *The Oracle,* boggled the mind as well as the eye, while the Diggers, a hippie group, started the Free Store—in which everything was free. Donations kept the Free Store open.

Although the word *hippie* had already been coined, it did not enter the mainstream until *Time* magazine ran a July 7, 1967, cover story entitled "The Hippies: The Philosophy of a Subculture." San Francisco was then truly submerged in a sea of hippies, along with the tourists who came to see them. At the same time, the Maharishi Mahesh Yogi greatly expanded his influence along with other Eastern sects present in San Francisco. George Harrison of the Beatles then paid a visit to the Haight. The Maharishi followed on August 7, in a visit that vastly increased the original fervor.

All of this finally overwhelmed the counterculture. By the end of the summer—after drug dealers begin murdering each other on the street— the residents of Haight-Ashbury held "The Death of Hip" parade— hoping that the tourists and the kids would just go away. The legendary *Oracle* newspaper stopped publishing. The famous Diggers Free Store and its free newspaper lasted for only another year, and then closed up shop, as many of the original hippies left for the woods, dispersing throughout the country in the beginning of the back-to-the land movement.

The significance of the summer of love in the history of the Organic Shift is crucial, for it was the first time that the organic worldview— long an underground phenomenon—had affected such a broad swath of the population. This entry into the consciousness of the mainstream was so powerful that almost the entire boomer generation was affected profoundly in that year of 1967; they sensed something new, something they wanted desperately to join. It was much more than the sex, drugs, and rock 'n' roll; it was a statement for peace, brotherly love, and the new way of thinking. Like Gandhi's unities, it was the coming together of a new global community, a new consciousness.

Although it may have been a young, immature generation at the time, it was an entirely new mind-set, and its new thinking has endured. Once a historical pattern is set in motion, there is no turning back, although it may take several decades to play itself out. Nearly an entire generation was affected by this early enlightenment and—even if they put activism aside later on—most were deeply transformed and would go on to view life through this new prism for the rest of their lives.

THE SOCIAL AND
SCIENTIFIC REVOLUTIONS UNITE

In 1966, Stewart Brand began a campaign asking NASA to take a picture of the whole Earth. This finally happened in November 1967. In showing the fragile and beautiful nature of our planet, this photo had an immediate and significant impact on the young organic generation. The whole Earth even became the unofficial flag of the counterculture, being flown on some flagpoles instead of the U.S. flag. Brand himself used the photo for the cover of his 1968 *Whole Earth Catalog,* which reviewed and sold tools, books, and all the odd things being created by the alternative society.

The whole Earth was an extremely powerful organizing image. By combining a desire to protect the environment of this fragile spaceship Earth with a near-religious set of beliefs about the virtual sanctity of the environment, an informal spirituality based on protecting nature and the biosphere took shape. Many in the young counterculture developed sophisticated beliefs about the world and reality itself, discovering the books and thoughts of the scientific and philosophical paradigm shift swirling all around them. These social/spiritual revolutionaries and the scientific revolutionaries of the Organic Shift would soon join forces in the 1970s and begin a new activism to protect the planet from the onslaught of mechanistic technology and the fossil-fueled industry.

ASSASSINATIONS AND RIOTS

The summer of love in 1967 was followed by the great turmoil of 1968, starting with the TET offensive, in which U.S. and South Vietnamese forces suffered nearly 2,700 killed or wounded. Although the combined losses of the North Vietnamese and Viet Cong were around 8,000, and it was a basic loss for the North, the ferocity and widespread nature of the TET offensive shocked America. It was now understood by most that Vietnam could never be held without a grave loss of life. By 1968, the United States had over 500,000 troops in Vietnam, and the casualty figures continued to mount throughout the year. LBJ declined to run in response to Eugene McCarthy's antiwar challenge in the primaries.

When Martin Luther King was assassinated on April 4, 1968, at least 125 riots erupted in the week that followed. It was madness. SDS meanwhile took over Columbia University to protest war-related research and the razing of a public park to build a school gym. The university called the police, who injured 200 students in retaking the school, arresting 712. Similar protests and occupations spread to some forty campuses around the country. Even high school students in New York and other cities walked out.

Then Robert Kennedy was assassinated in June. The two leading lights of the progressives were now suddenly gone. The Democratic convention in Chicago turned into a massive protest and then a police riot. A grisly convention scene, with tear gas inside the hall, left the Democratic Party fractured between the young progressives and the older Humphrey liberals. Nixon, saying he had a secret plan to end the war, beat Humphrey in a close election for president. It was not a good year—unless you were Richard Nixon.

If Nixon ever did have a secret plan, he took it with him to the grave, as he continued the war for years to come, escalating it even further with the massive B-52 carpet bombings of Hanoi and Haiphong. The stage was thus set for a showdown between the young, activist organic generation and the conservative, mechanistic, and ruthless Nixon.

Radical countercultures and many antiwar protests appeared around the developed world. Paris especially was rocked by student-led unrest in 1968. Students in Czechoslovakia meanwhile ignited the spring revolution that year, but the Soviets soon responded with tanks and a crackdown. Amsterdam was a center of the new counterculture, as was Berlin. There was a large radical movement in England and Japan as well. The new thinking had become a global phenomenon, although it was made up of multiple movements with multiple causes and was not a unified whole. All the counterculture communities grew as world events continued to radicalize nearly an entire generation.

Conservatives pushed back in different areas of the country. In

Berkeley, California, the University of California purchased land on Haste Street, where many progressives lived. The rationale given for buying the land was that new athletic fields were needed. People's homes were then torn down amid noisy protest, but no athletic fields were built. This drove the counterculture community over the edge. On Friday, April 18, 1969, an announcement appeared in *The Berkeley Barb,* an underground newspaper, stating that a park would be built on the site; construction would begin that Sunday. Two days later, large numbers of volunteers appeared with shrubs, trees, and tools. People's Park, a community park, was born.

After being confronted in this way for a while, the university struck back with bulldozers and threw a fence up around the property. Thousands of unruly protestors counterattacked, and the Berkeley police were soon in over their heads. Governor Ronald Reagan called out the National Guard, and with helicopters, tear gas, and full battle gear, they suppressed the riot. Many demonstrators were injured, one blinded for life, and one killed.

Similarly, in 1970, after Nixon invaded Cambodia, student protests and occupations of schools flared all over the United States. At Kent State, the National Guard shot and killed four students. By the fall, the size of the demonstrations, happening simultaneously in many cities, overwhelmed support for the war, and soon many more joined the protestors' ranks.

GETTING RADICAL

"Say it loud! I'm black, and I'm proud!"

BLACK PANTHER CHANT

In October 1966, Bobby Seale and Huey P. Newton formed the Black Panther Party for Self-Defense. Inspired by the radical ideas of Malcolm X, who had been assassinated the year before, the new revolutionaries released a newspaper, *The Black Panther,* in April 1967. Black pride and radical community action—including violence, if necessary—were the foundational tenets of the group. A month later, Bobby Seale and thirty fully armed Panthers protested at the state capitol in Sacramento against a law prohibiting the carrying of loaded weapons in public. They were all immediately arrested. Further confrontation with the authorities was quick, as the party grew rapidly in cities around America, particularly following Martin Luther King's assassination. When two black athletes won gold medals at the 1968 Olympics, they raised their closed fists in the black power salute, shocking middle America. The riots of 1967 and the assassinations and riots of 1968 drove over four thousand young people into the party.

In 1968, to finance the purchase of shotguns, the Black Panthers began selling Mao's little red book of quotations on college campuses. Ever alert, FBI director J. Edgar Hoover started the infamous COINTELPRO (counterintelligence programs) in response. Its mission: Arrest and kill the revolutionaries in the new radical groups and use covert methods, including psychological warfare, to disrupt all of their organizations. On April 6, 1968, just two days after the assassination of Martin Luther King, Oakland police gunned down seventeen-year-old Bobby Hutton. Several months later, in September 1968, Hoover declared the Panthers "the greatest threat to internal security in the country."

The Panthers then really got Hoover worried when they started a successful free breakfast program in January 1969, at which the kids learned to chant "Say it loud! I'm black and I'm proud!" along with their cereal and milk. By the end of the year, the Panthers were serving a free breakfast to ten thousand children every day before school.

In Chicago, Panthers leader Fred Hampton, only twenty-one years old, started five free breakfast programs, helped to open a free medical center, organized a door-to-door drive to test for sickle cell anemia, and began blood drives for Cook County Hospital. The Chicago Panthers persuaded local gangs to give up crime and join the "class war," which resulted in a much larger party in a very short period of time.

On December 4, 1969, on the tip of an FBI informant, a tactical unit composed in part of the Chicago police deparatment and the FBI raided Fred Hampton's apartment. He was murdered in a hail of ninety bullets. This signaled the start of a war by COINTELPRO against the Panthers. Within a couple of years, Huey P. Newton, Bobby Seale, and party chief of staff David Hilliard were all arrested, along with many other leaders who were accused of a plot to bomb public buildings and senate police officers. To disrupt the group, the FBI forged a clever series of letters from one imprisoned leader to another. This created suspicions among them, and the Black Panther Party split up, becoming a shell of its former self. The FBI and the police had reduced them to tatters. The decimated group nevertheless continued into the 1970s, and Bobby Seale even renounced the gun—running for mayor of Oakland in 1973 and garnering 40 percent of the vote.

Latino student organizations also became active at this time, starting with the Mexican American Student Association in 1967. As discussed earlier, many young Hispanics worked for the United Farm Workers and other causes. Then the more revolutionary Brown Berets developed in the *barrios* in 1968. Modeling themselves on the Black Panthers, though not as radical and violent, they sported brown rather than the black berets of the Panthers. The Brown Berets supported radical antiwar actions and allied themselves with all protests of a radical nature. Like the Panthers, they organized breakfast programs and other community actions, as well as printing their own newspaper, *La Causa*.

The Brown Berets were successful in bringing attention to police brutality against Latinos and the poor schools of the *barrio*. They organized a strong movement to restore interest in Hispanic culture and started the demand for bilingual education. Allying themselves with the antiwar movement, the Brown Berets organized contingents and speakers for all of the major antiwar rallies. In March 1968, in the "East L.A. Blowout," they incited high school youth to protest the Vietnam War, yelling "Walk out! Walk out!" Hoover soon turned the FBI loose on these young radicals as well, and by 1972, the organization was destroyed by infiltration and agent provocateurs.

In Texas in 1968, La Raza Unida party in Texas became proactive in seeking Latino control of communities in which they were the majority. Despite the radical excitement of the 1960s, organization of the Latino community was slow but gained speed later on when many individuals finally registered to vote in the 1980s and 1990s.

AMERICAN INDIAN MOVEMENT: ALCATRAZ AND THE PINE RIDGE WAR

The 1960s were also the time of the American Indian Movement (AIM), organizing in 1968 to stop police brutality against Native Americans. Like the Black Panthers and Brown Berets, AIM supported and organized cultural restoration and community organization. As in the Hispanic community, the revival of the culture of Native communities can be traced back to the radical AIM and the late 1960s. Inspired by the publication of *Black Elk Speaks,* which detailed the visions, sweats, and spirituality of the Native Americans, along with the story of resistance and eventual defeat, Indians in the late 1960s got radical like everyone else their age.

In 1969, AIM and a group of students declared the abandoned island of Alcatraz to be reclaimed Native land, and they peacefully took over the old prison with seventy-nine young Native American students. AIM became famous overnight. They stayed for nineteen months, negotiating all the while with the Nixon administration for several programs, which eventually led to the passage of major policy shifts and tens of thousands of acres being returned to reservations. In addition, Nixon quietly ended the horrific practice of termination, which had already relocated two hundred thousand from reservations to inner cities. The occupation of Alcatraz was thus one of the most successful protests of the entire 1960s. However, Nixon finally got fed up with the whole business, and sent in the U.S. marshals, who easily ejected the remaining fourteen Indians. AIM grew more and more radical as a result of that action.

For years, the Department of the Interior had wanted to lease a large piece of the Pine Ridge Reservation in South Dakota to mine for uranium and oil, and Chief Dick Wilson was determined to make the deal. On the reservation is Wounded Knee, the site of the notorious massacre by U.S. troops in 1890 in which three hundred Sioux had been killed, half of them women and children—all because of a revival of the Ghost Dance. Now the AIM returned there. On February 6, 1973, two hundred Indians marched on the Custer Courthouse to protest the suspended sentence of a murdered Native, Wesley Bad Heart Bull. The police descended on the crowd and dispersed them, arresting thirty.

On February 26, local Lakota who were against Wilson and the mining asked AIM to lead the protest movement. The next day, three hundred young radicals left for Wounded Knee Village. They were armed and proclaimed that they were ready for the government. Within hours of the occupation, police set up roadblocks and cordoned off the entire area, arresting all who attempted to leave. Nixon soon brought in reinforcements—in the form of armored personnel carriers, helicopters, 50- and 60-caliber machine guns, CS gas, and snipers. The outcome was inevitable.

Following the end of the occupation of Wounded Knee, a "war" broke out on the Pine Ridge Reservation, in which Wilson's Indian vigilantes, a ten-man BIA SWAT team, and FBI agents terrorized and murdered at least sixty-one people over the next few years. Two hundred dead may have actually been the real toll. Nixon and Hoover let it all happen. Finally, in 1975 the killing of two FBI agents led to the arrest of Leonard Peltier, who is still incarcerated for the shooting, despite many protests from human rights groups and AIM.

GAY LIBERATION

Along with the other minorities, gays and lesbians got radical at the same time. Walt Whitman, we saw, spoke of gay liberation in the later 1800s. It was just a dream then. Now in the 1960s, with the new freedom of the organic worldview influencing the mainstream, the homosexual community started their own crusade for equal rights—in the form of an ugly riot in New York City.

For years, police had busted gay bars and hauled off the customers to jail with no resistance. On June 27, 1969, however, all that changed forever. At Greenwich Village's Stonewall Inn, when a few police routinely arrested a bar steward, a lesbian, and three drag queens, the urge to fight back suddenly surfaced: "Then, without warning, Queen Power exploded with all the fury of a gay atomic bomb. . . . The lilies of the valley had become carnivorous jungle plants. Urged on by cries of 'C'mon girls, lets go get 'em,' the defenders of Stonewall launched an attack."[12]

Soon, four hundred were in the streets, throwing rocks and bottles at police trying to make the arrests. The police retreated into the inn as a defense and to await reinforcements. To pass the time, they beat and brutalized the customers still trapped inside. Radical students joined the crowd and someone set fire to the inn. The chant "gay power!" rang out for the first time. Although all were eventually rescued, feelings still ran hot.

The next night, four thousand turned out on the streets of Greenwich Village. Rioting and fighting ensued for three days. Ending the violence, a noisy but peaceful gay power parade finally marched up Sixth Avenue. The anniversary of Stonewall became Christopher Street Liberation Day, and between five thousand and ten thousand marched the next year in celebration. Even larger marches were held the next year in New York and Los Angeles, eventually becoming the annual gay pride parades.

As a result of the Stonewall Riot, the Gay Liberation Front (GLF) was formed. This group, based on the idea of coming out of the closet, openly confronted the old paradigm. To curb brutality and oppression, the GLF began "patrols" of the LA police. They picketed a whites-only gay bar in Houston and boycotted other bars in Chicago that did not allow men dancing with men.[13] The GLF struck at homophobia and racism wherever it appeared, even invading the 1970 convention of the American Psychiatric Association to protest the classification of homosexuality as a mental illness. Like the other civil rights movements of the time, it was all a direct challenge to the status quo and the cultural oppression of a minority.

WOODSTOCK NATION

Six weeks after the Stonewall Riot, on August 15 and 16, the fabled Woodstock concert was held, a festival of peace, music, and art. As before, the music was secondary for counterculture music fans. For most, making a statement about lifestyle or a protest against the war was just as or more important than seeing their favorite groups. Two million tried to make it to the concert, although the roads quickly clogged and only a half-million longhairs actually got to the site. When rains came and the event turned into a mud-filled debacle, the musicians and the counterculture celebration continued unperturbed. The music included "Freedom" by Richie Havens, "The Fixin' to Die in Vietnam Rag" by Country Joe and the Fish, and the Jimi Hendrix version of "The Star Spangled Banner," with the entire message of the concert being to question authority and change the world. Although the concert was a logistical disaster requiring a massive cleanup, it was a success in its goal of changing the world.

The impact of Woodstock, like the summer of love, was enormous. A love of music, along with the concern for world peace, the environment, and civil rights, widened the generation gap even further. The hippie counterculture was now political and determined to end the Vietnam War—and then build a new organic world to replace the old polluted one. The sense of community was so strong a name evolved for it: Woodstock Nation. No matter where a person might live in the world, whether they went to the concert or not, they could now *belong to something*—to Woodstock Nation.

THE 1970S AND WATERGATE

All we are saying is give peace a chance!
JOHN LENNON, "GIVE PEACE A CHANCE"

Following Woodstock, the movement against the Vietnam War grew larger than ever. The protests grew when Nixon invaded Cambodia.

Vietnam Veterans Against the War organized their own protests and held their own investigations of American operations against civilians. After becoming their leader, a young John Kerry testified before the Senate Armed Services Committee in April 1971, pleading that the war be ended and all troops withdrawn. After Kerry's testimony, many more senators turned against the war.

The next year, George McGovern and the boomer activist generation beat the Humphrey wing of the Democratic Party for the presidential nomination. Nixon, beside himself that McGovern might somehow defeat him, ordered the Watergate burglary. All that summer, the Watergate cover-up concealed White House culpability as the press buzzed about the connections of "the plumbers," the apprehended burglary ring. At the Democratic convention, the young progressive radicals meanwhile razzed the older Democrats, and McGovern ended up getting little help from the party leadership. Despite the scandal, Nixon trounced McGovern. Watergate then preoccupied America and much of the world, as Nixon fought desperately to stay in the White House. He finally resigned in August 1974, after Articles of Impeachment passed across party lines in the House.

THE PROTESTORS MOVE ON

The implications of the Watergate scandal were many, including the temporary reform of government in the form of the special prosecutor and other legislation, as well as the temporary return of power by the Democrats in 1974 and 1976. Yet most important, in terms of the paradigm shift, the resignation of Nixon so pleased the members of the counterculture that many simply left activism and protest altogether. Also the last compulsory U.S. military draft was in 1973, ending that burden on every man. By April 1975, the Vietnam War was over, and that issue, too, disappeared. Although fifty-eight thousand Americans died, the all-powerful nuclear bomb could never be used in Vietnam, and the most powerful military in the history of the world was forced to retreat by the determination of a small nation. It was a lesson learned—one that would later be forgotten in the conquest of Iraq.

As the protest movement evaporated, many in the counterculture left the cities and suburbs of their birth and began the back-to-the-land movement. Millions of young Americans moved to the country, bumping headlong into the conservative culture of rural life. After having to deal with a decade of war, scandal, and the draft, most escaped from politics and social change, got married, and had children. All the rest of the problems of the world would have to wait. The energy crisis, the deadly boring educational curriculum, the destruction of the environment, the power of the military-industrial complex, the oppression of women and minorities—all of it would have to wait.

Then, just as the protest movement burned out and began to transform itself, the old economy in the United States reached a peak around 1973. The standard of living for the average worker began to slip. Cheap gasoline and a world war that flattened the overseas industrial base had given America a free ride for more than twenty-five years. That free ride was now over, as the rest of the world industrialized and cheap labor in developing nations took away First World jobs forever, starting with manufacturing and steel. The rust belt emerged, growing larger and larger every year for decades to come, at the same time the protestors moved on with the rest of their lives.

The radical groups were meanwhile harassed and disrupted by the FBI into working within the system. Violent radicalism descended into a series of senseless, intermittent bombings by the leftovers of the SDS, the Weathermen, and other tiny groups based on violence such as the Symbionese Liberation Army (SLA). In 1974, the SLA kidnapped Patty Hearst, the daughter of newspaper mogul William Randolph Hearst. By the end of that adventure, which involved bank robberies and senseless killings, radical was no longer "in"; it was now "out." Even music lost its radical edge and was replaced in the late 1970s by nihilistic punk rock and mindless disco and pop. The stage was set for decades of drift and little activism by a new type known as generation X or the me generation.

Not all protest disappeared. A milder form came with the founding of Earth Day in 1970 and its celebration ever since. The environmental movement grew in leaps and bounds during the 1970s, as people took in the Love Canal disaster and the building of dozens of unsafe nuclear power plants in every region of the country. The 1973 oil embargo and resulting gasoline shortage made it clear how dependent the world had become on Mideast oil and on fossil fuel in general.

There had meanwhile been more ecodisasters, raising environmental awareness even higher. A 1976 explosion in Italy spread dioxin in the Milan suburb of Seveso, injuring thirty and giving three hundred schoolchildren chloracne, a skin condition. Nine million gallons of oil were spilled in the 1976 *Argo Merchant* wreck off Nantucket. And then two years later, the biggest oil spill of all: sixty-eight million gallons off the coast of France by the supertanker *Amoco Cadiz*.

A crucial point in the centuries-long paradigm shift now brought the change in worldviews to a new level. As in the eighteenth century, great minds began to explore the boundaries of the organic paradigm, and the Organic Shift in science accelerated even further. The holistic awareness of how multiple environmental crises are interconnected developed as a result of ecology becoming popular, prompting another look at the big picture in many disciplines.

ORGANIC-PARADIGM SCIENCE LOOKS FOR THE BIG PICTURE

In the enlightenment phase of the early 1700s, the Scientific Revolution gathered speed as the old scientists died off and the new thinkers replaced them. In the early years of the eighteenth century, new scientists wrote remarkable theories about the universe, seminal works like George Berkeley's *Treatise Concerning the Principles of Human Knowledge* in 1710 and Newton's *Optiks* in 1704, or the science-based history of Giambattista Vico in his *New Science*. Piece by piece, medieval dogma was replaced by the mechanistic laws of nature. The early 1700s was also a time when the new scientific "academies" and associations became stronger, less magical, and more respected. The early enlightenment phase brought a much wider audience to the new paradigm, and a large part of the population began reading the advances in thought as they were published.

In a similar way, the organic worldview started to mature as the new science was understood and discussed openly in the progressive community during the 1960s. Many important books became bestsellers. All replaced an obsolete and rigid mechanistic view with an organic theory, from biology and medicine to economics, history, and beyond. Kant's new order of sciences was now maturing.

First was Buckminster Fuller's *Operating Manual for Spaceship Earth,* published in 1963, which explained how society was overtaken centuries ago by the "Great Pirates": the European miners, buccaneers, and their bankers. The miners have, of course, created a civilization based on mining and drilling. Yet it doesn't have to be that way, said Fuller. It can be a civilization based on alternative means; we just need to be open to new thinking and find the solutions. In the following paragraph, he emphasized that we must figure out how best to maintain the life of our Spaceship Earth; we must search for the alternative.

> Now there is one outstandingly important fact regarding Spaceship Earth, and that is that no instruction book came with it. I think it's very significant that there is no instruction book for successfully operating our ship. In view of the infinite attention to all other details displayed by our ship, it must be taken as deliberate and purposeful that an instruction book was omitted. Lack of instruction has forced us to find that there are two kinds of berries—red berries that will kill us and red berries that will nourish us. And we had to find out ways of telling which-was-which red berry before we ate it or otherwise we would die. So we were forced, because of a lack of an instruction book, to use our intellect, which is our supreme faculty, to devise scientific experimental procedures and to interpret effectively the significance of the experimental findings. Thus, because the instruction manual was missing we are learning how we safely can anticipate the consequences of an increasing number of

alternative ways of extending our satisfactory survival and growth—both physical and metaphysical.[14]

Next on the reading list is Bertalanffy's *General System Theory,* published in 1968, which took holistic philosophy and earlier systems thinking and created a true science from it. No matter what the system was, Bertalanffy's principles now explained interrelations between the different parts of any whole. This advance, going well beyond Newtonian limitations, is the new foundation of the information and biological sciences:

> Compared to the analytical procedure of classical science with resolution into component elements and one-way or linear causality as basic category, the investigation of organized wholes of many variables requires new categories of interaction, transaction, organization, teleology. . . .
>
> These considerations lead to the postulate of a new scientific discipline, which we call general system theory. Its subject matter is formulation of principles that are valid for "systems" in general, whatever the nature of the component elements and the relations or "forces" between them. . . . General system theory, therefore, is a general science of "wholeness."[15]

Another book in 1968 addressed the metaphysical side of the Organic Shift: Joseph Campbell's *Creative Mythology,* revealing how myth evolved into modern religion, and how the roots of mythology are alive today in many traditions and privileges. Understanding how mythology evolved into our current institutions helps us understand ourselves and our society in a whole new way, making mythology a creative tool for personal and social development. Campbell, moreover, was responsible for making clear the deep nature of Jung's archetypes and their significance for personal growth and health.

In 1969, Abraham Maslow and Anthony Sutich started the *Journal of Transpersonal Psychology,* which integrated the spiritual traditions and mysticism of East and West into psychology. Theodore Rosak's *Making of a Counterculture* explained how the movements of the 1960s would be the start of an epoch-making transformation of thought and so the world. Charles Tart's *Altered States of Consciousness* explored the realm of inner space, giving a scientific look at the different types of consciousness that have been studied. Elisabeth Kübler-Ross's *On Death and Dying* revisited the taboo subject of death and brought the new paradigm to that area, too. This was a welcome enlightenment, eventually leading to the hospice movement of today.

THE 1970s: BIG PICTURE MEDICINE, HISTORY, AND PSYCHOLOGY

Holistic medicine emerged in 1971 when the Esalen Institute started its Program in Humanistic Medicine. Here the consciousness movement entered the healing arts, which had been in the hands of the allopaths since the defeat of homeopathy in America in the 1920s and 1930s. Much of the alternative knowledge was soon recovered, and holistic medicine grew around the world during the 1970s. In addition to acupuncture, acupressure, and shiatsu from the East, herbalism, homeopathy, and vitamin therapies became the norm among new-paradigm thinkers, who introduced the new methods to many others. Chiropractic science offered novel options for those with back pain, while Dr. Linus Pauling showed that vitamin C (ascorbic acid) was an effective cold remedy if taken in a specific megadose at or near the onset of the cold or flu. Though studies have debunked the idea that it is a "cure," newer studies (following the correct level and protocol) reported that overall flu and cold symptoms in the test group decreased 85 percent compared with the control group after the administration of megadose vitamin C in the form of ascorbate (ascorbic acid)—the form of vitamin C deemed most effective for viruses. This study as well as other studies confirmed that maintaining fairly high doses after the megadose significantly reduced the symptoms and shortened the duration of a cold or flu even when started after the infection has fully set-in.[16] Above all, alternative medicine was based on an initial holistic conversation with the patient in which the doctor listens—really listens—allowing a proper diagnosis. Just looking at physical ailments was not enough in the new medicine.

Nearly every field was by now feeling the influence of the Organic Shift.

The study of the past entered the big picture process with William Irwin Thompson's *At the Edge of History* in 1971. As the first holistic historian, Thompson brought the role of myth, mysticism, and culture into historical analysis. With the much broader perspective of the new worldview, his essays synthesized myth, culture, and history to explain the human condition and how we stand at the edge of history. In 1972, Thompson started Lindisfarne Association, originally in Southampton, New York, now located in Crestone, Colorado. He intended it as a model for a future planetary culture, a community in which artists, scientists, and mystics live together without chemical farming or fossil fuels.

Gregory Bateson, the husband of cultural anthropologist Margaret Mead, was another great thinker deserving mention. His *Steps to an Ecology of the Mind* showed the place of the human mind in the universe—as being part of a larger whole of society and culture. A grand synthesis of culture, anthropology, systems thinking biological models, ecology, and psychology, this book was a deep exploration of thought and its interrelation to everything else.

The 1970s continued, with Ervin Laszlo's *Introduction to Systems Philosophy* becoming a textbook on systems thinking for many sciences, from cybernetics to biology and beyond. John Lilly's autobiographical *Center of the Cyclone* meanwhile looked at the outer reaches of consciousness through sensory deprivation experiments, hallucinogens, and dolphin research. The film *Altered States,* with William Hurt as the researcher, was inspired by this book. *The Center of the Cyclone* was a journey through inner space that finally came to the conclusion that the mind creates itself and sets arbitrary limits for itself. It is the researcher's job to go beyond those mental barriers: "In the province of the mind, what one believes to be true either is true or becomes true within limits to be found experimentally and experientially. These beliefs are limits to be transcended."[17] A scientist must always be aware that the mind will unconsciously design experiments to find what it wants to prove. To compensate, experiments must be designed taking that into account, although one can never be absolutely sure one has transcended the final limit.

Robert Ornstein's *The Psychology of Consciousness* is a fascinating look at brain research and psychology, synthesizing the findings of mysticism with those of science. Ornstein showed, in particular, how the mind became so familiar within a paradigm that anomalies contradicting it are ignored or not noticed. These quick assumptions make mental limits comfortable but hold back new paradigms.

> One assumption of our society concerns the suits of playing cards. Through years of experience, we have learned that spades and clubs are black, hearts and diamonds are red. Normally, each deck of cards we see confirms this assumption. Bruner asked his observers to look at some cards through a tachistoscope—a device that flashes visual materials on a screen for a brief period of time. Intermixed with ordinary cards were several anomalous ones, for instance a red ace of spades and a black four of hearts. Many of the observers in this experiment did not see the unusual cards as they were but corrected them, reporting a red six of spades as a six of hearts. Assumptions can limit the scope of awareness. At one point in the experiment, it was suggested to the observers that although hearts are usually red, this does not logically imply that they will *always* be red. With this new idea extending their category system, some observers were quickly able to see what was in front of them.[18]

The experiment mentioned by Ornstein is very revealing. Once the limits are recognized as arbitrary, the brightest are able to transcend their assumptions and more consciously make a true observation of the facts. This is true for paradigm shifts as well as art, sports, and many other parts of our culture.

SMALL IS BEAUTIFUL

E. F. Schumacher's *Small Is Beautiful: Economics as if People Mattered* was another influential book, questioning economics and provoking a deeper understanding of the world. Schumacher added not only the environment to economics but also people, including the poor of developing countries. This was a real leap for the dismal science, as all mechanistic economic models ignored both the poor and ecology. "The whole point is to determine what constitutes progress," counseled Schumacher.

The definition of progress must include ecology, people, local cultures, the preservation of special sites, and many other factors, whereas mechanists only create theories that increase consumption through any kind of growth. He complained that "growth of GNP must be a good thing, irrespective of what has grown and who, if anyone, has benefited."[19] Theodore Roszak in his introduction to *Small Is Beautiful* notes that Schumacher believes "economists, for all their purported objectivity, are the most narrowly ethnocentric of people . . . since their world view is a cultural by-product of industrialism, they automatically endorse the ecological stupidity of industrial man and his love affair with the terrible simplicities of quantification."[20]

Schumacher saw the assumptions that misled the mechanists, especially the idea that high levels of growth and waste-ridden consumption can go on forever—despite Earth's finite resources. He warned that world peace would eventually be threatened by the current power structure and its dependence on oil: "what else could be the result but an intense struggle for oil supplies, even a violent struggle."[21] It's not a prophecy; it's simple economics.

He then explained the real failing of mechanistic economics. It assumes, quite conveniently, that nothing can be done to really help the poor, especially those in developing countries. This relieves corporations and governments of the responsibility they actually have to transform the lives of these people. As he said, "It is a strange phenomenon indeed that the conventional wisdom of present-day economics can do nothing to help the poor. Invariably it proves that only such policies are viable as have in fact the result of making those already rich and powerful, richer and more powerful. . . . It is always possible to create small ultra-modern islands in a pre-industrial society. But such islands will then have to be defended, like fortresses, and provisioned, as it were, by helicopter from far away."[22]

Schumacher concluded that the solution lies in developing the Third World and other places with sustainable means and appropriate technology. Small, human-scale village projects were deemed to be helpful, while industrializing on a large scale was destructive to the environment, culture, and people alike. Deciding what is or isn't appropriate technology can only be determined by meeting the local people and immersing oneself in the culture. And this technology must be sustainable, not based on resources like fossil fuel that one day will run out.

This perspective was very different from that of the mechanists, who funded inadequate top-down development programs designed to harness cheap labor for factories owned by foreign investors in a form of modern economic colonialism. Understanding the lessons within *Small Is Beautiful* is, quite simply, crucial to our creation of a workable future.

THE NEW PHYSICS
AND CONSCIOUSNESS

Another great book of the early enlightenment in the 1970s was Fritjof Capra's *The Tao of Physics*. This was one of the simplest explanations of the new physics and the many similarities it has to Eastern philosophies. *The Tao of Physics* not only was a clear explanation of the limits of mechanistic/Newtonian models, Capra also gave an easy-to-read description of the subatomic realm.

The Origin of Consciousness in the Breakdown of the Bicameral Mind was the long title of a book about history by Julian Jaynes, a Princeton psychology professor. Jaynes's theory is that early civilizations were ruled by both the left and right sides of the brain, meaning the right hemisphere had far more control in ancient and prehistoric times than previously believed. Since the right hemisphere is more intuitive and autohypnotic, bicameral consciousness would be very different from our own. For example, the bicameral mind is less apt to question religious authority because it is more easily influenced by and susceptible to suggestions from spiritual leaders. The community is all, and ego is not apparent.

Jaynes's theory emphasized that the right hemisphere can hallucinate voices. This threw new light on old texts, many of which assert that from the earliest times it was possible to speak to the ancestors and receive spiritual and practical guidance. Before reasoned analysis, this was how the bicameral mind made important decisions. Early humans asked their ancestors and heard the answer.

With the invention of writing, Jaynes believed that the use of the dominant right hand, which emphasized the left hemisphere and its specialization in language, tipped the balance. The bicameral mind began a millennia-long breakdown, which Jaynes found gloriously reflected in the writings of the time. Where the gods or God talked directly to mortals or prophets in early times, by the sixth century BCE, many agonized over why they ceased to speak.

Jaynes's theory helped make sense of much of history, as well as allowing us to better understand the thinking of remote tribes and peoples, such as the aborigines in Australia. Here, the bicameral mind can be directly observed; however, it is quite difficult for the modern person, whose brain is dominated by the left hemisphere, to submerge his or her ego enough to truly understand this other type of consciousness. Along with William Irwin Thompson, Julian Jaynes was one of the first to bring holistic awareness and the new worldview into the field of history.

ORGANIC AGRICULTURE BECOMES A SCIENCE

Just as scientific agriculture developed during the Enlightenment of the eighteenth century, an agriculture based on an organic paradigm came about during the early enlightenment phase of the 1960s and 1970s. We saw how the foundation for organic agriculture emerged in the 1940s and 1950s, a direct descendant of Goethe's search for nature's open secrets and Rudolf Steiner's invention of biodynamic gardening in the earlier twentieth century. During the 1970s, the Rodale family and many others made a great effort to turn the study of organic farming into a science, convincing some in academia and in the USDA that the subject deserved experimental research.

The expansion of organic farming itself led to the formation of regional groups that could certify the organic nature of the operation. The California Certified Organic Farmers, started in 1973, was the first certifying body in North America. Also in 1973, DDT use in the United States was finally banned. Some writers chart the beginning of the modern environmental movement to this event. Rachel Carson had finally won her argument—eleven years later.

Throughout the entire history of paradigm shifts, what was most striking was the irrational resistance put up by the old worldview to the new thinking—even after being confronted with the irrefutable facts. Corporate corruption of the governing process was part of the problem, but resistance to the new paradigm by the status quo was, by itself, substantial. This resistance delayed the banning of DDT for eleven years after Carson's bestseller exposed its hazards.

There are many other examples of this resistance to facts, from the expansion of nuclear power to the promotion of chemical farming—despite all the danger signals about both industries. Nonetheless, as more and more people learned the truth about chemical farming and the pesticides and herbicides it relies on, organic farming, organic food cooperatives, and health food stores expanded substantially during the rest of the 1970s.

Organic agriculture is based on the idea that the soil is alive, that numerous organisms and microorganisms create and maintain healthy soil. Each microorganism has a different function to perform, and so it is important to preserve all the many tiny forms of life. Chemical farming, focusing on the mere chemical nature of soil—the pH level, the nitrogen content, and so on—ignores this biological factor, destroying most of these minuscule beings with hot fertilizers and deadly pesticides.

Mechanists also ignore or dispute the clear evidence that chemical residues cause cancer and many other diseases, for accepting this evidence would mean acknowledging that chemical farming and all the pollution is a crime against humanity rather than a benefit. There are no known significant pollutants or diseases caused by properly managed organic farms.

Chemical farming, like the rest of mechanistic civilization, is based on consuming finite resources made mostly from fossil fuel, bringing them in from long distances with more fossil fuel, plowing and planting with fossil fuel, and then using still more fossil fuel to transport the food produced incredibly long distances. In contrast, organic agriculture recycles manure and other organic waste into fertilizer, removing the need to create and transport nitrogen inputs. It also typically tries to sell its food as locally as possible. It is, in short, sustainable, while chemical farming is a ticking time bomb in terms of disease, malformed gene pools, and pollution of the land, air, and water.

A most significant book that appeared at the end of the 1970s mirrored the organic farming view that soil is alive and populated by countless numbers of microorganisms. This book declared that not only is the atmosphere above us a living thing, *all of Earth is a living being*. As a living being, Earth deserved not only respect but also a name. This book gave it one.

NEGATIVE ENTROPY AND GAIA: A NEW LOOK AT LIFE ON EARTH

By showing how the atmosphere operates through a host of microorganisms interacting with gases and heat, in 1979 two organic scientists proved that Earth *is* alive. In that year, James Lovelock and Lynn Margulis released their long-awaited book, *Gaia: A New Look at Life on Earth*, a theory already widely known from the early 1970s. Just as organic soil science understood the complex levels of roles played by the tiny creatures in creating fertility and soil health, Lovelock and Margulis discovered an amazing array of microorganisms in the atmosphere covering the planet. In short, the air around and above us *is alive.* The atmosphere—and the biosphere—could thus be made sick and could even die.

To make their point, they named the biosphere *Gaia* (pronounced *guy*-uh), after the Greek earth goddess. The Gaia hypothesis is based on the principle that life is *negative entropy,* what Teilhard had called reverse entropy—the force that drives the unfolding of evolution in all its complexity. Goethe had first said evolution was an unfolding. Now the new paradigm began to quantify the concept. Life is the coming together of that which had been separated; it is the movement toward the synthesis of molecules. Buckminster Fuller called it *syntropy.* Negative entropy, reverse entropy, or syntropy, the concept now became the basis of biological science. For example, when life is looked for on Mars, the robotic explorers today conduct Lovelock's test for negative entropy.

Negative entropy is also the cornerstone of chaos theory, which mathematically shows how order (life) evolved out of the original chaos of early Earth. Ilya Prigogine, who won the 1977 Nobel Prize in biochemistry for this discovery, simply turned the entropy charts of physics on their heads, showing how life, evolution, and human culture self-organizes itself. With entropy, the bonds between molecules fall apart and the universe itself eventually burns out. This is true. Yet when a planet's atmosphere supports it, negative entropy can create life and all of evolution. Molecules synthesize with one another with ever-greater complexity, leading to sophisticated organisms such as dolphins and humans.

Rather than Newton's clockwork universe, chaos theory reveals a self-organizing universe. Prigogine's work was a milestone for many reasons. Not only was it the mathematical basis of many future biological models in many fields of science, the idea of the self-organizing universe is a revolution in the way we think about society and the future itself. Even though chaos theory is a scientific explanation for the beginning of life and thus contradicts the old religious myths; it nevertheless led one to a spiritual conclusion, for when an evolutionary change occurrs through negative entropy, things that were impossible suddenly become commonplace.

Imagine two halves of a rubber ball. Apart, they have little potential for movement, but glue them together and immediately the formerly separated halves have new properties. As one sphere, they can now roll or be thrown through the air, even hit with a bat. By synthesizing two separate things in a new arrangement, negative entropy unleashes incredible potentials. This analogy of the rubber ball helps explain the emergence of thought and the other abilities wrought by evolution. Like the ball's newfound ability to roll, significant properties keep emerging.

Every level of increasing organization in major evolutionary thresholds has been accompanied by a rise in brain size and consciousness—up to the reflective awareness of human beings. Yet as Teilhard's reverse entropy predicts, this process is not yet finished. Consciousness is about to make the next leap, to the next level of reverse entropy—a convergence of culture we are just beginning to imagine. Evolution means we will soon be living in a fully "planetized" human society, one that has a new potential unleashed: a *collective consciousness*. In other words, the human race is becoming conscious on a planetary scale, as technology provides the circuitry of a global brain and the paradigm shift raises awareness on many levels at once. If the self-organizing universe theory and Teilhard are correct, the human race wakes up—and in our near future.

Between Teilhard's futurism, systems thinking, chaos theory, alternative medicine, and all the other organic advances, the new worldview was poised by the late 1970s to have a huge impact on science and spirituality alike. Yet the maturing scientific revolution was also profoundly affecting technology at the same time.

RENEWED RENEWABLES:
MODERN WIND AND
SOLAR POWER

The dramatic entry of the organic worldview into the mainstream in the 1960s and 1970s soon led to rapid progress toward renewable energy. Yet without higher oil prices, alternative renewables could not come close to competing. This problem solved itself during the 1970s, however, as the era of very cheap oil ended in gas lines and interest in solar power took off. The U.S. Department of Energy was established and soon funded the Federal Photovoltaic Utilization Program, which installed and studied three thousand photovolatic systems.

Wind power also won research funding from the U.S. government at this time. Development centered around making much larger, interconnected machines, rather than the smaller models. This proved fruitful, with a fairly simple design being the most workable. Large multimegawatt turbines were now feasible. Meanwhile, the Danes made use of new advanced materials and fine-tuned the old Gedser Mill design, allowing larger turbines, as the Germans developed a lightweight, high-efficiency generator. These two advances would soon be combined in the large turbines to come. By the late 1970s, wind power was making real technological advances in Europe and America, and practical wind generators of several sizes finally became a technological reality.

At the same time that research pushed the envelope on large turbines, a counterculture wind power faction formed the Wind Energy Association. This group pushed through a small machine development program with money from the U.S. Department of Energy. This was successful in establishing basic technical research. By the end of the 1970s, both large and small renewable systems powered by wind, solar, or microhydro were being tested, and tens of thousands of small home units had been installed.

NUCLEAR POWER EXPOSES
THE MAIN FLAW OF MECHANISTIC THINKING

If just a fraction of the money spent to subsidize nuclear power would be put into researching renewable energy, the payoff would be tremendous. Our energy problems—accompanied by key lifestyle changes—would be solved (discussed further in chapter 10). Yet due to the power of long-standing mechanistic institutions in the government and industry, this has never taken place. Herein lies the crux of the mechanistic dilemma. Trying to address the paradox directly, through political action, is not adequate to the task. The institutions themselves are what needs changing—along with the minds of billions of individuals around the planet.

Maintaining control of the institutions with an iron grip, the old paradigm was able to perpetuate itself—even though it may have already lost the scientific and moral high ground. Remember that in the eighteenth century nothing really changed until a paradigm flip occurred in the general population, caused by the revelations of the new history of Voltaire and the writers of *The Encyclopedia*. This new story undermined the old worldview to such an extent that after 1760 a revolutionary attitude emerged. Today we must learn the truth about the old worldview and power structure—and how it came to be. If we are to work our way out of the dilemma created by the mechanistic worldview, everything that has been concealed or advocated by the mechanists should be reexamined in the light of the new thinking.

For example, to make up for the horror of the nuclear bomb and to improve the image of nuclear research, the mechanists in the 1950s decided to crack the atom and make electric power, no matter what the cost. Congress even passed an act to limit corporate liability in case of a meltdown, which might otherwise lead to lawsuits in the hundreds of billions of dollars. (In the case of Indian Point power plant, located only twenty-two miles from New York City, the property damages could be as much as several trillion dollars, clearly an impossible figure to pay. This deception about insurance continues to this day.)

This peacetime use of the atom was positioned as part of the package for entering the future—it was to be the Age of the Atom. The small antinuclear movement of the time thought it a terrible mistake. Science has always progressed by finding simpler, more elegant solutions to problems. Nuclear power production, however, tried to turn this truth on its head—and failed.

Making electricity with the atom, which was supposed to have been "too cheap to meter," was instead an extremely difficult task requiring miles of complex plumbing, pumps, and equipment. Additionally, the mechanists despicably left the problem of storing used radioactive waste to the future. History recorded the first nuclear power plant protest in Detroit at the experimental Fermi plant in 1956, following a partial meltdown. These early protestors were, of course, correct—the very next year one of the worst ecodisasters in history took place in the Ural Mountains when a radioactive explosion irradiated a whole region, which unfortunately was populated by 250,000 people.

In that same year of 1957, the Windscale plutonium reprocessing plant in England leaked copious amounts of radioactivity after three tons of uranium caught fire. At first, fans were used to try and control the disaster—but this only worsened the blaze. Carbon dioxide was then tried, also to no avail. Finally, despite fears of a hydrogen gas explosion, the pile was flooded with water and, after three days, this finally succeeded in putting out the blaze. The British government destroyed two million liters of milk and tried to suppress the true story of the radioactivity actually released. Safety was clearly sacrificed, this time for nuclear bomb construction.

Vast amounts of information showing the dangers of nuclear power were classified secret or simply never reported by the government and the corporate-owned media. Yet SANE and the rest of the progressive movement knew a great cover-up was underway and continued the fight against the building of plants. All to no avail, as the United States government was determined to push nuclear power through.

The United States narrowly escaped a full meltdown in 1955, when a breeder reactor near Idaho Falls, Idaho, began to melt because of "operator error," a nasty problem that will crop up again and again in the history of nuclear power. Six years later, the SL-1 Reactor at Idaho Falls blew a rupture in the core, a serious accident that emitted 500 rems per hour. And in 1966, the Fermi breeder reactor near Detroit partially melted down again, disabling it permanently. Despite all of these close calls, both the Northeast Blackout of 1965 and the Oil Embargo of 1973 were then used by industry and the government to push for the building of as many nuclear plants as possible. By the late seventies, more than one hundred plants were either built or scheduled for construction in the United States.

Then it happened. At 4:00 a.m. on March 28, 1979, due to equipment failure and operator error, a partial meltdown occurred at Three Mile Island Unit 2, near Harrisburg, Pennsylvania. Years later, it was realized how bad the meltdown really was and how close the world had come to a global catastrophe. For days, engineers struggled with the meltdown and were forced to vent radioactive gas that had built up. People in the local area could taste the radioactive iodine in their mouths from breathing the leaks. About 140,000 people were told to evacuate the Harrisburg area for fear of their developing cancer later on. President Jimmy Carter, himself a nuclear engineer in the navy, came to the control room of Three Mile Island to reassure the public—but it was too late. Public resistance to the building of new plants was absolute after Three Mile Island, and no new plants were given permits in the United States after that point. The cleanup of the disaster took years.

In addition to Three Mile Island, and, of course, all the chemical and atmospheric pollution, some of the worst environmental disasters during this time were sunken nuclear submarines. Today, they still leak radioactive cesium and plutonium into ocean currents *with radioactive levels many times that of Chernobyl.*[23] To protect the fishing industry, the government and the media ignore the sunken nuclear sub disasters. Even Norway is guilty of this; Knut Gussargard, director of the Norwegian Nuclear and Safety Institute, responded to news of sunken Russian subs by remarking, "If news of radioactively contaminated water and fish spread around, it would have disastrous economic results on the Norwegian Fishery and Fish Export Industry."[24] Nuclear power reflects the mechanistic dilemma in all its glowing glory. Mechanistic models of the world, combined with the corruption of crony capitalism, have proven themselves to be a disaster for the public good and for the environment.

On top of all these nuclear catastrophes came the realization that atmospheric ozone—protecting all life on Earth from UV radiation—was being seriously depleted by a gas called CFC (chlorofluorocarbons). Being an environmentalist was no longer a matter of simply caring about DDT and bird populations or worrying about the effect of dams on fish. Environmentalism was now a matter of planetary survival, of your own and your family's existence, as well. You could not trust science and technology anymore, for the mechanists continued to defend the old discredited technology, even in the face of multiple contradictions.

With Three Mile Island, the sunken subs, and the many other catastrophes, the mechanists have taken severe—but not mortal—hits to their credibility. So the dilemma continues to this day, with no real safety for the public, no comprehensive cleanup of corporate and government waste dumps, no solution for storage of radioactive waste, and no actual long-term planning for the future.

THE BIRTH OF DEEP ECOLOGY

From Thoreau to Haeckel, Adams, Ricketts, Leopold, and Snyder, the study of ecology had evolved from conservation and use management into deep ecology. Rather than the conservationist and the park service mechanistic mentality of wise use, which still allowed clear-cutting of parks and national forests, the deep ecologists put forth Leopold's paradigm B: do *not* manage resources; let them go back to nature and old-growth forests. The land ethic combined with a Buddhist-like respect and reverence for nature.

The first person to actually use the term *deep ecology* was the Norwegian philosopher Arne Naess in 1972. He said, "The essence of deep ecology is to ask deeper questions," specifically about humanity's relationship to ecosystems and the biosphere in general. All this fit in with the new Gaia hypothesis, which laid out the science that Earth is, for all intents and purposes, a living collective being. "In the movement instigated largely through the efforts of Rachel Carson and her friends, the 'unecological' policies of industrial nations were sharply criticized. The *foundation* of the criticism was *not* pollution, waste of resources and disharmony between population and production rate in non-industrialized nations. The foundation rested on answers to deeper questions of 'why?' and 'how?'"[25]

In particular, Naess detailed the fallacies of anthropocentric thinking and how everything from religion to economics and ecology itself was tainted by the belief that humanity is more important than other species. Although known to a very small minority in the 1970s, which soon included Gary Snyder and several professors and writers, the deep ecology movement was empowered by the publication of the Gaia hypothesis in 1979. Deep ecology would soon be incorporated into much of the burgeoning environmental movement to come. Anthropocentrism had been exposed by Naess, Snyder, and others, and now all of society's previous assumptions were open to reexamination.

THE 1960S AND 1970S
AS AN ORGANIC EARLY ENLIGHTENMENT

From the beginning of the Organic Shift in 1781 with Kant, and on through Goethe and the nineteenth and the first half of the twentieth centuries, the new paradigm had been limited to some philosophers, scientists, artists, poets, and the small countercultures. This low visibility of the young Organic Shift was similar to the mechanistic shift before it. During the Copernican Revolution, the general population knew little or nothing about the new worldview and the idea of a heliocentric solar system or a science-based democracy. Only when the Enlightenment philosophers explained in the early 1700s what Locke, Newton, and Copernicus had discovered did the new thinking enter the mainstream. Duplicating that early enlightenment phase, the organic worldview went mainstream in the 1960s and 1970s, creating most of the turmoil and change of the times.

In the 1960s, the spiral of history also took the definition of freedom to the next level by proclaiming the beginning of an age of peace, love, and brotherhood—when war and nationalism was envisioned to be replaced with international harmony. Just as the eighteenth-century public began to understand John Locke's theory that political power was derived from the people, in the 1960s the cry "power to the people!" energized many new thinkers.

Television connected the world and brought not just Haight-Ashbury but also the Vietnam War and the continuing civil rights struggle straight into everyone's living room. The new generation was stimulated to reflect on the state of the world—just as Teilhard de Chardin had predicted in the 1940s and 1950s. He foresaw how the closing of the electronic circuits of global communication would help to awaken universal feelings and actions on the part of a "planetized" humanity. The new ideas of the 1960s spread rapidly and quickly took on a life of their own.

Although there was no LSD or electric guitars in the early 1700s, hard liquor had been invented shortly before and brandy and other drinks were becoming very popular. Modern pianos and larger violin orchestras were prevalent, increasing the power of music. As we saw, the early eighteenth century was also known for the beginning of a new attitude toward sexuality and love, parallel to the free love of the 1960s, although not as direct. In eighteenth-century technology, early steam engines were inefficient and dangerous but already working to pump water out of mines, preparing the way for the Industrial Revolution. Renewable energy in the early enlightenment of the 1960s was similarly inefficient, yet it pointed the way to a future green energy revolution. Organic agriculture and the implementation of scientific agriculture in the early 1700s were still another parallel.

Another key parallel was that both early enlightenments witnessed a loss of faith in the old beliefs and the rapid acceptance of a new science. In the 1960s, this new worldview came complete with a different basis for spirituality in (1) deep ecology and (2) the Teilhardian view of evolution as a great unfolding of complexity and a great infolding of consciousness. Although there are great differences, in that the old science was based on machine models and the other on biological models, the similarity of the overall pattern of change is unmistakable. In short, in the 1960s and 1970s, society clearly entered an early enlightenment phase, the first stage of the decades-long social transformation pattern of social and political change from one worldview to another.

Yet the hippie movement of the 1960s lacked true substance, making up for it by resorting to mystical mythologies. Although it was a time of protest, exciting music, and shocking new ideas, there was little change of the world's institutions—trends very similar to the early enlightenment phase of the eighteenth century. Despite the conservative political scandals such as Watergate, institutional change would have to wait, as the new social movements mostly devolved into a series of insubstantial fashions and mere fads.

And all that change, all that seemingly irresistible force, was about to meet the theoretically immovable object: the still-powerful institutions and traditions of the old worldview. For just as the eighteenth century had its conservative backlash between 1726 and 1740, so the activists and thinkers of the 1960s and the 1970s were set back in the 1980s by a similar reversal, a powerful counterreaction against change. Only this time the backlash phase was televised—and powered by electricity, oil, and the military-industrial complex. It was a dramatic swing of the historical pendulum, all the way from the new back to the old. To the organic progressives, it felt as though time suddenly started running backward. Being an organic process, the progression is not apparently consistent or smooth but more like a surging-and-falling-back wave of change.

Ilya Prigogine

9

THE CONSERVATIVE BACKLASH

1976–1990

After Big Media, U.S. colleges and universities are the biggest enemies of the values of red-state Americans.

PHYLLIS SCHLAFLY

THE PARALLELS BETWEEN
1726–1740 AND 1976–1990

As we have seen before, history is like a spiral and there are distinct similarities between periods of time. There are, of course, always new characteristics added to the complexities of society, creating the spiral effect. We must always keep these new layers of complexity in mind when we compare one time period to another. So we have democracy and protest demonstrations in our time, whereas the early eighteenth century did not. Our culture today has always been able to enjoy affordable books, whereas book publishing was still a young business at the beginning of the Enlightenment. Although there are great parallels, the great differences must be remembered as well.

All of nature has a rhythm—patterns that can be observed and understood. The social transformation pattern is no different. Events may swing toward a progressive outlook and then back when a regressive conservative era attempts to undo the progress. In examining the similarities between 1726–1740 and 1976–1990, we first need to examine in-depth the previous social transformation. From 1726–1740, a strong conservative backlash fought all the secular change of the previous decades, battling back with fundamentalist religion and regressive policies. This slowed the pace of the Enlightenment and influenced an entire generation that came of age during the backlash.

Progress was replaced by regress as the first fundamentalist, fire-and-brimstone preachers appeared. Huge revival meetings were held, and many swooned and talked in tongues, confirming to all present the truth of the preacher and his sermon. The point was to terrify all who would listen on the horrors of hell awaiting the unrepentant sinner. In his most well-known sermon, "Sinners in the Hands of an Angry God," Jonathan Edwards told rapt crowds that sinners were but loathsome spiders held aloft by a mere thread over the fire and brimstone.

As their popularity soared, revivals start attracting several thousand people in some towns, truly a large number for colonial America. In the end, however, sectarian rivalry and competition between the preachers themselves put the movement into a slow decline, especially when the revivalists started accusing each other of being agents of Satan.

Politically, conservative reactionaries had near-absolute power between 1725 and 1740. In France, the treasury was looted repeatedly at this time by corrupt royal officials, blowing a hole in the national budget. Meanwhile, the aristocracy in France and England took up the call for liberty and arranged to have more of it—*but just for themselves.* And after winning huge tax breaks by adopting some of the radical language of the philosophers, the nobility end up paying *nothing* in taxes, beginning a decades-long financial squeeze on the small middle class.

Throughout the entire setback, however, the scientists and writers of the Enlightenment kept exploring and fleshing out the new

*Fig. 9.1. Renowned preacher of the eighteenth century
George Whitefield ministers to this revival crowd in the 1730s.*

mechanistic worldview of the day. Seminal works were written and new theories proposed that had enormous influence later on, such as David Hume's *A Treatise of Human Nature* in 1739, in which he promotes skepticism as a philosophy and thoroughly thrashes medieval scholasticism. Such a shocking notion earned Hume a healthy dose of backlash, and he laments in the treatise: "I have expos'd myself to the enmity of all metaphysicians, logicians, mathematicians, and even theologians; and can I wonder at the insults I must suffer? . . . Can I be sure that in leaving all establish'd opinions I am following the truth?"[1]

Despite the conservative backlash stage, the new worldview thus kept developing, storing up ideas that would soon be deployed to thoroughly undermine the entire establishment of the time, the *ancien regime*. All of this was similar to our own recent backlash phase—to an uncanny degree. Just as the early enlightenment of 1700 to 1725 was followed by a severe fundamentalist religious backlash from 1726 to 1740, the challenge of the 1960s and 1970s gave rise to a similar counterreaction from 1976 to 1990. *Matching up of the two distinct backlash periods tells us exactly where we are today in the social transformation pattern.* Keep in mind that, because this is an organic process, the edges are flexible and there will probably be a continuation of backlash or certain elements of backlash going into the future.

THE FUNDAMENTALIST RISE

There have been many conservative backlashes in history, yet these two particular counterreactions are clear responses to the paradigm shift entering the mainstream. As in the eighteenth-century backlash, fundamentalist religious revivals are the major trend in the years between 1976 and 1990. Both fundamentalist trends succeeded in temporarily holding back even more secular change, displaying a striking number of parallels despite the centuries between them. These similarities include not only the success of the two revivals but also their effect upon the politics of the time. In a generational braking process, the intense new sects swayed large numbers of young people. The 1980s generation X, dubbed the nihilistic me generation, did turn out more conservative, becoming today's most conservative part of the population (people born between 1962 and 1982). Nevertheless, there are many progressives among generation X, and they basically think with the new paradigm, as do their boomer forebears.

Even the way the fundamentalist movement began to break down was similar. In the eighteenth-century backlash, the revival preachers had begun to call each other Satan, which repeated itself in the mid-1980s in the Jimmy Swaggart and Jim Bakker TV preacher scandals. Just as the revivalists of the 1720s and 1730s began to self-destruct, major fundamentalist preachers in the 1980s fell to the temptations of money, power, and sex.

Although fundamentalism remains a powerful political force

today, the TV preacher scandals of the 1980s—by showing that such a faith was no guarantee against the failings and materialism of human nature—slowed the rapid growth of the backward-looking movement.

THE REGRESSIVE REAGAN REVOLUTION

Following the rapid paradigm change of the 1960s and 1970s, as in the eighteenth-century Enlightenment, a powerful political counterreaction swept the world as conservatives and fundamentalists joined to roll back the many accomplishments of the progressives and liberals. It was the rise of the New Right.

The *Washington Post,* the *New York Times, Time* magazine, *Salon,* and the *New York Observer* all identify one man as being crucial in the rise of the New Right: Richard Mellon Scaife. Scaife, a recluse billionaire, funded Nixon, Reagan, and most of the new conservative advocacy groups such as the Heritage Foundation, American Enterprise Institute, Hoover Institution, Cato Institute, and, beyond, to Newt Gingrich's GOPAC, which resulted in the Republican takeover of the House in 1994. Then, in the mid-1990s, Scaife funded the *American Spectator* magazine and put $2.4 million into the Arkansas Project to investigate and whip up a case against Democratic President Bill Clinton.

Scaife's efforts, and those of other conservatives such as California's Howard Jarvis, began to pay off in 1978. In that year, the election of the regressive Margaret Thatcher, as well as the passing of Proposition 13 in California, signaled that the conservative backlash phase had begun. Ronald Reagan's election to the U.S. presidency in 1980 was the culmination of Richard Mellon Scaife's drive to bring about a conservative counterreaction in the United States. Just to be sure no one in the South could mistake the difference between him and President Carter, Ronald Reagan kicked off his 1980 campaign for president in Philadelphia, Mississippi, where the three civil rights workers had been murdered in 1964. Using the code words *states rights* as a bulwark against the increasing effect of affirmative action laws, and with the help of money from Republican millionaires like Scaife, by election day Reagan was able to woo white Southern voters away from the Democratic president from Georgia.

On April 30, 1984, when a student at Shanghai's University of Fudan asked Reagan which life experiences best prepared him for being president, he replied: "You'd be surprised how much being a good actor pays off." Just as Reagan knew how to play a cowboy, he could also play the role of president, helping him to mouth lines that he knew to be mere rhetoric or outright lies and make them sound appealing.

Liberals were trashed regularly in the 1980s—and with good reason. Mechanistic liberalism had lost its humanity in large part. In particular, after the welfare system was established, no innovative approaches to get people off the dole were tried. Although most were on welfare for only short periods, people were given incentives to stay unmarried and not find a job. The system sucked them in, and many stayed permanently. Reagan told untrue stories about welfare queens driving Cadillacs and selling drugs, which struck a chord with "Reagan Democrats." These voters also felt good when Reagan bashed the antiwar liberals and progressives, portraying them as negative and anti-American. Reagan pounded home a strong patriotic message, and the middle was swayed by the feel-good politics, after so many years of Vietnam and antiwar sentiment. The administration took a hard line toward the tottering Soviets and pushed for a huge buildup of weapons and deployment of untested missile defense systems.

Nuclear war fears surged, however, when members of the administration tried to explain it was possible to survive a nuclear attack by digging a hole in the ground and covering yourself with a door. One million people demonstrated and pleaded for nuclear disarmament in New York City in 1982.

KETCHUP BECOMES A VEGETABLE

Although always mindful to take his daily nap, Reagan nevertheless was able to dismantle what he saw as the country's biggest enemy. No, not the Communists; the U.S. government. This he did with astonishing efficiency, cutting needed programs in any way he could. David Stockman later reported that the general strategy of the ultra-conservatives was to "starve the beast," to create tremendous deficits on purpose through huge tax breaks to corporations and the wealthy. Then the cry of "no money" could be raised during budget negotiations on everything from Medicare to education—a strategy that had continued to erode the health and welfare of the country. Rather than cut what is really needed, the burden is placed on the ones least able and least deserving of the cut. For example, to cut the school lunch program, Reagan's minions even had ketchup reclassified as a vegetable, just to save money on vegetable portions.

Nearly all the new tax cuts went to corporations and the rich, who were supposed to invest it and create new jobs. The top individual income tax bracket was lowered from 70 percent, which was certainly too high, but Reagan brought it down all the way to 28 percent. Most of the untaxed cash ended up in yachts and luxury condos. The deficit ballooned, and even more red ink was spilled when $3 trillion in military spending was handed out to supposedly defeat the "evil empire" of the Soviet Union. Although mortgaging the future this way does help the economy in the short term—and helped buy Reagan the 1984 election—we are still paying the bill today in debt service that now totals over $200 billion a year.

It was all very similar to the economic shenanigans and fiascos of the conservative politics of the Enlightenment period from the 1720s until the 1770s. Favorites were allowed to loot the treasury in both eras—while the middle and lower classes were stuck with the bill. To Reagan, government was evil, but the free market was infallible—and so all he did was justified in his mind.

While the tax cuts helped the rich and the upper middle class, a Reagan increase in the payroll tax hit the working class. The payroll tax is probably the most regressive tax that we have because it removes dollars from a family's income before they get to invest or spend those dollars—suppressing the economy. It also makes hiring or creating new jobs more expensive for privately owned businesses that do not have corporate loopholes and to some extent greatly hampers even corporations' ability to hire more workers. With the amount of money that the employer must pay for each employee on payroll, the ability to hire entry-level jobs was reduced. For every midlevel to high-level executive, the payroll deduction eliminated funds that could create at least one and often several entry-level positions.

Midlevel management became overworked and undersupported, creating a drag on the success of the entire enterprise. We can increase our productivity, stimulate the economy, and create new jobs if we find an alternative to the regressive payroll tax that continues to steal away money and divert it into tax breaks for the wealthy; subsidies for big oil, the nuclear industries, and agribusiness; stock market capital gains; and funding for the ongoing wars.

Reagan also tried to destroy the union movement by throwing out the air traffic controllers and fighting labor wherever he could. Even worse, the jobs Reaganomics did create were lower paying work, as manufacturing jobs begin to hemorrhage to Mexico and other countries. In 1973, median pay for working men was more than $10 an hour, but by 1987 it had fallen to $8.87. Where an average worker without an advanced degree had an annual income in the 1970s of $24,000, by the end of the second Reagan term, that same person was only bringing in $18,000 a year. The superrich on the other hand, did quite well. In 1970, the top 1 percent of the population owned 20 percent of the nation's wealth. By 1990, that figure had doubled to 40 percent. Remember that at the height of the robber barons, at least 45 percent was owned by 1 percent, a figure that coincides with today when 1 percent of the wealthiest Americans own 42 percent of the country's wealth.

The spending cuts in federal programs were truly monumental and have put the nation into disrepair ever since. From 1980 to 1990, transportation and infrastructure were down 68 percent, education and training slashed 60 percent, and law enforcement and government expenditures cut 58 percent. Meanwhile, $909 billion in national debt soared to $2.6 trillion by the time Reagan left office, most of it for the arms buildup. In hindsight, the spending was unnecessary, as the Soviet

Union would have folded anyhow. The "borrow and spend, starve the beast" policies did, however, nearly bankrupt the United States, starting our country down the slippery financial slope it finds itself on today. Under Reaganomics, deficits exceeding $200 billion would go on "as far as the eye could see," as White House numbers man David Stockman put it. And they did.

THE SCANDALS PILE UP

With Reagan's election, the new laissez-faire economics so loosened banking industry regulation and enforcement that hundreds of billions of dollars were quickly fleeced from depositors in the notorious savings and loan scandal. The S&L scandal eventually cost $400 billion[2] and was actually much more with interest. Reagan allowed total corruption, first letting the S&L scandal go uncontested and then, when forced, setting up a system that sold off the foreclosed assets at bargain basement prices to insiders. New York governor Mario Cuomo at the 1984 Democratic convention called the S&L bailout "the biggest bank heist in the history of the world."

The scandals in the Reagan administration actually began *before* election day, as a secret deal was struck with Iranian leader Ayatollah Khomeni to release the fifty-two American hostages in return for arms sales. At the very moment Reagan was inaugurated, at noon, the hostages were released to carry out the agreement. Arms sales then continued for years, with the money being transferred to illegal accounts established to help the contras in Nicaragua, as well as right-wing dictatorships in El Salvador, Honduras, and Guatemala.

Reagan had a whole secret government in the White House basement, run by Vice President George H. W. Bush and Oliver North. When the "arms sales for hostages" deal was exposed in November 1986, Reagan went on television and denied it. He took it back a week later, but still denied the sales were for hostages. This, too, was finally retracted in March 1987 when the White House finally confessed. An ABC News/Washington Post poll at the time found that 62 percent of the country thought Reagan had lied about Iran-Contra. Thanks to an investigation led by then freshman Senator John Kerry, convictions were finally made, but the biggest perpetrators, including Weinberger, Reagan, and George Bush Sr. (who was later elected president in 1989 and served four years), either got off scot-free or were given presidential pardons.

In other foreign policy debacles: Reagan resisted all efforts to end South African apartheid and encourage free elections, truly a reprehensible move. Reagan can be held responsible for the deaths of tens of thousands of innocents in El Salvador and the other Central American conflicts. Eighteen thousand died when Reagan gave the green light for the Israeli invasion of Lebanon in 1982 and even sent U.S. "peacekeepers" in to help Israeli Prime Minister Begin maintain his hold on the occupied country. In 1983, 241 U.S. Marines died when a massive truck bomb went off in Beirut, the biggest single-day loss of soldiers since the TET offensive.

Reagan also supplied weapons and funding to the Islamic rebels in Afghanistan. As part of this effort, the CIA enlisted Osama bin Laden to help them fight the Russians. Thanks to the United States, the radical Islamists, known to hate all modern cultures, were trained in the most sophisticated techniques of modern espionage and sabotage. The Russians were sent packing, but soon thereafter, Bin Laden and the Islamist radicals—in a massive blowback—turned their full attention to the United States, Israel, and Saudi Arabia. Bin Laden and al-Qaeda were Reagan's and Bush's Frankenstein—a monster of their own making. This would lay the foundation for the terrorist attacks on the World Trade Center in New York on September 11, 2001.

Another monster of the day created by Reagan and Bush Sr. was Saddam Hussein. Urged to fight the Iranian mullahs, Reagan and Bush Sr. armed Saddam—*even though they were secretly arming Iran at the same time.*[3]

The arming of Saddam Hussein is a fact that must not be forgotten. Happiest of all were Bush and Reagan's defense contractor friends, who supplied both Iran and Iraq and then gave big checks to conservative reelection campaigns. When Saddam gassed over ten thousand Kurds in 1988, killing up to five thousand, Defense Secretary Rumsfeld personally went to Baghdad to receive the results of this rare field test of the American weapon. Rumsfeld was reportedly pleased about weapon effectiveness when Saddam informed him that many more died than had been predicted.[4] In 1990, President Bush gave Saddam the green light to invade Kuwait, telling him America would not interfere in territorial disputes between Arab nations. In 1999, an article in the *Christian Science Monitor* reported the following:

> Eight days before his Aug. 2, 1990, invasion of Kuwait, Saddam Hussein met with April Glaspie, then America's ambassador to Iraq. It was the last high-level contact between the two countries before Iraq went to war. From a translation of Iraq's transcript of the meeting, released that September, press and pundits concluded that Ms. Glaspie had (in effect) given Saddam a green light to invade. "We have no opinion on your Arab-Arab conflicts," the transcript reports Glaspie saying, "such as your dispute with Kuwait. Secretary [of State James] Baker has directed me to emphasize the instruction . . . that Kuwait is not associated with America."[5]

Bush then reversed himself, and the Gulf War was the result.

Reagan always supported dictators like Duvalier in Haiti and Ferdinand Marcos in the Philippines, who was an especially close friend of the Reagans for decades. Thousands died and were oppressed because of the friendly cowboy's love of dictatorship in those two countries alone. Reagan continued to prop up Marcos until the People Power Revolution in 1986 forced the horror of the Marcoses out—but not before they stole at least $10 billion more from the Philippine treasury, which Reagan let them keep.

Setting the record for what was then the most scandalous administration in history, 138 Reagan administration officials were investigated, indicted, or convicted of misconduct and/or criminal activities. This hall of shame included such names as Elliott Abrams, Richard

Allen, Richard Beggs, Anne M. Gorsuch, Carlos Campbell, Duane R. Clarridge, Michael Deaver, John Fedders, Alan D. Fiers, Guy Flake, Clair George, Louis Glutfrida, Edwin Gray, Arthur Hayes, J. Lynn Helms, Max Hugel, Rita Lavelle, Robert C. McFarlane, Marjory Mecklenburg, Robert Nimmo, Oliver North, J. William Petro, John Poindexter, Thomas C. Reed, Emanuel Savas, Richard Secord, James Watt, E. Bob Wallach, Casper Weinberger, and Charles Wick. In his 1988 summation of the Reagan legacy, *Corruptions of Empire,* Alexander Cockburn revealed the true spirit of "Reaganism": "Reaganism is shorthand for a particular culture of consumption, a reverie of militarism, of violence redeemed; of a manic, corrupted and malevolent idealism. The priorities of this culture at the directly political level were simple enough: the transfer of income from the poor to the rich, the expansion of war production and an 'activist' foreign policy, traditional in many ways but as Noam Chomsky has said, 'at an extreme end of the spectrum: intervention, subversion, aggression, international terrorism and general gangsterism and lawlessness, the essential content of the Reagan Doctrine.'"[6]

AIDS AND WOMEN'S RIGHTS IGNORED

Many people also died when Reagan ignored the global and national AIDS crisis, resisting pleas for funding for years. Fundamentalist supporters such as Jerry Falwell told him the disease was divine retribution to homosexuals for sinning against nature. Although AIDS was discovered in 1981 and became widely known in 1983, Reagan never even mentioned the word until September 1985, when a reporter asked him about the need for more funding. Reagan replied that half a billion dollars has been allocated since he took office—yet that was a ludicrously small amount for a disease that threatened the entire planet with early death.

He continued to underfund the research whenever he could. Later, critics charged that because of this resistance, the discovery of protease inhibitor treatments that have helped patients stay alive after 1996

Fig. 9.2. In the mid-1980s, fundamentalist TV preacher scandals signaled the beginning of the end of the modern conservative backlash. Here Jimmy Swaggart is confessing his sins after being caught with a prostitute.

would have occurred earlier. According to Larry Kessler of the AIDS Action Committee, hundreds of thousands of lives could have been saved, but Reagan blocked funding.

Reagan also became the most antiwoman president in American history. He first nixed support for the Equal Rights Amendment (ERA) in the Republican 1980 platform. After the defeat of the ERA, Reagan began loading the Supreme Court with regressive conservatives like Antonin Scalia. The overriding impulse was to overturn *Roe v. Wade* and make abortion illegal again—to please the moral majority and the Christian Far Right.

Reagan made a mistake, however, when he nominated Robert Bork for the Supreme Court. Bork had such far-right views of the Constitution that he even negated the right to privacy. The partisan battle over Bork, which caused the GOP to lose the Senate, eventually became a tit-for-tat game between the Democrats and Republicans. It resulted later in another fight over Clarence Thomas, nominated for the Supreme Court and accused of sexual harassment by Anita Hill, and then the retaliation of the Clinton impeachment.

THE ENVIRONMENTAL
DESTRUCTION AGENCY

Hiding behind the guise of the friendly cowboy, throughout the 1980s Reagan also tried to undo the early work of the environmental protection movement. Famous for saying "Trees cause more pollution than automobiles do" and "If you've seen one tree you've seen them all," Reagan was now the fox in charge of the henhouse. He put known anti-environment advocates in charge of regulation and enforcement: James G. Watt as secretary of the Interior and Anne Gorusch as EPA director. Watt, the former director of the Mountain Legal Foundation, a corporate and mining land-use advocacy group, quickly became infamous.

The irrational backlash against environmental science and the new awareness was nothing short of astonishing. In articles like "Ours Is the Earth" in the *Saturday Evening Post,* Watt expressed his fundamentalist anthropocentric view, that the world is "merely a temporary way station on the road to eternal life. . . . The earth was put here by the Lord for His people to subdue and to use for profitable purposes on their way to the hereafter."[7] Watt proceeded to dismantle his part of the government and let industry return to laissez-faire controls, meaning no controls at all. Gorsuch also got to work, trying to gut the Clean Air Act. It took Congress two years to defeat her attempts to weaken pollution standards on nearly all manufacturers, from cars to furniture and electronics and so on. Meanwhile, corruption became so rampant at the Superfund Program that Director Rita Lavelle was thrown into jail for lying to Congress about the stalled cleanup.

Conservative advisors and technocrats constantly surrounded the president and gave him possible strategies in his quest to unravel the government and turn the country over to the miners and the corporations. Gorsuch proposed a 39 percent cut in the 1983 EPA budget, even though she knew the new laws on hazardous waste would theoretically double the enforcement workload. Despite the need to carry out new laws created from the Love Canal disaster, EPA enforcement became a joke, as violations filed by regional offices dropped 79 percent in the first year of the Reagan administration. Refusing to accept the science in front of his eyes, Reagan joked about acid rain, which had already killed much plant and animal life in northern areas. Time after time, proven environmental protection laws were turned into voluntary corporate programs that accomplished little or nothing.

Watt was busy as well. Without preparing environmental impact reports as required by law, Watt attempted to rewrite the environmental protection laws relating to strip mining—effectively making the regulations moot. He also proposed to more than double the amount of public timber cut, even though the forest service warned him of lower future yields and loss of old-growth forests. Watt tried repeatedly to open up wilderness areas to oil and gas development, despite the illogic of developing a wilderness area in the first place.

"Never has America seen two more intensely controversial and blatantly anti-environmental political appointees than Watt and Gorsuch," said Greg Wetstone of the Natural Resources Defense Council in 2004.[8] Gorsuch was forced to resign in 1983 when documents revealed major misconduct within the EPA. Watt, meanwhile, sold a billion tons of coal from federal lands in Wyoming to mining companies. When questioned about this clearly illegal action, he implied, on September 21, 1983, before the U.S. Chamber of Commerce, that his decision could not be criticized because his coal advisory panel included "a black . . . a woman, two Jews, and a cripple." In the ensuing uproar, Watt resigned his post eighteen days later.

The Reagan onslaught against the environment nevertheless went on, a mindless rejection of good environmental science for the sake of corporate profit and to help win the war against the new paradigm. Efforts to gut the Clean Air Act eventually reached a climax in 1987 when Reagan vetoed reauthorization. Congress fortunately overrode the veto.

Solar energy was, meanwhile, set back for at least a decade as Reagan cut off successful solar tax credit programs and ended many promising research programs. European companies ended up making money off the earlier wind and solar development, rather than American ones.

Who knows where we would be today if Reagan had not ended the push for solar in the United States?

THE END OF THE FREE PRESS

To help in his quest to move America and the world backward, in 1987 Reagan finally pushed through the end of the Fairness Doctrine in media publishing and broadcasting. It was the repeal of this policy that removed any sort of balance to the way news and especially opinion were offered to the public by the corporate owners of the mass media. One Congressman warned that with the end of the Fairness Doctrine, "Candidates would lose the right to reply, parties out of power would not be able to respond, radio stations could allow supporters of one candidate to dominate the news, and local and state ballot issues could no longer be covered." Another Congressman said, "I am concerned that . . . broadcasters could use the public airwaves as their bully pulpit. They could every day pound away at their point of view, with absolute, total disregard to the other point of view." Twice the Congress passed the Fairness Doctrine into law and twice Reagan vetoed it.[9]

The predictions about the demise of the Fairness Doctrine have, unfortunately, come to pass. Abandoning its responsibility to investigate stories that might harm the regressive Republican/corporatist cause, the country has never been the same as the mass media now regularly holds back newsworthy items and has been taken over by the conservatives. Even the *New York Times* and the *Washington Post* have succumbed, while Fox News and the *Washington Times* regularly omit key news and put outrageous slants on nearly all stories with gross impunity. It can all be traced back to the Reagan era.

All the while, mechanism ruled the world's institutions, blocking any deep change. The Reagan administration now allowed corporations to write their own regulations and create enormous tax loopholes for business. Education grew even worse; environmental protection and alternative energy were actively fought and a conservative corporate-media oligarchy grew stronger and even more ambitious. Meanwhile the World Bank and IMF spent trillions of dollars on macroeconomic, international development programs that often bankrupted the countries they were trying to help. Over the fifty years of this top-down funding, in which aid was given to governments rather than villagers, an estimated $100 billion was not accounted for, either lost in the system or stolen.

As for his supposed triumph over the evil empire, Ronald Reagan did not defeat communism, which was already on its last legs. Although he deserved credit for listening to protestors and Mikhail Gorbachev and agreeing to important disarmament talks, Reagan took on the Soviets by throwing the United States into massive debt for arms that would never be used. Fortunately for Reagan, the Soviet Union collapsed of its own impossibility shortly after his two terms. The corporate media, of course, gave him the credit—despite the fact it would have fallen no matter who had been president. To sum up, Ronald Reagan was a disaster for the world and for the progress of the new paradigm—the perfect embodiment of the meaning of the second phase of the social transformation pattern, the conservative backlash.

CHERNOBYL, BHOPAL, AND THE *EXXON VALDEZ*: THE MECHANISTIC DILEMMA DEEPENS

During the height of the conservative backlash phase, from 1976 to 1990, it seems as though progressive change was brought to a halt. Although there were substantial groups and centers for alternative reform, there was little traction for the new paradigm to deeply transform anything. On the other hand, mechanistic institutions reached a zenith of power, yet they also had nowhere to really go—except to further corrupt science for profit and increase their hold on the media.

The limits of mechanistic thinking were fully revealed in the 1986 Chernobyl nuclear reactor accident that terrified the world. Contaminating thousands of square miles and killing or maiming several thousand firefighters, emergency workers, and locals through high doses of radiation and early deaths—as well as a huge global toll—the inadequacy of the old worldview was never more apparent. Estimates vary from the International Atomic Energy Agency from 4,000 future cancers to numbers using dose-effect figures. Dose figures from the 2006 Torch Report estimate the world-wide collective dose of 600,000 sieverts will result in 30,000 to 60,000 excess cancer deaths, seven to fifteen times greater than those given in the 2005 figures from the IAEA. According to the official pronuclear International Commission on Radiation Protection (ICRP) report, a global death total of 50,000 to 70,000 would be closer to the truth. Radiation expert John Gofman, who believes the risk is six times worse than the ICRP numbers, puts the global toll at 317,000 to 475,000.[10]

Even today, the Chernobyl reactor is dangerous, covered by a make-shift housing that is crumbling around the old reactor core. The world will likely hear of Chernobyl once again when the housing collapses and rearranges the radioactive mess—it's not over yet.

Three years before Chernobyl, the world had been horrified to learn of the Union Carbide chemical disaster that killed at least 3,000, wounding 50,000, in Bhopal, India. A year after Chernobyl came the tragic *Exxon Valdez* oil spill of 11 million gallons, despoiling Prince William Sound in Alaska. Then there was Saddam Hussein's invasion of Kuwait and his setting of 500 simultaneous oil well fires in retreat—with pollution spreading thousands of miles.

Mechanistic science and technology, such as nuclear power, chemical plants, and fossil-fuel systems, had already been exposed as irrational hazards. Each disaster now increased the pressure for change and led to a burgeoning environmental movement, as well as resulting action such as the Superfund Program (1980) and the CFC reduction agreements (1987). Yet the conservatives and corporations fought each effort to the very end, worsening the dilemma, as lost time led to greater and greater catastrophes.

Nonetheless, the new environmental consciousness was powerful. One billion people in 140 nations took part in the twentieth anniversary of Earth Day in 1990—the single biggest demonstration in history.

Although Ronald Reagan and Margaret Thatcher ruled America and England and the media downplayed the Earth Day demonstration and omitted or marginalized other environmental efforts, there was much activity on the part of the environmental movement.

There was also great activity and many important discoveries on the part of new-paradigm scientists and thinkers. In fact, several seminal theories vital to the future were advanced during the backlash phase. Try as they might, regressive conservatives could not stop the mighty forward motion of the Organic Shift.

THE DEEP ECOLOGY MOVEMENT

> *I pledge allegiance . . .*
> *I pledge allegiance to the soil*
> *of Turtle Island*
> *and to the beings who thereon dwell*
> *one ecosystem*
> *in diversity*
> *under the sun*
> *With joyful interpenetration for all.*
>
> GARY SNYDER, "FOR ALL"

When ecology is looked at in a shallow way, as the conservation movement often did by seeing only the utilitarian value of a unique ecosystem, it loses sight of what Thoreau saw in nature, that all life is equally important and has an inherent value beyond the utilitarian. Rather than humanity being apart from nature, we must see how we are rather a part of nature—that we have an obligation to think not only of humanity but also of every other species on Earth. We must feel in our hearts for any life-form as we feel for our own human brethren. It's all rather Buddhist.

In 1982, Michael Soulé (a conservationist, biologist, and Buddhist), Gary Snyder, and Robert Aitken put on one of the first deep ecology conferences in Los Angeles. Arne Naess, Bill Devall (a sociologist, philosopher, and environmental activist), and the philosopher George Sessions were in attendance, along with several Buddhist scholars. It is this combination of spirituality and ecology that underlies the philosophy of the future. Both Paul Ehrlich and Frijof Capra see deep ecology as the necessary path for humanity. As Paul Ehrlich said, "The main hope for changing humanity's present course may lie . . . in the development of a world view drawn partly from ecological principles—in the so-called deep ecology movement."[11]

Fritjof Capra said, "Shallow ecology is anthropocentric, or human-centered. It views humans as above or outside of nature, as the source of all value, and ascribes only instrumental, or 'use,' value to nature. Deep ecology does not separate humans—or anything else—from the natural environment. It does see the world not as a collection of isolated objects but as a network of phenomena that are fundamentally interconnected and interdependent. Deep ecology recognizes the intrinsic value of all living beings and views human beings as just one particular strand in the web of life."[12]

Inherent in deep ecology is the obligation to protect nature when it is under siege. So Snyder and Naess and other deep ecologists like Sessions emphasized community involvement in social and political movements for change. In 1981, in Norway, Naess chained himself, along with several hundred other protestors and native Lapps, to demonstrate against the construction of an enormous hydroelectric dam. They were all arrested.

Gary Snyder also emphasized that science and society in general need to become more aware of the regional nature of things. He saw natural bioregional connections being more important than the artificial and arbitrary human-imposed borders of states and nations. Future social solutions lie in recognizing that regional and bioregional ties could be leveraged to create a better world, a world grounded in its region and the spiritual ecology of resident or ancient native cultures. We can learn a lot by learning how to think regionally, explained Snyder. "Bioregional awareness teaches us in *specific* ways. It is not enough to just 'love nature' or to want to 'be in harmony with Gaia.' Our relation to the natural world takes place in a *place,* and it must be grounded in information and experience."[13]

Besides starting a rural zendo near his home in the Sierra Nevada foothills of California, Snyder was one of the founders of the Yuba Watershed Institute, a bioregional nonprofit devoted to the study and support of the local Yuba River Watershed, involving itself in everything from topography and ecology to cultural history and social action. The social activism of the deep ecologists, their combination of a scientific spirituality and bioregional awareness, inspired the leading edge of the environmental movement, and the new paradigm in general.

THE MANY PATHS OF ORGANIC SCIENCE

Despite the power of the conservative backlash phase, new-paradigm science not only continued down its path of discovery, it now realized several fundamental truths about the world. In 1980, Ilya Prigogine's *From Being to Becoming* laid down the fundamental science behind chaos theory. By transferring the physics of thermodynamics to biology—in the concept of negative entropy—Prigogine made possible the quantification of life and evolution itself, a dramatic advance in science. Rather than the steady gradual evolution assumed by Darwin, the fossil record instead shows long periods of stability interrupted by sudden leaps to more complex organisms. Chaos theory and negative entropy explain this "punctuated equilibrium." Time after time, "staircase charts" of evolutionary or social transformation appear in the data, with all life and even human history displaying this clear pattern.

Chaos theory is also crucial to understanding the very dynamics of change itself. Chaos theory and negative entropy thus explained organizational change and development, subjects which will be crucial in the upcoming intensive and transformational phases as the planet is forced to quickly transition off fossil fuels. Such a great change requires that we fully comprehend how society and personal lifestyles can be transformed through community development and consciousness raising on a planetary scale. By showing us how change works, chaos theory helps us create a plan for the future, a common agenda and strategy. We will soon need all the new science we can get—after the paradigm flip.

Building on the theories of Prigogine, as well as those of James Lovelock and Lynn Margulis and Chilean biologists and philosophers Francisco Varela and Humberto Maturana, Eric Jantsch released *The Self-Organizing Universe,* a synthesis of the negative entropy theories behind the Gaia hypothesis and chaos theory, in that same year (1980). Showing how cultural and technological change also displayed the characteristic staircase chart of chaos theory, Jantsch posited an organic self-organizing universe, in contrast to the mechanistic clockwork universe. It is the very nature of life and evolution to self-organize, powered by the incredible force of negative entropy.

The Self-Organizing Universe and *From Being to Becoming* in many ways quantified the earlier concepts of Henri-Louis Bergson, J. C. Smuts, and Teilhard. It was Teilhard who first saw how reverse entropy could create future turns of the spiral of evolution, leading eventually to the rise of a new science as well as the awakened collective, noospheric awareness. Both Prigogine and Jantsch acknowledged Teilhard's contribution.

Although mention cannot be made of every important book and theory of this time, the following list covers some of the advances made during the conservative backlash phase:

- 1980 David Bohm, *Wholeness and the Implicate Order,* created holographic model of universe in which quantum mechanics leads directly to "implicate order," supporting the concept of the self-organizing universe.
- 1980 Marilyn Ferguson, *The Aquarian Conspiracy,* a synthesis of New Age thinking and science up until the late 1970s. Out of all thinkers, Teilhard de Chardin was found by Ferguson to be the most influential.
- 1981 Rupert Sheldrake, *A New Science of Life,* a new theory of how information is transmitted through "morphic resonance" in a collective manner. These morphogenetic fields help us understand

how life not only takes new forms but also how new ideas can be thought of simultaneously.

- 1981 Duane Elgin, *Voluntary Simplicity,* a seminal work on how one can be "inwardly rich" and "live a life that is outwardly simple." Great explanation of the changes individuals and society itself must make for civilization to survive in the near future. Voluntary simplicity will be the rage of the upcoming millennial generation.

- 1981 Ken Wilber, *Up from Eden,* a study of evolution from the Paleolithic and the cultural evolution of human history up to the present and beyond. Not understanding the almost certain chance of transcendence and transformation in paradigm shifts, however, Wilber reaches a pessimistic conclusion: that the human race will *not* make it.

- 1982 Fritjof Capra, *The Turning Point,* examines paradigm shifts and the differences between the two worldviews, one based on mechanical models, one on biological analogies. Predicted massive change in the future by showing how the Newtonian worldview, and the society based on it, is coming to an end.

- 1983 Peter Russell, *The Global Brain,* an easy-to-read explanation of the modern concept of the collective consciousness. Russell explained how the brain reached reflective awareness when the brain accumulated 10 billion neurons. He predicted the planet will acquire a collective mind when the population levels off at 10 billion people—with each person like a neuron in a global brain.

- 1985 Charlene Spretnak and Fritjof Capra, *Green Politics,* covers the rise and contributions, as well as some of the faults, of the Green Party movement. Based on the organic worldview, Greens won Parliamentary and local seats around the world.

- 1988 Thomas Berry, *The Dream of the Earth,* a powerfully written plea and guide on how to save ourselves from the fix we find ourselves in today. Covers the massive changes individuals and societies need to make for the revival of the mythic hope that has always driven civilization.

- 1989 Deepak Chopra, *Quantum Healing,* popularized the concept of the mind affecting the body, becoming a great influence in health and therapies of various kinds.

- 1992 Brian Swimme and Thomas Berry, *The Universe Story: From the Primordial Flaring Forth to the Ecozoic Era,* primarily the evolutionary part of the new story; this book retells evolution as a great epic, ending in the rise of today's new ecozoic era, in which humanity becomes ecologically aware.

We can see from this list that despite the conservative counterreaction, the new science charged forward in many different fields. All of these advances were significant indeed, but one researcher in particular discovered a whole new branch of science, one critical to finding the workable, sustainable future.

REGENERATIVE ECONOMICS, AGRICULTURE, AND THE ZONE METHOD

As discussed earlier, when the world runs out of cheap oil and gas in the early twenty-first century, the planet will be forced to scramble for alternative energy. This is a large part of the coming paradigm flip, as the world is forced by raw economics and supply to make the switch. Changing over to renewables is fortunately very doable, with wind power already competitive with oil—even according to the skewed mechanistic economics. Beyond the clean energy technology that directly produces electricity, however, we must make use of the world's most efficient solar power converter ever discovered: dirt, or more correctly, healthy soil growing plants, which takes solar energy and water and converts it into food, fiber, and biomass crops.

If the world can regenerate enough of the 70 percent of the world's farmlands that contain depleted or dead soil, we can till and compost our way to a workable future. We now know how to build innovative agricultural systems that require no chemicals or fossil fuel and produce large volumes of organic food—we just need the will and the way to teach that to hundreds of millions of farmers around the world. One person stands out in the drive to understand how nature and the growing of plants actually works: Robert Rodale.

In the 1980s, Rodale not only helped to make soil science and organic agriculture full courses of study at universities around the United States and the world, he also conducted his own research at the Rodale experimental farm, showing that organic fields of corn raised side by side with chemical-input corn would match the conventional yields. This was proven over fifteen years. At the same time, soil science took several leaps forward during the 1980s and 1990s, oftentimes with Rodale's help, as the role of many previously unstudied microorganisms and micronutrients were revealed.

These accomplishments, however, are not why I have singled him out, for Robert Rodale discovered something even more vital, a new vision even more encompassing. Goethe had said that nature holds open secrets, secrets that are not hidden; we merely need to study and intuit them and we shall understand them. Robert Rodale uncovered many of those open secrets in the 1980s. Going beyond mere organic agriculture, Rodale founded a new order of science: the regenerative sciences. These included not only a new, ecological economics and system of agriculture but also a regional economic development method. It was all based on regenerative philosophy.

Regenerative means the ability to restore, the opposite of degenerative, the ability to destroy. Whether something is regenerative or degenerative is the question we must ask ourselves in all our futures to come. Rodale knew his ideas deserved a test, and so he dreamed up the Cornucopia Project, designed to reinvigorate the Lehigh Valley region surrounding the Rodale Institute Experimental Farm in Pennsylvania. With the help of several others, he next devised the regenerative zone method, which involves the creation and implementation of a zone development plan. This was somewhat like the farm transition plan an organic farmer must file today with the government to receive certified organic status. Everything is planned out, with local resources and regenerative potentials identified and strategies put in place to reach and maintain chemical-free status.

During the 1980s and 1990s, universities everywhere established courses to study and research organic and regenerative agriculture. The Russians, in particular, became interested in regenerative farming. Unfortunately, on September 20, 1990, while teaching and inspecting Soviet farm operations, Robert Rodale was killed in a horrific road accident. Although regenerative agriculture and agro-forestry were already widely known, regenerative economics, the regenerative index, and the zone development method all, unfortunately, fell into disuse upon his death. Chapters 10, 11, and 12 more fully describe how Rodale's discovery of many key open secrets could be deployed to create a workable future.

While the conservative backlash was trying to turn back the clock, the new paradigm thus continued to establish powerful theories, including the founding of the regenerative sciences by Robert Rodale. The mechanistic conservatives and the mechanistic liberals were always looking backward and never got past the illusion of the world as a machine. The organic progressives were meanwhile always looking to the future and dealing with the actual realities—such as the problem of nuclear waste, the unsustainable nature of society's addiction to oil, and the outrageous pollution of agriculture and industry, to name just a few.

What happens next is perhaps the most interesting part of the whole paradigm shift: the intensive phase. For it is in the intensive that new forces arise and suddenly change the dynamics, bringing society to the tipping point, to the paradigm flip. It is in the intensive that deep change finally begins, where the anomalies of the mechanistic dilemma are directly exposed in the light of day and finally confronted by a much larger part of the general public. And it is in the intensive phase where the new worldview finally finds a strong public voice.

Vandana Shiva

10

THE INTENSIVE PHASE AND THE COMING PARADIGM FLIP

1991–2011

May you be cursed to live in interesting times.

CONFUCIUS

CHANGING THE EQUATION

Progressive movements have come before but have always been beaten back in conservative counterreactions. Why would this time be any different? Conservatives have turned back progressives in the 1920s, 1950s, 1980s, 1990s, and more recently during the early years of the new millennium. The difference this time would be that we are not in the earlier parts of the Organic Shift, we are approaching the paradigm flip of a new intensive phase. In the eighteenth century, the intensive phase changed the equation through a number of new factors, giving the new paradigm a far better chance of finally defeating the old power structure. The most likely scenario today finds us approaching a similar paradigm flip, created by a young activist generation, a wave

318

of new thinkers, and a new history, a new story of the world.

The mechanistic worldview can never solve the problems of its own making. This is not to say that the organic paradigm will definitely succeed in every way, that it will be instant utopia. Yet if the world works its way out of the mechanist/nationalist dilemma over the next forty to fifty years and shifts to a new form of global society based on peace, renewable energy, and regenerative science, the odds are far greater for success.

We have lived through only fifteen years of the present intensive phase, and the events so far seem to be living up to its name: the tied 2000 election, the World Trade Center attack and the dot-com crash of 2001, the continuing U.S. mortgage crisis that started with the housing crash of 2008, the corruption of Wall Street and the U.S. bank bailouts, the confirmation of the greenhouse effect, the quagmire in Iraq and the Abu Ghraib prison scandal, the ongoing war in Afghanistan, and now the European Union financial crisis—all expose the short-comings of our current society, its institutions, and the old mechanistic worldview. Yet these earth-shaking events are only the beginning.

At the same time, great advances have been made by the new paradigm in science and technology and in communicating their views.

In the general flow of events, conservatives will continue their backlash against the new worldview at all costs, even as they accelerate their drive to take over much of the world's remaining oil supply. Just as Louis XV and mad King George III fought all sorts of reform—and let their rich favorites loot the treasuries of France and England as they started one war after another for world domination—we may expect to see the current round of conservative power brokers to do the same.

THE CLINTON YEARS:
A REPRIEVE

Following the terms of Reagan and Bush Sr., Clinton ran and so did Ross Perot, harping on the size of the deficit until it became an issue. Perot and Bush together equaled 56 percent of the election day vote, splitting it 37 percent to 19 percent. Clinton was thus elected. He promptly began to fix the deficit hole, raising the top rate for the rich back to 36 percent and 39 percent for the superrich. A transitional figure, part progressive and part moderate liberal, Clinton started addressing many problems. Yet the mechanistic dilemma grew worse as the basic problems of corruption, environmental protection, and the subsidies to chemical farming and polluting industries, including nuclear power, continued.

Newt Gingrich and the conservatives in Congress roundly attacked Clinton when he raised the "gays in the army" issue as his first reform. This controversy and the attempt to have Hillary Clinton head up health-care reform whipped up such a conservative storm against President Clinton that in the 1994 election, Newt Gingrich and the Republicans won a stunning seventy-five additional seats in Congress. Both the backlash of 1726–1740 and that of 1980–1995 thus burned hot for fifteen years—and then slowly but steadily began a decline over the next twenty years. After fifteen years of success, as in the eighteenth century, the conservative backlash reversed, for it was not really a movement with any kind of coherent direction but simply a reaction against all the secular change of an enlightenment period. Once it succeeded, there was nowhere else new to go, so it simply went further down the "starve the beast" road—especially in the Bush II administration to come.

THE REGRESSIVE PRESIDENCY
OF GEORGE W. BUSH

The George W. Bush presidency became far more than a rerun of the Reagan disaster. It created a situation far more serious than the Great Depression because it eroded the very foundation of our democracy and financial infrastructure of our country. The media—emasculated by the repeal of the Fairness Doctrine—reveals very little about the true damage done by the complete corporate and Republican takeover of the U.S. government. Although Bush did not try to reclassify ketchup as a vegetable, he did try to hide the outsourcing crisis by attempting to recategorize burger flipping as a manufacturing job. His war on the environment was unprecedented, even worse than James Watt.

From A to Z, Bush was such a regressive conservative that progressives, liberals, and moderates started to miss Nixon. Just as the wealthiest 2 percent of France paid virtually no tax at all under the reign of Louis XV, under George W. Bush the wealthy paid so little tax after loopholes that some billionaires asked that their taxes *be raised*. Parallels between the two intensive phases are fascinating—and rather comprehensive—as the aristocracy/corporate power-grab develops. The middle class is squeezed, and then squeezed again and again. Today, thanks to tax deferment loopholes, the top corporations, especially those shipping jobs overseas, typically pay no tax at all. They even receive refunds while reporting record profits. Expect the squeezing of the middle class to soon reach even more outrageous levels as the intensive phase unfolds.

CORPORATE WORLD DOMINATION

Yet it is within the neocon hidden agenda that the real story of intensive phase politics lies. Coming out of the corporatist big-money crowd of cronies, the U.S. neocons have apparently implemented a plan of world domination of the global oil supply and major economic spheres. Even water and the very gene pool itself are under the threat of economic takeover by the multinationals in the form of privatization, perhaps the biggest damn lie of them all.

Genetic engineering and the patenting of seed crops is very ominous. Jeremy Smith, consumer advocate, writes:

The American Academy of Environmental Medicine (AAEM) urges doctors to prescribe non-GMO diets for all patients. They cite animal studies showing organ damage, gastrointestinal and immune system disorders, accelerated aging, and infertility. Human studies show how genetically modified (GM) food can leave material behind inside us, possibly causing long-term problems. Genes inserted into

GM soy, for example, can transfer into the DNA of bacteria living inside us, and that the toxic insecticide produced by GM corn was found in the blood of pregnant women and their unborn fetuses.[1]

The genetic tampering that mechanistic science allows—with our food supply and other life-forms—may be the deadliest mistake of all.

Russia, China, and Europe, of course, have their own strategies to dominate as much as they can of the fast-approaching global end game. It is this dangerous battle for the planet's remaining fossil fuel and resources that will drive international politics for decades. This coming war over old-paradigm resources is a major part of the mechanistic dilemma, an unnecessary and wasteful tangent taken by the neocons and their counterparts in Russia and China. It is all very reminiscent of the eighteenth- and nineteenth-century struggles between the European powers for colonies and raw materials.

Even though the war in Iraq was obviously about its massive oil fields, the corporate mass media does not report this truth, always portraying it as a fight for freedom or to find weapons of mass destruction—a war against terror. So the neocon wars to come will also be falsely justified as a battle for freedom and democracy, never mentioning the oil resources that just happen to be in countries like Iran, Uzbekistan, Tajikistan, and the Caspian Sea countries, nor the pipelines planned for Afghanistan and Pakistan. Television's incredible power to propagandize is living up to George Orwell's predictions in his book *1984*. The neocons and their cronies currently enjoy the benefit of the media. (Yet Orwell may not have the last word, as the historical pattern says all that changes with the paradigm flip.)

PARALLELS IN TIME: HOW TO TRANSCEND AN OLD WORLDVIEW

Another interesting parallel between the two intensive phases is enlightening in itself. Old paradigms delay for as long as possible any admission that their science might be incorrect and slanted. For example, it took until 1751 for the Catholic Church to admit it was wrong, that Earth does indeed move around the sun. Compare that long-delayed correction to August 25, 2004, when the Bush administration, the EPA director, and cabinet secretaries finally conceded that global warming is at least partially man-made (and thus a real danger)—something known since 1909.

Even though the mechanistic dilemma is reaching a zenith, so too is nationalist military and economic power. By the end of the intensive phase, however, and despite the censoring of the news by the corporate press, the bankruptcy of the corrupt old worldview will likely be known to all—and the approach to the paradigm flip will begin. For despite all its military and institutional might, the new paradigm has something the old worldview can never possess: the way out of the dilemma.

No one alive has ever seen anything like the paradigm flip that swept the West from 1755 to 1760. What would a twenty-first-century version be like?

As bad as the second Bush presidency was—with the upending of the fiscal applecart by tax cuts to the rich and the invasion of Iraq to primarily gain control of 115 billion barrels of oil—Bush and the neo-con wars will not likely be seen as the most significant event of this time. Nor will the fact that the world is running out of cheap oil. The most significant factor of the decade between 2005–2015 will be the appearance of powerful new trends, intensive phase trends. Only these new currents of change have the ability to transcend the present world-view, its technology, and its politics—and that will be far more significant than the Iraq War or 9/11.

By making their appearance right on schedule, these intensive trends are setting in motion the run-up to the paradigm flip and the comprehensive social revolution that follows. These are huge, visible Organic Shift trends, landing right in the middle of the old world and proclaiming a radical progressive message. As in the intensive phase of the Enlightenment, these new factors completely change the equation. If the Enlightenment pattern continues, social transformation, already at a rapid rate, would accelerate to a breathtaking pace.

In the intensive years of the Enlightenment (1741–1760), historians agreed that certain new factors emerged, factors that would eventually prove decisive in the overthrow of the old power structure and the development of scientific democracies. These include:

- A wave of transformational books and plays by the mechanistic *philosophes,* which exposed the lies and contradictions of the medieval world.
- The rise of an activist generation that only thought with the new mechanistic worldview and thus despised the old structure they were trapped in—without freedom, without true science. George Washington and Thomas Jefferson were members of this new generation.
- The older *philosophes* taught the new generation, guiding them in how to shape a different future, based on science, democracy, and the right to vote. A common agenda was reached, greatly increasing the pressure for change.
- New histories were written by Voltaire and the *philosophes,* creating a mechanistic story of the universe, human history, and the progress to come. This new story radicalized about one-quarter of the general public in the paradigm flip of the mechanistic shift, around the years 1755 to 1760, and the younger activists began to lead the nascent revolutions that would soon arrive.

During the intensive phase from 1741 to 1760, book after book, by philosopher after philosopher, used the new mechanistic science to show how the medieval worldview was a fraud and an intellectual dead end, exposing all its glaring faults and contradictions. Each book suggested new scientific solutions to the medieval dilemma, a decades-long onslaught of thought that finally undid the entire paradigm. Within a short time, the new philosophers eventually decided upon a common agenda of change, a plan of global reform that reinvented everything from agriculture to medicine, from spirituality to politics.

Then Voltaire delivered not only his new history of Europe—

debunking the divinity of royal birth—but also a new history of the world. The upper and middle classes were astonished by revelation after revelation from the new philosophers. All together, the *philosophes* had created a new story of humanity and what its potential really was.

It was radical. And it swept the West. Just before the year 1760, approximately a quarter of the general population passed through the paradigm flip. By understanding this new story of the past, the general public and a whole new activist generation became radicalized in the few short paradigm-flip years. No longer believing in the myths supporting the divine rights of the king, the priests, and the aristocracy, a large part of the population was suddenly getting mentally prepared to overthrow the whole structure in the American and French revolutions.

Now, in our own time, as the four stages of social transformation apparently repeat themselves, the end game of the Organic Shift—the Global Awakening—draws near. Following the early enlightenment phase of the 1960s and 1970s and the conservative backlash phase between 1976 and 1990, the world has now entered the intensive phase of the new enlightenment. It is here that the extraordinary becomes an everyday occurrence.

THE RAPIDLY GROWING ECOCRISIS

The first extraordinary occurrence we must discuss is the rapidly growing potential of environmental catastrophe. In the 1980s, it became apparent that the very future of civilization on Earth was at stake, as carbon emissions continued to rise and many studies predicted a catastrophic rise in global temperatures. By the late 1990s, far more sophisticated studies and computer models of the atmosphere confirmed all these fears. Just as frightening, as noted earlier, the planet has been growing rapidly darker from all the pollution, enough to affect crop yields. This is the madness of the old fossil-fuel economy. These new findings must redouble the global effort to stop and then reverse all CO_2 emissions. In addition, chemical farming actually wears the soil out and is polluting large tracts of our seas, causing massive dead zones. The early conservationists and deep ecologists—Thoreau, Adams, Ricketts, Leopold—were worried about losing various parts of nature to industrialization and sprawl. Today we are in danger of threatening the future of civilization and nature itself. What kind of future will our children inherit?

The 1999 UN Environmental Program study on global warming helped to make clear the urgency of this very real ecocrisis. The pre-industrial carbon dioxide level of Earth's atmosphere was around 240 to 270 parts per million (ppm). In the 1950s, CO_2 levels from the Industrial Revolution reached 315 ppm. As noted earlier, in 2011, we were at 388.92 ppm, and in 2012, we are at 395 ppm, thanks to *just 56 more years of fossil fuelishness.* According to the atmospheric calculations, once we reach 450 ppm the devastation from storms, drought, and rising seas to low-lying areas, especially Third World coastlines, will overwhelm the ability of the global economy to recover. Civilization and world trade will enter an irreversible spiral downward. In short, this study says the world has about five years to start introducing nonpolluting power sources and dramatically decrease emissions growth. Though there was the first slight decrease measured in 2010, we must act *now* to build and invest in alternative fuels. In addition, we should initiate a petroleum austerity and waste reduction campaign to reduce levels more quickly. The window of opportunity to save the planet from climate meltdown is closing fast.

As noted earlier, we're burning up oil at an ever-increasing rate. Industrialize China and the Third World with old fossil-fuel technology, and we would need *at least two more planets.* Another recent study showed that the world *was due to remain in a period of fifteen thousand years of stable, temperate climate.*[2] Global warming will undo this delicate balance. The bottom line? We must stop using fossil fuels within fifteen to twenty years to stabilize CO_2 and then bring it down to around 300 ppm, the point at which it becomes healthy again.

In November 2011, the International Energy Agency (IEA) released its *World Energy Outlook,* the most thorough analysis yet of world energy infrastructure. A UN Wire reporting on the analysis stated, "The last real chance to combat climate change could be lost forever within the next five years if countries continue to build power plants, factories and buildings dependent on fossil fuels," and went on to say "that the entire global budget—in which global emissions of carbon dioxide must be held to no more than 450 parts per million in order to stay below 2 degrees Celsius of warming—will be used up by 2017."[3]

James Hansen of NASA and many other scientists and climate experts say this target is too low and that the target should be below 350 ppm. (350 is the true maximum because otherwise the momentum of the current conditions will overwhelm the reduction . . . it will be too late to stop that momentum. Time is a critical factor, and so it is not just the figure of 350, it is also the time factor that must be taken into account.

> If humanity wishes to preserve a planet similar to that on which civilization developed and to which life on Earth is adapted, paleo-climate evidence and ongoing climate change suggest that CO_2 will need to be reduced from its current 395 ppm to *at most* 350 ppm. The largest uncertainty in the target arises from possible changes of non-CO_2 forcings. An initial minus 350 ppm CO_2 target may be achievable by phasing out coal use except where CO_2 is captured and adopting agricultural and forestry practices that sequester carbon. If the present overshoot of this target CO_2 is not brief, there is a possibility of seeding irreversible catastrophic effects.[4]

So the 450 ppm figure doesn't allow for the time needed to lower emissions enough to actually have a healing effect on the atmosphere; it doesn't take into account the momentum that climate change has or the unknown factors such as the amount of time it will take us to convert to renewable energy and fuels nor the fact that the corporations and the people in power who hold stock in the oil, nuclear, mining, timber, and war industries will obstruct and fight the changes needed to save the planet.

Bill McKibben, founder and head of 350.org, agrees and says we must lower the level to 350 ppm or below to avoid global crisis. He created this worldwide educational and activist campaign to implement a plan to reach this target and stop global climate change from reaching this point of no return. He travels the world with hordes of new environmental activists who have joined his effort. If, he says, we can bring it down to 350 ppm, the larger goal of reversing it to 300 ppm in two decades is reachable, and if we do not bring the level down to 350 ppm, it will be too late to reach safe levels (the 300 ppm level)—there just will not be enough time. "Accelerating arctic warming and other early climate impacts have led scientists to conclude that we are already above the safe zone at our current 392 ppm [395 ppm by the time this book went to press], and that unless we are able to rapidly return to *below 350* ppm this century, we risk reaching tipping points and irreversible impacts such as the melting of the Greenland ice sheet and major methane releases from increased permafrost melt."[5] Lester R. Brown of the Earth Policy Institute has written the most extensive and authoritative book on how to stop climate change. Entitled *Plan B 4.0: Mobilizing to Save Civilization*, the book is an invaluable guide to anyone wanting to learn how and what to do to effect change. Brown has made it available as a free download at the earth-policy.org website. He not only provides us with a clear and detailed plan, he has shown us how to be a star-quality change agent by giving his book over the Internet. McKibben and Brown are prime examples of how to be effective change agents for the Organic Shift.

It is unimportant that various analyses and studies may vary somewhat in what they report on climate change and levels of CO_2, and data will probably change by the time this book reaches print. We have no time to become distracted. What is important is to focus on what is agreed are bad ideas, such as building more coal plants and fracking, which will greatly *accelerate* pollution and CO_2 emissions as well as deadly pollution. What is important is that we act *now* to lower CO_2 emissions and avoid *any* so-called clean solution that is petroleum based.

The scientists know it, and most of the population now understands this; yet the regressive conservatives and their corporate contributors continue to run the world, and so the problem grows worse, not better. Of course, this situation is what will help drive the consciousness raising of the coming paradigm flip, so in a way that's good.

IMMINENT ICE AGE

A 2004 study, by a consortium including the National Oceanic and Atmospheric Administration (NOAA) and the U.S. Department of Energy (DOE), shows that the oceans—previously assumed to be able to absorb enough carbon dioxide to allow time to transition to alternative fuels—are themselves being destroyed through carbon overload. The CO_2, which creates a more acid ocean and a lower pH, actually retards shell growth, dissolving shells, disrupting growth and reproduction, and destroying the delicate balance in the environment necessary to sustain marine life. In addition, "ocean acidification" changes the way the planet's entire carbon cycle works, accelerating global weather changes. A 2010 *Scientific American* article reports on this and the impact on global warming and climate change. Johnathan Havenhand, Swedish ocean scientist, explains:

> The ocean's interaction of CO_2 mitigates some climate effects of the gas. The atmospheric CO_2 concentration is almost 390 parts per million (ppm), but it would be even higher if the oceans didn't soak up 30 million tons of the gas every day. The world's seas have absorbed roughly one third of all CO_2 released by human activities. This "sink" reduces global warming—but at the expense of acidifying the sea.
>
> Marine life has not experienced such a rapid shift in millions of years. And paleontology studies show that comparable changes in the past were linked to widespread loss of sea life. It appears that massive volcanic eruptions and methane releases around 250 million years ago may have as much as doubled atmospheric CO_2, leading to the largest mass extinction ever. More than 90 percent of all marine

species vanished. A completely different ocean persisted for four million to five million years, which contained relatively few species.

Ultimately, the solution to ocean acidification lies in a new energy economy. In light of recent lethal coal mine and offshore drilling explosions and the catastrophic Gulf of Mexico oil spill, the U.S. has more reason than ever to forge a safer energy strategy for the planet. Only a dramatic reduction in fossil fuel use can prevent further CO_2 emissions from contaminating the seas. An explicit plan to shift from finite, dangerous energy sources to renewable, clean energy sources offers nations a more secure path forward. And it offers the planet, especially the oceans, a chance for a healthy future.[6]

Our seas are actually more vulnerable than the atmosphere.

In addition, new scientific proof indicates that the great ocean currents are slowing down due to a decline in ocean salinity—a phenomenon predicted to trigger a new ice age for Europe and Northeast America. Almost all oceanographers, from Woods Hole to the University at Potsdam, now believe global warming will first trigger an ice age. A great ocean conveyor powers Earth's heat transfer mechanism, making Europe and Northeast America livable—*and global warming is about to shut it off.*

This slowly moving conveyor belt of water, one hundred times the volume of the Amazon, takes one thousand years to complete a circuit from the North Atlantic to the North Pacific, creeping along at about three inches per second. Running on the surface up toward England, the powerful current channels the warm waters of the Gulf Stream across the Atlantic toward the British Isles. That is why Britain is as warm as it is, despite being so far north. After the great conveyor passes the British Isles, it submerges to the bottom of the North Atlantic, thanks to the denser nature of saltier, colder water. Continuing along the bottom from South America and Antarctica, the great conveyor surfaces again in the Pacific before finally turning back for the Atlantic—picking up surface heat from the equator all along the way.

Powered only by salinity and the energy of the sun, the conveyor's

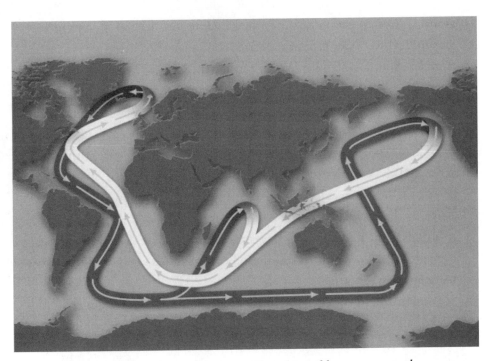

Fig. 10.1. The Great Ocean Conveyor. The white ribbon represents the warm surface current of the ocean, the darker ribbon is the deep water current. (Image courtesy Woods Hold Oceanographic Institution.)

plunge to the bottom of the North Atlantic *has already slowed 20 percent since the 1970s,* with global warming adding much freshwater from melting glaciers and ice sheets. In short, the warmer it gets, the more freshwater dilutes the ocean, and the slower the conveyor goes. As extra freshwater continues to dilute the salty ocean current, the conveyor reaches a point where the thermohaline circulation suddenly stops (*thermo* means "heat" and *halos* means "salt"). Winter temperatures would fall dramatically, an average of up to ten to fifteen degrees Fahrenheit.

Research shows a temporary conveyor shutdown was a contributing factor in the little ice age, which lasted in Europe and the North Atlantic from 1300 to 1850.[7] A Pentagon study in 2004 predicted the same possible scenario, saying England could very quickly revert to a climate like that of Siberia. President George W. Bush meanwhile had the global warming section of the annual EPA report deleted, a convenient way to deal with the problem. Nearer comes the day, therefore, when a complete shutdown could plunge Europe into a new ice age, giving the northeast United States the same climate as present-day central Canada. A shutoff could develop quickly, within three to ten years—and could last *five hundred to one thousand years.*

These studies are nothing less than a scientific declaration that we are facing the end of the world as we know it. Yet, instead of a ragged bearded fellow on a sidewalk holding a beat-up, hand-lettered sign proclaiming the end is near, these are NOAA/DOE and UN studies—this is Woods Hole and the world's oceanographers. The corporate media, dependent on car commercials and oil company ads, are not allowed to report on the seriousness of it all.

According to the enlightenment pattern, it is this ultimate crisis of science and democracy, along with 9/11 and the Iraq War, which will ignite the new generation of organic-paradigm youth in the intensive phase, turning them—for the rest of their lives—against the unfairness of corporatism, crony capitalism, and industrial pollution. In a very real way, the failures of the Bush administration in Iraq and in general are precisely what the shift to the new paradigm needed in the intensive phase, something that would truly energize the young. For just as the eighteenth-century founding fathers were enraged by the mistakes of King George III, a new generation will soon see the conservatives make clear the utter failure of the entire old worldview. At that point, the conservatives *and* the liberals will have both been completely discredited, leaving the political world wide open for the progressives and their organic worldview. New leadership might then bring the young millennial generation together with the progressive boomers, an unstoppable force that could change the world.

THE ACTIVIST MILLENNIAL GENERATION

In the eighteenth century, an activist generation—one who had been raised from birth with the mechanistic scientific paradigm—arose and collectively decided that they were going to change the world. Today, the mostly nihilistic generation X is waking up and becoming active and beginning to join older and younger generations working for change. The next generation, called the millennials by many demographers, should be similar to the activist generation of 250 years ago—to people like George Washington and Thomas Jefferson. According to the stages of social transformation, most children born during and after the 1980s will answer to a higher calling. They will be aspiring *homo noumenons*, always thinking from the outside in, respecting the unknowable and revering nature as a being.

In their seminal 1992 book *Generations,* William Strauss and Neil Howe predicted the return of activism and large street demonstrations upon the arrival of the millennials—who came of age during the Iraq antiwar movement and were an essential part of the 2004 presidential campaign as the "Dean generation." These children of the 1960s boomer generation think *only* with the organic analogy and are passionate environmentalists. In fact, the broken windows and street fighting during the battle in Seattle in 1999 and the 2001 G-8 riots in Genoa were clear indications that the millennials are indeed the predicted activist generation. One young protestor in Genoa, Carlo Guiliani, actually died for his beliefs. If so, these large protests over the Iraq War, the direction of globalization, and the increase in corporate power are just the beginning. As the millennials come of age, their protests will likely draw millions, dwarfing the tens and hundreds of thousands that demonstrate today.

The Occupy Wall Street Movement (OWS) and related movements created a global progressive movement to end corporate corruption of governments and the environment and restore the people's rights, freedoms, equality, and quality of life through peaceful occupation of public spaces. These movements feature peaceful assembly, demonstrations, marches, nonviolent civil disobedience, and general assemblies for "direct democracy." The motto of the OWS—we are the 99 percent— identified the members of the movement of the overwhelming majority. At the same time, the motto delineated the stark reality of gross inequality of wealth and power in the United States, created over the past few decades as a direct consequence of corruption and abuses of power. The statement of the first general assembly of OWS in Liberty Plaza, called the Declaration of the Occupation of New York City, reads:

> As one people, united, we acknowledge the reality: that the future of the human race requires the cooperation of its members; that our system must protect our rights, and upon corruption of that system, it is up to the individuals to protect their own rights, and those of their neighbors; that a democratic government derives its just power from the people, but corporations do not seek consent to extract wealth from the people and the Earth; and that no true democracy is attainable when the process is determined by economic power. We come to you at a time when corporations, which place profit over people, self-interest over justice, and oppression over equality, run our governments. We have peaceably assembled here, as is our right, to let these facts be known.[8]

In language reminiscent of the Declaration of Independence, the statement goes on to list examples of corruption and abuse by the powerful 1 percent, government, and corporations. The statement then ends with: "To the people of the world, We, the New York City General Assembly occupying Wall Street in Liberty Square, urge you to assert your power. Exercise your right to peaceably assemble; occupy public space; create a process to address the problems we face, and generate solutions accessible to everyone. To all communities that take action and form groups in the spirit of direct democracy, we offer support, documentation, and all of the resources at our disposal. Join us and make your voices heard!"[9]

Is this the first of massive global awakenings? The words "create a process to address problems" and "generate solutions accessible to everyone" state the mission and intent of a *collective* awareness and power to act. Is this the first of massive global awakenings? The assertion to exercise our rights of self-determination and the rights of all life on the planet and the offering to join with others in collective sharing and action on a global scale is unprecedented. If this were accomplished, it just might signal the beginning of the organic flip.

History, according to the new paradigm, has already appeared—specifically Thom Hartmann's *Unequal Protection* and John Stauber and Sheldon Rampton's *Toxic Sludge Is Good for You.* While Unequal Protection explains how the Supreme Court granted all the rights of an individual to public corporations and thus set the stage for the abuses that followed, Stauber and Rampton's book details how the lies of corporate propaganda are perpetrated on the public by phony institutes and phony studies touting the benefits of new products. It was new history like this that undid the aristocrats and tyrants of the eighteenth century, and it will be new history that unravels the validity of the corporate power structure.

Seminal books in this period are:

- 1992 Margaret Wheatley, *Leadership and the New Science,* brought new paradigm and human potential concepts into the field of organizational development in corporations. This began a very important trend to raise the consciousness and thinking processes of people at work and within corporations as a whole.
- 1995 Daniel Goleman, *Emotional Intelligence,* revealed that our emotional quotient (EQ) is as important as our intelligence quotient (IQ). Many intelligent people are otherwise immature due to low EQ. EQ can be more useful than IQ in determining successful leadership and in generally having to cope with the stress and challenges of modern life. Unlike IQ, EQ can be raised with personal work and discipline.
- 1995 Peter Senge, *Fifth Discipline Guide,* a seminal work that showed exactly how to bring systems theory and new-paradigm concepts to the business world, dramatically raising productivity by unlocking the human potential of organizations, teams, and individuals.

Taught by the new books of the new-paradigm authors and by transformational entertainment, such as *Lord of the Rings, Fahrenheit 9/11, Avatar,* and *The Manchurian Candidate,* as well as the billions of pages to be found on every subject on the Internet, the millennials in the intensive phase will slowly reach a common agenda with their boomer elders. Similar to the common agenda of the Enlightenment and the revolutions, which valued scientific democracies over theocratic monarchies, the millennial agenda would be to end mechanism and corporatism and move society as fast as possible to the Organic Shift, true democracy, reforms of all kinds, as well as regenerative science and development.

A NEW STORY FOR
THE MILLENNIAL PEACE MOVEMENT

The millennials, as a mass movement of *homo noumenon,* are more aware than any other previous generation. They will likely start a global peace movement that would dwarf all others. If the terrorist attacks become even more horrific than 9/11, the millennials will resist patriotic jingoism and grow even more determined to bring about world peace, to create the everlasting peace first spoken of by Immanuel Kant at the very beginning of the Organic Shift.

The millennials could learn and make use of the newly revealed story of the Peacemaker, the young Indian shaman who brought peace to the Iroquois people and formed the League of Five Nations around a thousand years ago.[10] Generations of horrific war had ravaged the Iroquois when a mere teenager received a vision of a great peace among all the nations. After twenty-five years of traveling to various Native nations and teaching peace, the young Peacemaker and his allies finally forged a lasting peace, which endured for hundreds of years. The special personal development method of the Peacemaker, his negotiation strategies, and his Condolence Ceremony, which was designed to relieve the mental trauma caused by war, could inspire the millennials—as they yearn to be peacemakers themselves.

The Native American peacemaking and negotiation process could be used as a new initiative to forge a real commitment to change and to live peacefully with former enemies. To forge a lasting peace, the Peacemaker called a grand council of all the tribes and clans in the Northeast. The Condolence Ceremony was a critical part of the whole process. In this ritual, the trauma caused by the horrors of war and hatred were healed, first the trauma of the eyes, of what was seen, then that of the ears, of what was heard, and then the throat, to cleanse it of the hateful things that were spoken. Through the Condolence Ceremony, in which your enemy took part and offered his condolences for your losses, hate was transformed into brotherhood; the participants now saw and heard the world anew. They could talk to their former enemies without animosity in their voices. Upon reaching the peace, the chiefs buried their weapons of war beneath a great white pine with an eagle's nest in it—the great tree of peace.

This collective act of apology, compassion, and healing was essential to creating a new dynamic. A League of Five Nations was then possible, made up of allies instead of eternal enemies—all thanks to the efforts of the Peacemaker. The method of the Peacemaker could add an entirely new level of sophistication to peace negotiations. This new approach, and the horror of terrorists finally using weapons of mass destruction, would at some point make the new global peace movement far stronger and more determined than all the pacifist movements of the past combined.

THE POWER OF THE INTERNET

The Internet is another large factor in the paradigm flip. Suddenly, the control of the news flow has passed from the corporate media to individuals and virtual communities. Through the Internet, people can read all the newspapers in the world, as well as radical and progressive blogs and nonprofit sites. Every time an Internet connection is made, millennials—and their elders—become aware of the true history and news that has always been hidden from them. The inequality of corporatism and crony capitalism are exposed to one and all, but especially to the young. According to the Pew Research Center, 29 percent of Americans regularly go online for their news. In 2010, 54 percent of all U.S. adults and 73 percent of adult Internet users went online for political and election news during the midterm elections.

This trend began in 2004 with the birth of Air America Radio across the United States. That same year former vice-president and Nobel Laureate Al Gore and Joel Hyatt cofounded Current TV, a youth/progressive cable news channel. Gore went on to create the Climate Reality Project, a center for climate activitism and climate news and blog, that now has five million members and supporters worldwide. Progressive bloggers, now reported upon daily by cable news, will increasingly affect the politics of the intensive phase. For example, the otherwise certain nomination of John Bolton as UN ambassador was recently stopped when Melody Townsel went on the DailyKOS website and revealed Bolton as unhinged and volatile by blogging about it for a couple of days. Unable to win Senate confirmation, he resigned in December 2006 when his recess appointment would have ended.

Regressive conservative blogs cannot match the progressive ones, as millions of progressives speak the truth about a world that no longer works. It's the perfect situation for the "shotgun approach," as millions of progressives come together over the Internet and discuss tens of thousands of problems and their potential solutions. The regressive blogs can merely repeat the talking points of the neocons and the corporatists, adding only a little bit to the echo chamber. In contrast, by shaming them into real reporting, the progressive blogs are starting to drive some of the news cycles of the mainstream media. Here the divide between the two worldviews is clear and distinct. The progressive blogs even like to say they live in the reality-based community, while the media, the government, and the corporations have all "drunk the Kool-Aid"—they believe their own propaganda even though it is clearly suicidal to do so. As uncensored reporting of the unfolding environmental, military, and fiscal disasters awaken one and all to the mechanistic dilemma and as more new-paradigm books make the multiple crises visible to every one, the United States—and the world—will approach the paradigm flip.

THE BEGINNING OF THE REGENERATION REVOLUTION: RENEWABLE ENERGY

Equal in scope to the Industrial Revolution, in the intensive phase we should see the beginning of the regeneration revolution. This transformation of industry and whole economies will start with many studies and pilot programs in the intensive phase and then move on to planet-wide efforts in the next phase. George Washington Carver's chemurgy solutions, which substitute plant oil for crude oil, will soar in use, especially as cheap oil runs out and prices of crude skyrocket. We must turn to his inventions and the work of many others since him, to take what is now a niche oil market and make plants the primary oil source for industry. Here, some sort of legislation and/or tax policy may be necessary to force the issue—but this will likely accelerate in the intensive, with broader programs going into effect.

Renewable energy and conservation will make amazing leaps forward, adding substantially to the energy mix, and the widespread use of solar photovoltaics in construction materials, along with minihydro, conservation, and other advances, have turned renewables from being a pipe dream to replacing the gas and oil pipelines of today. According to a new global wind map, based on the several thousand locations measured, wind power could produce 72 trillion watts of electric power.[11]

In 2011, the world wind energy production rose to 238,000 megawatts. China is the leader in wind energy, with 63,000 megawatts, with the United States coming in second at 47,000 megawatts. Denmark leads in Europe, with 25 percent of its electricity generated by wind as of 2012, and the European region as a whole currently has a wind power capacity of about 100,000 megawatts.[12]

Offshore wind resources are far more powerful than on land, but few major offshore wind projects exist. Tremendous wind power exists off shore on both the East and West coasts of the United States as well as the Great Lakes areas, South Dakota, Colorado, North Dakota, and Montana, and wind projects have not been built into most of the best areas. Little effort has been made to install wind power in regions that could provide the most power even though we have detailed studies and know exactly where the best wind resources are. Because wind projects do not yet exist yet in the best areas, transmission lines have not been built that would transmit the energy. Wind remains a vastly underdeveloped renewable power resource worldwide. Meanwhile, low-tech solar energy machines are now practical—especially for developing countries.

ELECTRIC CARS AND BIOFUELS

Right now an electric car is entirely feasible. On August 15, 2005, CNN News reported that Energy CS in Monrovia, California, created from a citizen-based research and development movement, would start converting Toyota Prius hybrid cars to plug-in hybrids. This means that you can plug them into an ordinary wall outlet, recharge the lithium ion batteries, and go all electric for commuting, equaling a miles per gallon rate of 230! The car, moreover, is charged in the middle of the night, using electricity only at off-peak hours. This off-peak electricity is essentially wasted every night between midnight and 6 a.m., as boilers take a good three days to power up, so they simply are left on all night. The voltage delivered to the grid is turned down to even out the loads. Since 85 percent of all passenger miles are for commuting, charging plug-in hybrids at night could replace electricity for gasoline for almost all commuting, and we would not need to add a single power plant or buy any additional fuel. That's right.

Since passenger cars produce two-thirds of all pollution and consume two-thirds of all gasoline, if nearly 85 percent of all passenger miles go electric, the United States would no longer require oil from Saudi Arabia and Venezuela. *The United States could lower its oil use by more than half if all passenger cars went plug-in.* Now that's conservation of energy!

Yet it took a citizen-based movement of tinkerers to shift the industry, as the car makers dragged their feet for over three decades. Toyota took them seriously and recognized the advances that have been accomplished. The Electric Auto Association geared up to spread the Prius conversion service all around the United States by the end of 2006 and since then has been a prime resource for electric car conversion. Energy CS provided the superior lithium batteries, which will last the lifetime of the car and should never need replacement. As we will see, this was the wave of the future, of new-paradigm groups that gather their resources and make change happen on their own. In comparison, in 2005, the $13.5 billion the Bush administration granted to the oil, coal, and nuclear industries for further research, development, and insurance for potential nuclear disasters was just more money down the unsustainable rathole.

With the new electric cars like the Tesla Model S, Ford Focus, Nissan Leaf, Chevy Volt, Le Car, and Tango (to name a few), plus using green hydrogen for planes and second-generation biofuels or biodiesel derived from grasses, woody materials, algae, and used cooking oil from restaurants for trucks (instead of less efficient soy or corn), the science is there for society to transition rapidly to a non-fossil-fuel future. The renewable fuel energy is there—if we are clever enough to learn how best to harvest it. Although it may have a long way to go, the regeneration revolution has clearly begun in the area of renewable energy and conservation.

REGIONAL PLANNING

Another area in the near future that will theoretically transform itself is regional planning. Regenerative regional planning pilots could be conducted in the most progressive regions in the United States, Europe, and Canada. The process of these pilot regenerative zones would be to identify not only local resources and potentials but also to eliminate pollution points by replacing old technology with organic substitutes. Then a regional regeneration plan is envisioned by the community, as first espoused by Robert Rodale in the 1980s. Professional regenerative consultants and firms would assist the region in transitioning, for example, from the spraying of massive amounts of herbicide and pesticide in urban parks and commercial landscaping to using the clean organic management approach. Other consultants could help the community switch from crude oil industrial products to plant oil or from fossil fuel and nuclear to renewable energy or from conventional farms to organic. The whole thrust would be to make the entire region as green as possible, with the multiple regenerative sources in a single zone all reinforcing one another and creating a powerful synergistic effect.

Pilots testing the regenerative zone development method of Robert Rodale would also prove the effectiveness of bottom-up development to the international development community, which would lead to the freeing up of billions of dollars in aid later on. Pilot regenerative zones in the Caribbean and Africa are the most likely places to deploy the new bottom-up approach of regenerative economics and development. In developing countries, the regenerative zone method would accomplish all of the following within a specified geographic region:

- Regenerative soil and crop scientists find the most profitable basket of crops, fish, and livestock that can be grown for each region and develop a live and web-based e-course to teach it to local growers. As each regional regeneration plan is implemented, depleted or even dead soil is made healthy and capable of producing abundant crops.
- Farmers are taught regenerative agriculture and/or aquaculture. Crops are certified organic and, while filling local food needs, can also be exported at greater profit. Regional Organic Growers Cooperatives can help guide, consult, market, process, and sell the food produced. Malnutrition and vitamin deficiencies are no longer a life-threatening problem for two billion as food production soars.
- The unemployed, especially youth and women, are taught green

businesses they can start and are given microcredit loans for initial capitalization.

- Energy production is switched from expensive imported oil to local renewables. Farmers and others with appropriate land are connected with wind-farm investors where feasible.
- Men, youth, and women are given instruction and e-courses on general health, parenting, nutrition, and AIDS. AIDS treatment is provided free of charge to keep people healthy enough to work and support their families.
- Community development and improvement are undertaken to build leadership as well as infrastructure, including natural sewage treatment and clean water delivery.
- The digital divide is closed with regional or village computer set-ups, web mobiles, or other delivery systems such as cybercafés. Program participants are taught HTML and basic computer skills.

Each source of regeneration reinforces the other. For example, organic farming improves local health and nutrition while boosting family farm incomes, all at the same time. Chemicals, usually used in an unsafe manner, are eliminated. This cleans up the local environment, especially the water, lowering human and wildlife contamination and disease. AIDS courses help prevent the further spread of the disease and treatment programs keep the labor pool up. Renewable energy production reduces carbon emissions for the whole planet, while within the zone, far less money is lost to external oil companies, meaning more money remains locally to circulate.

All of the regenerative sources work together to increase regional self-sufficiency in food, energy, business needs, etc. The synergy created will give development in each zone a life of its own, which will be able to continue with little or no future funding. The Sun will then become the primary funding mechanism for the region, so to speak, as regenerative agriculture turns the soil of the zone into the most efficient solar power converter we know.

THE PILOT REGENERATIVE ZONE
ON THE ISLAND OF DOMINICA

The tiny island nation of Dominica in the eastern Caribbean would be a perfect place to show regeneration in action. Dominica is the easiest place in the world to quickly show the effectiveness of the zone method. By switching hundreds of farmers to organic food production, it is estimated the island, which has seen banana exports plummet, can again make a profit raising bananas and from new items, like fish from inland fisheries. The potential of Dominica is so enormous, for both organic exports and from renewable energy transition, that lowering the poverty rate from 30 percent to the lower 20 percent within five years is an attainable goal. The little island has 365 rivers and is already 47 percent hydropowered, while the wind-farm potential on the east side of Dominica is far more than will ever be needed.

Within five years, Dominica could be on its way to tens of millions of dollars in new annual exports of food, while renewable energy use will rise from the current 47 percent to the mid-50 percent mark and be headed strongly upward as wind-farms are leased. A success on Dominica would serve as a model for switching to regenerative development all over the Caribbean and the world. The spread of AIDS will have been stopped cold, while economic self-sufficiency of all kinds will raise the money circulation rates by at least 15 percent. These would be the main goals of the pilot zone. Once the zone method proves itself in Dominica and other pilots in Africa, the international development community could turn to it in the years that follow, ending the degenerative top-down approaches of the International Monetary Fund and the World Bank.

THE NEED FOR PROGRESSIVE
GLOBAL EDUCATION

Education delivered over the web is another crucial part of the regeneration revolution and regenerative development. The crisis in education is a prime example of the mechanistic paradigm's inability to solve the disasters of its own making. The structure of our high school curriculum *was actually created in the 1820s.*[13] This old, Herbartian curriculum, based on the pedagogical theories of German philosopher Johann Friedrich Herbart (1776–1841), is the rot at the core of what is wrong with education, leading to the demotivation of students and teachers alike. A progressive curriculum is what is needed, one based on global studies, one that includes the regenerative sciences, the new scientific method of Goethe, philosophy, and so on. There needs to be progressive curriculum pilots and models created: the web can be used to bring it to the entire planet.

Imagine an on-site school day in a late intensive phase pilot of the new curriculum. Students start each day with a group circle and go over how they will cooperate with each other in various learning projects throughout the day or maybe discuss something bothering them or that is giving them difficulty. The circle figures out how to help them, maybe with mutual instruction, where an advanced student might teach the one needing assistance. Then it's on to class, where a typical day might mean learning about regenerative philosophy or regenerative economics, studying the East and West together in global studies, learning soil science or regenerative design, along with integrated mathematics strands, physical education, living skills, self-development, parenting, health, foreign languages, arts, and music.

There would be a holistic integration of the curriculum, with students learning one main subject in several linked classes. There would be more generalization and less specialization, especially in grammar, chemistry, and physics. The emphasis will not be to supply a nationalist military with rocket scientists or the chemical industry with chemists but to regenerate the environment, science, and culture of our planet. There are many other aspects to a new-paradigm education that cannot be covered here.

In contrast, conservative governments have meanwhile cut education spending back severely in the last two decades, and the West never did fund a true global education program. We saw how Kant himself decried the war budgets robbing the world of the education it needed to move forward. The same struggle continues two hundred years later. He predicted, "But as long as states apply all their resources to their vain and violent schemes of expansion, thus incessantly obstructing the slow and laborious efforts of their citizens to cultivate their minds . . . no progress in this direction can be expected."[14]

While inner city schools in the United States suffer the most, in the Third World 125 million children don't go to school at all—unless they go to Islamic fundamentalist schools where they often learn hatred and terrorism at an early age. After September 11, the world ignores the state of world education at its own peril. If there is no real change in this key area, there is no way out of the dilemma. Yet the mechanists, who long ago lost their ability to think outside the box, continue to tinker around the edges rather than go global and update the curriculum itself.

The push for global education could also see innovative new-paradigm pilot programs started, especially in international development and on the web, with progressive e-courses delivered over the Internet. Video clips and e-courses in different languages can teach an entire curriculum from one central global website. In the intensive, there will likely be a global high school on the web pilot, where students from around the world can learn global studies lessons together and discuss the new knowledge in online groups. This use of the web in education, along with the many other innovations of progressive education, would revolutionize learning in the world and make a reality the rapid change that is needed.

INTEGRATIVE MEDICINE

Integrative medicine, combining the best of old and new, will likely replace much of mechanistic or Western medicine, which, according to the National Institute of Medicine, kills up to one hundred thousand in U.S. hospitals every year. Ninety thousand more catch fatal infections from poor sanitary controls in health-care facilities.[15] And *another* hundred thousand die annually from preventable adverse reactions from misprescribed drugs *at home*.[16] Most of these deaths would be eliminated under integrative medicine, saving not only lives but also billions of dollars in lost productivity, government deficits, and insurance premiums. Integrative medicine, as we have previously discussed, reaches a new level of effectiveness in care. As alternative approaches reduce the use of drugs and surgery and their deadly side effects, overall health would improve, increasing life span as well as quality of life—another notch along with individual economic output.

In the intensive phase, medicine is profoundly affected by the new paradigm. Alternative practitioners, such as Dr. Andrew Weil, have started schools teaching doctors the new methods. Integrative medicine will no doubt become the norm in the decades to come. We have seen the bitter battle between the AMA and its sworn enemy, homeopathy. Yet today, alternative health already accounts for one-half of all medical office visits in the United States and is continuing to increase by leaps and bounds. As many alternative methods and herbs are less expensive and more accessible in developing countries, distributing a mix of basic health information and proven alternative health remedies will be a major component in the regeneration revolution. Integrative medical research must be funded to provide scientific proof of its effectiveness in different areas.

Regenerative medicine now exists. The term is being used for cloning bioidentical cells and tissues; whole new organs, such as a bladder, can be grown from a person's own cells. New research in brain plasticity is finding ways to regenerate lost abilities by making the brain reorganize and actually grow new connections. The new studies show that injured brains can learn, reversing the long-held belief that lost brain functions due to brain damage is irreversible. Nootropics and antiaging drugs and supplements have been found to slow the progress of aging, Alzheimer's, and other neurological diseases as well as the ability to improve cognition, memory, and performance for anyone wishing to enhance these functions. As revealed in Stephen Larsen's *The Healing Power of Neurofeedback: The Revolutionary LENS Technique for Restoring Optimal Brain Function,* the new neurofeedback treatment can now stimulate the brain to regenerate and make new connections after brain injury, stroke, and neurological disease. Genetic medicine is now developing brain stem cell treatments and bioidentical cloned implants. Implant surgery, though still in the early stages of development and with some risks, has now produced significant improvements and sometimes dramatic recoveries.

Integrative medicine is being adopted more and more into ordinary health insurance plans. Chiropractory and acupuncture are now widely supported, as are many other alternative traditions. Expensive and debilitating back surgery is now routinely avoided by going to the chiropractor. Alternative "escharotic" skin cancer treatments are saving the lives of thousands of people every year, even though the FDA tries to suppress it wherever it can. St. John's wort is not only more effective than Paxil as an antidepressant, many studies have also found there are virtually no side effects with the herbal remedy. Medical students are even being tested for bedside manner at some medical schools—a true turn away from the mechanistic and toward the organic.

Conservatives and corporations from time to time try to regulate and ban many vitamins and herbal remedies on a global scale, leading to the formation of grassroots organizations to stop them. Future struggles loom, as the pharmaceutical industry plans to take over or eliminate competing alternative products and therapies. Yet integrative medicine and alternative health will likely weather these storms and continue to be merged with medicine as a whole.

As the proof of all the organic solutions mount—and the mechanistic crises build—the tension between the two worldviews will be laid bare for all to see. The old paradigm will know its days are numbered, that its "cover story" is now totally blown. The much-vaunted spin machine will have a massive breakdown. The only question is: What will the corporatists do at that critical moment? Will corporations and conservatives go on an even bigger PR defensive campaign? Will they try to institute an Orwellian police state to maintain power? Or will they finally throw in the towel and really start following an ecological and regenerative philosophy, seeing that there's money to be made that way, too? It will probably be the former, while they pretend to implement the latter.

According to the historical pattern, it will be too little, too late.

THE PARADIGM FLIP ARRIVES

If the social transformation pattern keeps repeating itself in the exacting manner we have seen, with literally dozens of corresponding trends and time frames, and as the millennial generation comes of age, it is estimated the Organic Shift will reach the paradigm flip sometime after 2011. Catalyzed by the new story and the rapid revelations of the whole intensive phase, paradigm flips are very short—lasting just a few years. On one side of the paradigm flip, the world's decision makers will still be predominantly mechanistic, as will the population. Progressive and mass media pieces on the books of the new philosophers and scientists will quickly raise consciousness, however, with many more people becoming fully aware of the new thinking.

The power of the new story and the broader view of the past would then spread rapidly, invalidating the mechanistic worldview in the minds of wide swaths of the population, especially the young millennials. More members of the mass media would become aware of the paradigm shift and begin reporting it as such. Crony capitalism, and its reliance on fossil fuels, would be put on the defensive all over the world.

The voting power of the millenials has been increasing since 2004 and affecting the political landscape. In the 2004 U.S. presidential campaign, some fifteen million millennials were suddenly eligible to vote for the first time. That year, there was a huge shift to radical progressivism among those born in the 1980s, going two-to-one against the Iraq War and Bush. By the 2008 election, another fifteen million joined the original fifteen million—truly large numbers to consider. By 2012, out of a seventy-four million millennial population, sixty-four million mostly progressive millennials will be eligible to vote in the U.S. presidential election; fifteen million millennials turn eighteen every four-year presidential cycle.[17] In addition, there are eighty-two million boomers. Just the sheer size of this group will change the political equation in the United States, as the progressive boomers and the radical progressive millennials form a "generation sandwich," overwhelming the mostly conservative or apolitical generation X in the middle.

As in the Enlightenment, the millennials think solely with the new worldview, in this case understanding the ecological crisis, as well as the basic inequality of a world set up to always increase the rule of corporations and the rich. Assuming the millennials maintain their progressive leanings, or unless the right wing is able to use the fear of terrorism to declare martial law, censor the progressive press, or somehow steal or forgo national elections, it looks as though the burgeoning corporatism of today will be reversed in the not-too-distant future. It is all very similar to the struggle against the king and the aristocracy 250 years ago, with the corporations and superrich playing the role of the eighteenth-century aristocrats.

Intensive phases are the most crucial part of every paradigm shift. The paradigm flip at the end of the intensive starts the change, but it is only the beginning of deep change. That comes in the transformational phase, a reinvention of the world far beyond anything experienced in the 1960s. What people do during the first few decades of the twenty-first century either brings about the Organic Shift to the paradigm flip and the Global Awakening or puts it so far off in the future that it loses momentum—and never stops the CO_2 from going to 450 ppm and precipitating global collapse.

Within a very short time, between three to five years, the paradigm flip would be over, as more people and the young millennials awake. A far larger portion of the population will then be aware and thinking with the organic analogy. Assuming that we do "activate"—and that is by far the most likely scenario—the organic tipping point will be directly followed by the heavy lifting of the transformational phase. This phase has always been a battle to the end between the two paradigms, as the old worldview attempts to put off its inevitable defeat for as long as possible.

Yet this transformational phase will be like no other. This will be the first on a global level—a great transformation disseminated through books, movies, plays, musicals, television, radio, and the Internet. It will be a transformational phase that already possesses democracy and the ballot box. What would this be like? How far could we go? Could the skeptics be wrong? Could the paradigm flip, progressive politics, and the regeneration revolution really occur and shift our society's course? According to the historical pattern of the Enlightenment, the most likely scenario is therefore something quite unexpected by the world today: a great transformation moving the world out of the mechanistic dilemma and into the sustainable future. The metaphysics of regeneration and hope could then replace the metaphysics of despair.

11

AFTER THE PARADIGM FLIP

The Transformational Phase: 2012–2050

We shall not cease from exploration
And the end of all our exploring
Will be to arrive where we started
And know the place for the first time.

> T. S. ELLIOT, STANZA IV,
> "LITTLE GIDDING," FROM
> "FOUR QUARTETS"

TRANSCENDING
THE MECHANISTIC DILEMMA

To transcend the mechanistic dilemma, we only need to have the social and political equivalent of the American Revolution and equal the all-encompassing scope of the Industrial Revolution with new technology. That's all. But we can do it. It won't be easy, for the old power structure will want to maintain its rule, every bit as much as King George III wanted to keep his cherished American colonies. Skeptics will scoff at this notion, saying the corporate power structure is too strong in the age of multinationals. Progressives are too small a portion of the population, they will say, and traditional mechanistic science and medicine control all the major institutions of the world. But if our ancestors in the Enlightenment were able to defeat King George and the most powerful force in the world, so can we. Here's how.

In the Enlightenment of the eighteenth century, we saw how the paradigm flip of the social transformation was reached around 1755 to 1760, how the activist generation came of age and allied themselves with the older philosophers. About one-quarter to one-third of the European American population became new-paradigm thinkers during those paradigm-flip years, and a revolutionary atmosphere developed soon afterward. Fueled by the new history and a common agenda of liberty, equality, and fraternity, continued outrages committed by the monarchy slowly built up the courage of the young radicals. Within twenty to thirty years of the paradigm flip in the late 1750s, the American, French, and Industrial revolutions had reinvented the Western world.

What does this time tell us about our own near future? If we continue to duplicate the historical pattern of the Enlightenment in exact years, our own paradigm flip would probably arrive sometime after 2010, and a dramatic increase in the acceptance of the organic worldview would occur over just a few years. This does not mean the dates of 2011 to 2020 for the paradigm flip are hard and set; the flip could come later and the completion of the transformation will need more time, at least until 2050. But in terms of generational change, which is what the stages of social transformation primarily are, those are the most likely years for the flip.

As the mechanistic system bankrupts itself and breaks down in a minicollapse or total collapse, a vast reinvention of individuals, corporations, and society alike would sweep the globe over the following decades. A new history and common progressive agenda is essential to creating this future, as that is what builds the outrage among the millennial generation, while providing an organizing image for the future. The essential questions of who are we, how did we get here, and where are we going must be reanswered and delivered to the general population for there to be deep change.

To create a new sustainable civilization, many people would expand their roles according to the new organic/biological model and regenerative economics, all possessing a profound collective will to transform and reform the planet. The mechanistic dilemma, bit by bit, problem by problem, would then be eliminated or reduced by organic solutions.

Of course, switching to the new paradigm and changing to new technology and institutions will not instantly solve all our crises at once. There will still be polluted places to clean up, sunken nuclear subs, old nuke and toxic waste, and many pockets of poverty, but we would be well on the way to a regenerated global society.

Yet all this will happen *only if enough people become active.*

Fortunately, it appears that the millennial generation is the activist generation it was predicted to be—fitting the intensive phase timing. Here are the numbers.[1] By 2016, 60 million millennials will be eligible to vote in the U.S. election. Four years later, the number increases to 75 million. Combined with the 80-some million of the boomer generation plus an awakened generation X, they could form a progressive voting block that would give a newly progressive Democratic Party an electoral lock for decades to come. Thanks to the ability of the Internet to easily collect small contributions, the Democrats will be able to match the Republicans dollar for dollar, as they did in the 2008 election. Nearly every advantage the conservatives now enjoy would thus be reversed in the upcoming transformational phase.

Outside of a complete coup or perhaps a split of young and old progressives, the political will for change in the United States is going to theoretically be overwhelming, and the world will, of course, join in the drive for global regeneration and the end of crony capitalism. Would this change be violent like the American and French revolutions? Or on this turn of the spiral—given that we now have a real democracy—would it be nonviolent evolution? Hopefully, it shall be the latter, along the lines of Teilhard's collective awakening—a wonderful thing.

How would the new paradigm fully activate the young millennials in the transformational phase? Probably entertainment—transformational entertainment on a scale and in mediums we cannot begin to imagine. Virtual reality enlightenment and immersion on the Internet perhaps, combining consciousness raising with an engaging 3-D interactive format. Get to meet all the greatest thinkers, artists, and writers in the world. Just put on the goggles and go meet them in the virtual enlightenment salon—ask them anything. We will likely go far beyond the beginning films of this trend, such as *The Lord of the Rings, Fahrenheit 9/11,* and *Avatar* in the intensive phase. In the transformational phase, the combination of film or theater, music, visionary visuals, spirituality, and the new paradigm, itself, would enlighten one audience after another—leaving them profoundly changed and inspired.

Millennials could also theoretically become a more literate generation and read books, lots of books—each one charting out the new society. The conservative mechanists, as in the Enlightenment, will not be able to compete on the scientific level of the organic theorists. They will have to resort to old dogma, mount PR campaigns, and attack the data as bad science, as they always have. In the mid-eighteenth century, there were authors who spoke out against science and the deism of the mechanists, but they had little success, suffering from the same lack of creativity that Toynbee found in every dominant establishment he studied. As nature's open secrets are revealed and the organic solutions mount, expect the mechanistic rhetoric against them to also increase.

Today, the progressive organics are the creative minority. They are writing and producing the groundbreaking books, films, music, and art. Creativity and talent will win this battle, as bad films and books will not reach near the audience of great ones—and that's what the transformational phase will be all about. Great, profound masterpieces will transform a person, make him or her reflect collectively on the survival of this planet and the future of the children. It would not be a level playing field if the Organic Shift goes Hollywood.

The new paradigm will also presumably be a regular guest on an enormous and planet-wide progressive network of cable TV and Internet programming. Here regenerative economists and ecologists would detail—with great production value—exactly how the corporatism of the late twentieth and early twenty-first centuries compromised the biosphere and the entire global economy. They will then lay out all the regenerative solutions, telling it all as a good news story, day after day. This would be a great contrast to the corporate media and the news reports that are only 5 percent hard news—and then that 5 percent is primarily tailored to support polluting forms of technology or the old worldview. Future viewers of the news, especially millennial viewers, will always want the real reports over the typical celebrity trial/shark attack/sick baby/terror alert news we have today. The change in the media will perhaps be the biggest part of the deep changes to come in the transformational phase.

Again, the advantage would go to the organic paradigm.

THE REGENERATION REVOLUTION
BRINGS IN A NEW ERA

Just as the Industrial Revolution swept all before it, the most likely transformational phase scenario predicts that the regeneration revolution will reinvent technology, economics, and a host of other fields. Powered by regenerative solutions and the new economics, technological/social innovation would replace the failed macroeconomic machine models of today. It will be an economics that counts environmental protection along with cultural protection, community development, and self-sufficiency—all of it part of the fundamental equation. Think of it as the totally ecological and humanized version of all industry and all our institutions. What would that be like? If the enlightenment pattern continues, expect several major components to this great wave of change:

- *Regional Regenerative Planning*—Each local region would envision a regional regeneration zone plan, identifying the pollution points and a course of action to eliminate the pollution. The zone plan will also carry out the regenerative zone method in each region. Agribusiness is broken up and restored to regenerative family farms by a progressive president, like Teddy Roosevelt busting the big money trusts. Carbon emissions are turned around by renewable energy, clean vehicles, and plant oil solutions for industry. Offshore wind energy, new solar power breakthroughs, and subsidized home and office systems change the energy picture forever. The dreaded level of 450 ppm of atmospheric CO_2 is never reached.

- *Global Regeneration Plan for Developing Countries*—Using regenerative economics and Rodale's zone method, ten thousand zones transition developing countries to renewable energy and regenerative development. The worst of world poverty is first cut dramatically and then eliminated by the midcentury as depleted soil tilth is replaced by healthy, fertile ground. Renewable energy replaces unaffordable fossil fuel, along with alternative vehicles and the new plant-oil-based industry.

- *Government Regenerative Planning*—Zero-based regenerative audits of every government budget would eliminate wasteful mechanistic programs and replace them with far less expensive new-paradigm programs that work.

- *Corporate Sustainability Movement*—George Washington Carver's chemurgy or the new algal biofuel made from algae is applied on a global scale, transitioning industrial manufacturing from crude oil-based products to plant oil like the biofuel made from algae. Each corporation voluntarily undergoes a sustainability audit by environmental experts. Low audit numbers would quickly bring ruin to companies, as a much larger percentage of the population will buy from their competitors. Competition and cooperation will drive the regeneration revolution.

- *Institutional Sustainability Movement*—Corporations and organizations of all kinds will join the sustainable corporation movement. Medical corruption by big pharma will also be reformed, as drugs and surgery are replaced by less expensive alternative treatments wherever possible and truly independent testing and stricter standards of acceptable side effects and risks are mandated. Education, nonprofits, NGOs, and so on all become part of a new widespread drive for regeneration and then sustainability.[2]

- *Regenerative Agriculture Movement*—Organic agriculture becomes widespread, and a new category, certified regenerative, comes into being. This means that the product was produced in a regenerative agricultural system rather than a mere organic farm. As regenerative farming replaces chemical farming, yields increase around the world, agro-pollution is eliminated, and the ocean dead zones are restored. Cancer rates lower as pesticide and herbicide food residues plunge and food additives, trans fats, artificial sweeteners, and other toxic food additives, toxic pharmaceuticals, and household chemicals are banned. Old chemical companies could be sued out of business.

- *Individual Regenerative Lifestyle Changes*—It will be essential that the attempt be made to comprehensively educate every person on Earth in the following: (1) awareness of ecological connections and (2) awareness that as individuals we are the biggest polluters of all, not industry. If we really want to avoid destabilizing the environment through global warming, if we really want to end terrorism, if we really want to attain true peace in this world, we must go through our own personal change and growth. We must find out how to use and waste less in our personal lives. We must become active and use our skills to try and shift the paradigm. This time it's personal.

- *Integrative Medicine Replaces Traditional Surgery and Drugs*—The many millions killed by the crude drugs and surgery of mechanistic medicine will survive in the transformed future. In addition, health-care costs will lower as less expensive alternative methods are used, far fewer doctors are needed, and corporations are prohibited from ever operating health care treatment centers again. Pharmaceuticals are reregulated and made subject to market. Government approves alternative treatments. The American Medical Association is replaced by the Association for Integrative Medicine.

- *Holistic/Regenerative Education Established*—Textbook publishers would be forced to compete with new, progressive web schools teaching the new-paradigm education in conjunction with on-the-ground public and private facilities. These paperless schools will likely be based on global studies, teaching every subject through history, and will impart the new scientific method, the history of the Organic Shift, and global consciousness and the regenerative sciences at an early age. Graduates of the new system will far surpass today's students in terms of academic understanding and critical thinking skills and become the regenerative leaders and scientists of the future. They will be change agents and peacemakers of the first order.

The regeneration revolution will make the pilot projects of the intensive phase, regenerative regional planning, and the zone method the blueprint for a global sustainable future. Is it regen or degen? Is it regenerative or degenerative? Is a coal plant degenerative and an offshore wind farm regenerative? Implement the regenerative solution. It's pretty simple.

This is almost the same division in worldviews that Aldo Leopold found. Is the practice sustainable or nonsustainable? But with regen versus degen, the argument goes one step further and asks whether the practice not only sustains but actually restores that which has been lost and then perpetuates through future generations. Does it regenerate what has been destroyed or diminished? Nature allows a starfish to regrow a lost limb. A wind turbine reduces carbon emissions and thus regenerates the atmosphere. A recycling program helps clean up and restore the environment. Is it regen or degen? In essence, this is what Robert Rodale was asking us to think about whenever we do anything. Applying this philosophy to farming and forestry, we saw how Rodale developed regenerative agriculture, in which a farm is integrated with brush and wildlands near the farm, primarily to provide beneficial spiders and birds, the next step beyond organic agriculture.

With regenerative farming, there is typically no waste and, of course, no chemical runoff into the environment. Regenerative agronomists design whole systems *with brush and wildlands as part of the design.* The natural environment is restored, while farmers increase their income. All it takes is training and lots of planning. When a farmer converts from conventional chemical farming to organic and especially regenerative methods, a virtuous cycle is set in motion.[3] Chemical farming is a vicious cycle, in which more and more fertilizers and pesticides are needed to maintain yields, and the farmer, his family, and customers are all exposed to increasing amounts of cancer-causing and neurotoxic residues.

The virtuous cycle of regenerative farming is the opposite. It is regen rather than degen. By switching to compost from local sources for fertilizer and using natural methods to manage pests and weeds, the expense of buying the chemicals is eliminated, while nearby organic matter is reused before it becomes waste contaminating the environment. The farmer, his family, and neighbors are also not exposing themselves to toxic substances and will thus live healthier, happier lives. The same is true of the nearby communities who—eating the locally produced organic food—will also grow healthier. Local consumption of farm products meanwhile raises money circulation and strengthens the regional economy. Add renewable energy, and you have a system that requires no chemical inputs and no fossil fuels. The farm can cycle forever with no pollution, given a stable climate and proper regenerative management.

All this works incredibly well in the real world, as the Rodale Institute and many university studies have proved.[4] If a farm can be planned out beforehand and managed in a regenerative way, why not a whole region? In short, why can't regenerative philosophy be applied to economics and social policy? Then apply it to the whole country, to the whole world. So Rodale took what he had learned in agriculture and applied it to the dismal science of economics, yanking it out of the ivory tower and bringing it literally down to earth.

As noted earlier, this was a momentous step in the history of economics, for Rodale created the first holistic way to measure economic growth that is regenerative, a system that restores the health of a region economically, agriculturally, socially, and ecologically. By adding ecology and humanity to economics, Robert Rodale humanized economics, and in a very concrete manner. Like E. F. Schumacher, it's economics as if people mattered, nothing like today's corrupt and skewed mechanistic economics and the resulting top-down international development plans. The old economics—taking neither people nor the environment into account—is a major reason the world is in the mess it is today. By *reversing* the usual thinking through a system that *works with nature, rather than against it,* regenerative economics transcends the failed policies of the mechanistic dilemma.

In common with Gary Snyder and the deep ecologists, Rodale had a strong regional awareness. Rather than the national, macroeconomic view, economics should focus on the region. This regional, bottom-up approach would first examine why a specific depressed region has lost its economic and human vitality, and what areas need regeneration to restore lost money circulation and self-sufficiency, or even human health. Going beyond employment rates and income levels, Rodale measured local ecology, regional energy use, health, community strength—all items ignored and artificially separated from economics by the mechanists.

Instead of brooding over macroeconomic statistics for the country as a whole, regenerative economics makes a diagnosis of regional economic ills and then prescribes a plan to reach full health of that single area (just as Rodale would bring depleted soil in a specific field back to full health). Self-sufficiency in all areas except a specialized few is the general goal. The idea is to develop each region's regenerative sources in a single zone, unleashing multiple beneficial synergies. This will increase money circulation, as well as food and economic self-sufficiency.

Rodale created a regenerative index to track data in several key areas—essential to understanding the whole process as it unfolds in the real world. The regenerative index describes an ideal regenerated state, such as full employment (100 percent employment), compared to current levels of local employment. The same goes for health, energy, food, and so on. Although 100 percent levels can, of course, never be reached, an area that totals below 50 to 60 percent on the regenerative index is unhealthy, while 30 to 40 percent is poor, and under 30 percent is destitute, true for many regions in developing countries. Table 11.1 explains the measurements contained in the regenerative index. Having to import energy from out of the local region, for example, can lower self-sufficiency and money circulation, while local renewable energy production regenerates the environment and keeps the energy purchase local.

What you are looking at in table 11.1 is a whole new way to look at economics—from the bottom up rather than the top down. This new perspective focuses on regional health as a whole, self-sufficiency, money circulation, energy use, and many other factors. It applies to developed economies as well as those of the developing nations. And it is there that regeneration will have its greatest and most important use.

Over three billion young people around the world will need jobs in the coming decade. Only the regeneration revolution can supply them, as continuation of the mechanistic dilemma continues to create multiple global economic crises. Most will be jobs related to regenerative agriculture or aquaculture, some will be in renewable energy, new vehicles, or some other green business. It will be a true parallel to the Industrial Revolution in scope and size. Regenerative agriculture will be the largest part of the coming revolution, one fully supported by the new science.[5]

Regenerative economics will be the base of the new society. Once the environment is accounted for, the public can begin to understand how their future is really for sale in so many hidden ways. Solutions can then be created. Developing countries can expand their economies without having to slash and burn their forests, pollute the atmosphere, or use up other precious resources.

Yet the world doesn't change simply because a government issues a proclamation. The Netherlands is an example of this truth. Despite industry regulation, information dissemination on the needs to reduce pollution, and even a threat to institute a green fuel tax, the growing popularity of cars in the Netherlands led to increased emissions year after year. Besides the need to switch over to low-emission vehicles, the Dutch encountered the real problem: how to reduce pollution at the personal level.

TABLE 11.1. THE REGENERATIVE INDEX FROM ROBERT RODALE

AREA	DEFINITION	MEASUREMENT	MONEY LOST TO ZONE
Full Employment	100 percent of those wanting jobs are employed	Employment levels	
		Years lost in zone due to under- or unemployment	Amount of money lost to lack of jobs below 100 percent employment
Full Health	100 percent health	Number of people sick	
		Years lost in zone due to under- or unemployment	Amount of money lost to illness below 100 percent health
Full Energy	100 percent energy self-sufficiency	Percent of energy produced in zone	Amount of money lost to out-of-zone energy
Full Food	100 percent agricultural self-sufficiency	Percent of consumed food produced in zone	Amount of money lost to out-of-zone food purchases
Full Materials	100 percent raw materials self-sufficiency	Percent of raw materials produced in zone	Amount of money lost to out-of-zone materials
Full Industrial Production	100 percent industrial self-sufficiency	Percent of manufactured goods produced in zone	Amount of money lost to out-of-zone industry
Full Capital	100 percent local capital reinvestment	Percent of local capital reinvested in zone	Amount of money lost to out-of-zone insurance, health care, etc.

To do this, you somehow have to reach people and make them feel it is their responsibility to reduce their personal share of CO_2 waste, water use, and so on. Fortunately, the new paradigm has a tool that has proven successful in doing this, increasing conservation from the 2 percent range to 20 percent. This was done by having a plan that approached people on the community level. As one community started to see the benefits of the program, it *communicated* that success through its local media and word of mouth. Other communities wanted to follow that community's example, and the ideas, experience, and knowledge base then spread, along with a new organizing image. A kind of community self-image is contagious and can spread to other communities.

The way to reach the individual is through his or her loyalty to his or her community. The regenerative method works for poor and low-income communities and areas of any country, not just for developing countries. It is a universal method that can be used in places ravaged by drought or famine or devastated by natural disasters such as hurricanes or earthquakes, as well as whole nations and regions destroyed by war. So for example, in this method, the art and music of the indigenous culture, or the roots of the culture are restored or preserved. The elders of the group, rather than being marginalized and forgotten about, participate in teaching the unique wisdom, history, and culture to the children, giving them a sense of continuity and a sense of belonging and community, as well as a historical perspective or consciousness.

When we forget our histories, we develop a disregard for lessons of the past—even the recent past. We can then develop a pervasive and chronic historical amnesia. Not only is this a very sad thing, it can be a very *dangerous* thing. With this new way of thinking, the new—the progressive and innovative—is welcomed but does not destroy or *degenerate* the good from the past; it does not impose a false culture but embraces a holistic one where past, present, and future are all part of an never-ending unfolding.

COMMUNITY DEVELOPMENT IGNITES REGIONAL REGENERATION

All over the world, cities and towns in the transformational phase will likely discover a whole new level of personal involvement in the community. Under a current pilot program called Livable Neighborhoods, many blocks in Kansas City, Missouri, have been organized by block leaders. Created by David Gershon and Global Action Plan, the Livable Neighborhoods program has the block team decide upon and organize up to thirty-five actions that make their block more livable. This fundamental approach can be used to bring community development and personal responsibility for the environment around the globe.[6]

In this simple but powerful method, residents meet each other for the first time and immediately form close relationships as they carry out action after action, initiating effective recycling and conservation programs, creating bartering and interblock commerce, forming crisis response teams, and so on. Block, neighborhood, and regional vision-planning sessions help create a sustainable local future. A loyalty to the block and a powerful new sense of belonging replaces the emptiness that lies within the hearts of many urban dwellers, who typically live alone among a multitude of neighbors. Loyalty and belonging are two of the most fundamental human impulses. Community development fulfills these deep needs, giving it a real power to influence.

Community development has huge potential. It could very well become the basis of a new grassroots politics—one that has quality of life and environmental protection as its primary focus. This would cause huge shifts in many elections around the world. There is no doubt that the great potential of community development must be unlocked to bring about the deep changes that are needed.

THE GLOBAL REGENERATION PLAN

Leading the charge of the transformational phase will be a new Marshall Plan for developing countries based on regenerative development.

Building on the regenerative zone pilots of the intensive phase, a global regeneration plan would go well beyond the current millennium development program. Much of the world would transition to organic agriculture and to industries running on plant oil and a renewable energy base. The main parts of this global plan reinforce each other on a region-by-region basis, as shown in the regeneration revolution plan, where the interconnections become positive feedback loops. All the nodes synergize with each other: global education, regenerative agriculture, integrative medicine, democratic peace, low-impact technology, renewable energy, and community and personal development. Within a couple of decades, the global regeneration plan would take on a life of its own and not need additional financing.

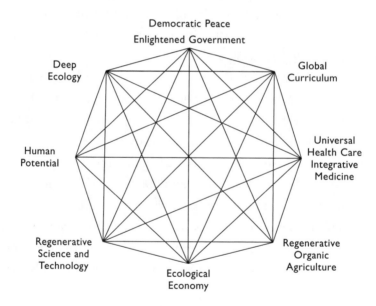

Fig. 11.1. The global regeneration plan wherein
everything is connected

After five years of the global regeneration plan, the impact will be noticeable in many countries, as two billion abject poor are given the chance to work their way out of hunger and unemployment. After ten to twenty years, the impact of a concerted worldwide effort to regenerate the planet would be enormous. The increased yields, the jump in Third World incomes, the drop in disease (especially those caused by malnutrition), the cleanup of the environment, the development of whole communities, the leap in education—all would clearly give us a much different future than the one the mechanists have in store for us. Terrorism in particular will decline, as regenerative development eliminates poverty and political oppression—its primary causes. Free trade will flourish, for even though regenerative development and planning dramatically increase local self-sufficiency, they will also rapidly increase markets in developing nations.

Twenty years is all it would take for a $750 billion global regeneration plan to change the world forever. It would happen quickly because agricultural systems mature within only seven to ten years, and the renewable energy switch is instant upon installation. Once the organic farms and the renewable production are set up, with proper maintenance they can theoretically run indefinitely—never using up the fertility of the soil or the power of the wind and the sun. What a bargain! We get rid of terrorism, much chemical pollution, the worst poverty, and most diseases and reduce and avoid global warming and the primary cause of recent wars—all for a mere $250 billion over ten years in developing countries and in the First World. The time frame for planet regeneration would be twenty years, or about one generation. Around a trillion dollars for the United States over twenty years would make it 100 percent renewable, or $50 billion a year. With the vast wealth released by the use of renewable energy, the ending of expensive wars, and with having the supperrich, the top 1 or 2 percent, finally paying their fair share, we could do this—and the *new* wealth that would be *created* would pay for it in the *long run*.

Through tax credits and tax revenues from new regenerative technologies and a greater number of employed citizens with regenerative jobs—millennials and other taxpayers—the government would end up spending very little of that sum. "Sustainability Image Drives" among newly sensitive corporations will make the new more profitable wind farms and other renewable technologies a *huge* investment vehicle in the United States and

Europe, with other countries—India, African nations, China—to follow. After the second decade or by the next generation, the global regeneration plan would have fully matured and accomplished *all* its goals. (Please see appendix 1 for more on global regeneration.)

Combine all of the above with dedicated regenerative regional planning in the industrialized world, and we really *could* clean up our planet and not only survive, but *thrive*.

MECHANISTIC INSTITUTIONS
ARE CLOSED OR TRANSFORMED

If the transformational phase develops, the mechanistic institutions of the world would be challenged by the political might of the new paradigm to either change or close. For example, the AMA will likely be shut down for having little credibility left, for having been a outright tool of tobacco companies, pharmaceutical companies, and the HMOs and not pushing for mandatory continuing education and testing for doctors. An integrative medicine association of some sort could theoretically replace it. Traditional medicine will not completely disappear, however, as there would be many components that will survive because of their validity and utility. The pharmaceutical industry, however, could be forced to give up most medicine based on chemicals and poisons and turn to, at least initially, less-profitable herbs, enzymes, and natural methods. Publicity on the fatal side effects of traditional drugs, and a consequent dwindling market for them, would accelerate this trend. Chemistry would not disappear but become a more sophisticated science, one that takes into account natural processes and the dangers of unintended consequences. Decoding the human genetic map will lead to whole new levels of understanding in health. New technologies will, of course, be invented, but with the organic paradigm providing guidelines, these new methods will not be monstrous creations but rather beautiful organically harmonized medicines and solutions. Tempered by ethics and the big picture view, regenerative medicine— which already exists, such as cloning an individual's own cells to grow a genetically identical organ transplant—will revolutionize medicine in an *organically principled* way.

Medicine as a whole, reformed along with the rest of science and technology, will likely be going increasingly down the organic path as the transformational phase develops. Costs would drop dramatically, as proven, less expensive organic methods that cure replace the traditional drugs that merely treat symptoms. Deaths would theoretically drop as well, as the hundreds of thousands killed annually through misprescribed medication would decline dramatically in the United States. Fewer drugs, more natural remedies, and better continuing education will be the answer.

Medicine's top-to-bottom reform will be mirrored in many other mechanistic institutions, especially the various industry associations and government departments and agencies. They will all either have to change radically or be entirely replaced by another organization. In the transformational phase, the public and a progressive Congress will no longer stand by while today's "kinder and gentler" corporatism and the mechanistic corruption of science cause the deaths of millions around the world.

The transformational phase will be a very different world from the one we live in today. Industrial associations formed solely to reverse the regeneration revolution would no longer be possible. The new science would, institution by institution, deploy Goethe's new scientific method to all research and operations—a zero-based analysis of all the processes of society interacting with nature. The environmental impact reports on whole industries in the transformational phase will have a real impact on practice as the progressive media report the damning data. Millennial boycotts against corporate polluters are already easily started over the Internet. The successful Internet boycott against Sinclair Broadcasting in the 2004 presidential election proved that we are in a new era, where the grassroots is activated and organized through the web. Skeptics will dismiss the very idea of all this—but according to the four stages of social transformation, this is the most likely scenario.

CLEAN TECHNOLOGY

A large part of the regeneration revolution will be clean energy. The renewable energy machines making that power will be more efficient than ever before, and they are already competitive today. Offshore wind towers will be larger than today's 197 gigawatt wind power generators and will produce more power at lower wind speeds. Costs are estimated to fall from its current four cents a kilowatt hour (kWh) to one or two cents per kWh with true mass production, according to energy consultant Harry Braun.[7] General Electric is already the biggest maker of wind turbines in the world, selling over $1 billion dollars a year. In 2005, Clipper Windpower, a wind turbine manufacturer, engineered a breakthrough turbine with variable speeds. With its multiple gears and four different generators in one easily replaceable unit, this wind turbine gets 25 percent more energy from slower wind speeds, allowing it to be placed closer to the homes and factories it serves.[8]

Solar panels will meanwhile be far cheaper than today and cover outside walls and roofs all around the world, while home wind and micro-hydro systems will be quite common. Even automobiles will be clean and regenerative. Cheap solar and wind energy makes possible the creation of hydrogen gas from water, also known as green hydrogen. Green hydrogen is made by using a water electrolyzer, powered by either wind or solar energy, that splits water into hydrogen and oxygen. It is an elegant solution showing how different kinds of alternative energy can be combined.

By the transformational phase, alternative vehicles like the straight electric or hydrogen fuel cell cars or perhaps another type of car will have replaced most of the old gas guzzlers, which will likely only appear at classic auto shows. Even NASCAR would go to pure ethanol or Tesla electric cars in this future.

GLOBAL PROGRESSIVE EDUCATION

Assuming that intensive phase pilot projects delivering learning over the web to those who can't afford tuition and the Third World are successful, global education would take the next leap in the transformational phase. Websites would support total online curriculums in different languages for both high school and four-year colleges, finally bringing education to the entire planet. And it could all be based on the new paradigm, especially in the teaching of the regenerative sciences. New holistic methods and new progressive curriculums would also be used in district after district instead of today's system. High school graduates would seem more like the college graduates of today, ensuring the rapid spread of the new education within ten years—as happened in the Herbartian wave of the 1890s, which established the current curriculum and our familiar "factory schools." Like the rest of the world, education will be reinvented and transformed, developing the full human potential of students and preparing them for the great task of regeneration they must accomplish. As education finally becomes fully relevant to the times in which they live, students and teachers will be motivated from a far deeper level than today, changing the dynamics of school completely.

By the time the transformational phase comes to a close at midcentury, global progressive education will have truly changed large parts of the world, although there will still be much more work to be done. Yet the future will have been set and a great sustainable foundation laid for the centuries of teaching and learning to come. The environmental ethic and the living response of the collective world will appear as a spiritual yet scientific experience to those raised in it, for these children will respect Earth, and thus their schooling and their teachers, in ways unimagined today.

POWER TO THE NEXT GENERATION: THE MILLENNIALS IN POWER

In the end, the success of all these progressive organic solutions will finally end the mechanistic dilemma. The benefits of the regeneration revolution, the efficiency of renewable energy, the common sense of global education over the web, and the realistic elimination of chemicals from industry and agriculture will consign the old science and technology to the oddities of history file. All the while, the progressive media would be spreading the truth of the new situation to an enlightened and increasingly radical populace. Raised on the new education, schoolchildren will be agog over the absurdity of nuclear bombs and the crude science of nuclear power, such as the ridiculous nuclear-powered plane, as well as countless other features of corporate domination and self-destructiveness of the mechanistic age.

From 2025 to 2050, the millennial generation will mature and come into power in its own right. Just as the reigns of mad King George III, Louis the XVI, and the power of the aristocracy all came to an ignominious end in the eighteenth century, today's corporatism and crony capitalism will finally fall—if the enlightenment pattern continues to its obvious conclusion. And just as the aristocrats lost their privileges in the French and American revolutions, so the corporations will have to give up their right of personhood, granted to them in the United States in an 1886 mistake by a court clerk and then passed and made into law by the Supreme Court ruling in 2010. No longer will companies be allowed to donate to political campaigns, nor lobbyists be allowed to throw lavish parties and other perks to legislators or to manipulate or unfairly influence any other government worker. Cronyism would end, and a properly regulated and regionalized capitalism could then thrive.

The power of the oil industry, the pharmaceuticals, the tobacco companies, and the banks financing it all would theoretically be pushed back by the unbeatable oratory and political muscle of the millennial leaders. Even if the millennials are at first defeated in their quest for equity and true democracy, they will rise up again and again—until the mechanists and the military-industrial complex are finally put in their place. Nothing less, after all, can save our fragile biosphere and atmosphere.

THE GREAT TRANSFORMATION
AND PERPETUAL PEACE

War, militarism, and terrorism are a constant threat to life. Sunken nuclear submarines will soon be leaking massive amounts of radioactive materials. Although most radioactive particles are heavy and will not leave the area, neptunium and other by-products could be carried by currents around the world. Nuclear power plants operate every single day, even though just one full meltdown could contaminate Earth's entire atmosphere, while making unlivable vast tracts of the planet's surface. Chemical pollution from industry and agriculture is now so widespread it threatens entire bioregions and the oceans themselves. Meanwhile, mechanistic institutions, such as schools, hospitals, and corporations, chew up people and whole communities when there is no rational reason to do so.

There is much to change. We cannot afford to waste any more time and money on war and killing ourselves over things like oil, religious differences, and millennia-old grudges. War causes global warming because it creates one of the largest carbon footprints in the world. The carbon emissions from firing rockets, exploding bombs, and machinery production for the war machine are enormous. Military budgets are black holes, sucking up tax dollars on the military budget to produce weapons and products we never intend to use, while consuming vast resources, human potential, and wealth. We need to finally end *all* war and realize that we have a common enemy—*ourselves, in the form of the old thinking*. We must throw out the old and ring in the new, a world without war, a regenerative world. A collectively awakened and ecologically aligned world, determined to turn old-paradigm civilization into an organic, global society, is what it will take to transcend the mechanistic dilemma.

In 1949, Teilhard believed that after the scientific revolution of his day, the new science would enter the mainstream and eventually be accepted as the next foundation for civilization. After evolving the hardwiring of the planet in communication and computer systems, a special global circuit would then close, and collective reflection and action could finally become a reality. The global mind, the noosphere, would awaken. In this next threshold of consciousness, thought converges or "infolds" into a single collective point of *reflexion*. This is global awakening on an unprecedented scale, something new under the sun. And if the historical pattern of the enlightenment continues, it would be coming very soon, perhaps somewhere around the midcentury.

Certainly many people will not change and will remain as they are today. Yet enough of the population and most of the decision makers will, theoretically, be swept up in the planetary catharsis, setting in motion a number of interconnected events. How does the paradigm flip lead to the great transformation? If Teilhard is right, how exactly do we get from here to there? What exactly would this new world look like?

Teilhard compared the planetary awakening to the realization of the first Stone Age person who thought *I am,* a new power of reflection that within a few hundred thousand years gave us all of civilization. This ability to reflect also brought with it the understanding of death—that one day you will die. Much of spirituality comes from trying to handle this realization.

In the future, because of the compression of humanity and information, Teilhard predicts the noosphere will be able to *reflect* upon its *global* being and realize *we are.* At that moment, the collective consciousness would understand that *we, the human race, and the entire planet could die,* triggering the collective impulse to survive. It is the explosion of information and the population together, trapped within the finite limits of the globe, which catalyzes the gathering change of state. Now, the *new* realization is that humanity has the capability to kill Earth and *that everything that we do is critical to the life of the planet.* And as we witness the suffering of souls, massive population die-offs, and the extinction of many of the exquisite and irreplaceable forms of life on the planet, the simple reality of all of it will finally be utterly clear to every single human being on the planet, and then we will *transform.*

No doubt the Internet—connecting the planet at an increasingly rapid pace—will play a large part in the global mind awakening. Just as the philosophers of the eighteenth century sold their words in inexpensive books, so the new enlightenment philosophers of tomorrow will delight their readers with inexpensive *free-speech* e-books, iPad apps, and interactive sites on the web. Today, people can publish their work for free and have the whole world reading it overnight. The Internet is truly an acceleration in what Teilhard called the "little-noted speed of thought"—likely to the point where the noosphere breaks through to the third threshold of consciousness, the awakening of the collective mind.

In short, the world will finally be brought to the point of truly reflecting on the human condition. It is then that we will think as a species and start addressing the thorniest question of all: war. On the political stage, world government would simultaneously become more influential than the remaining pockets of ultranationalism. The urge for world peace would then be overwhelming, and the millennials will galvanize the planet in the largest street demonstrations of all time. Yet as the Peacemaker taught, it is not enough to bring an end to the fighting. The wars will only start again—unless a strong structure is put in place to guard against breakdown.

Like the founding of the Iroquois League and the great peace among Native American tribes, a global grand council could be the model for our own time, as would the Peacemaker's great law of peace, which would be studied for its teaching on how to control the impulse to war. A global peace conference, assembling negotiators and those on the sides of all conflicts, could use the Condolence Ceremony to wipe away the trauma of war and the deaths of loved ones. Then all can negotiate with an open heart instead of a closed heart. Bring all the conflicts of the planet together in one collective conference and have all follow the same innovative peacemaking process, which would presumably involve years of intensive discussions. We would then finally have a real chance to wind down the insanely bloated war machine.

THE PARADIGM WAR ENDS TOO

Supporting all this deep change would be the new curriculum in education. Delivered over the web to the entire planet, students would learn from the earliest grades the fundamentals of global studies and the regenerative sciences. By using world history as a foundation, the curriculum teaches appreciation of other cultures, promoting peace and global unity. The Organic Shift would also be taught, to properly present the new science. As we saw, Kant advocated the teaching of a universal history as the permanent foundation for perpetual peace. This is still true, and the transformational phase will therefore see the beginning of this momentous change in education.

A new global peace in tandem with global regeneration would be the culmination of the whole Organic Shift. The transformational phase would thus be a leap to a whole new level of global awareness. Imagine the potential that positive *collective thought* and *collective action* might unlock. If history is any indicator, once the paradigm flip arrives it would all happen very quickly—within twenty to thirty-five years.

The infolding of the collective consciousness, the compression of information, the secular spirituality of deep ecology, the coming together of peace-minded communities (Gandhi's unities), the regeneration revolution and an invigorated Peace Movement responding to the horrors of war—all of these many movements coalesce into a single push for change. It is a common agenda that will be resisted for as long as possible by the old paradigm. Once begun, the environmental/ economic regeneration revolution, powered by 75 million millennials in the United States, and billions more progressives around the world, will be unstoppable. America will be transformed by its millennials, and then it would finally become a part *of* the world, rather than being apart *from* the world, as it is today.

And with that step, the paradigm war will end, the future would be won. The crises of our present-day world would be reversed by global regeneration, the desire for peace and the dismantling of the massive war machine. A new millennial generation leadership would then lead us not to utopia, the perfect place, but to melitopia (from ameliorate), a better place, the future that works.[9] There's only one thing that can stop it.

You.

Mahatma Gandhi

12

THE NEXT STEP

What You Can Do

If we could change ourselves, the tendencies in the world would also change. As a man changes his own nature, so does the attitude of the world change towards him.

MAHATMA GANDHI

IT'S ALL UP TO YOU

By transforming the world, that initial "small group of thoughtful, concerned citizens" who began the Organic Shift will have proved Margaret Mead right. Starting with Kant, Goethe, Wollstonecraft, and Blake, and on to Gandhi, Prigogine, Teilhard, and all the others, it was never a large percentage of the population that saw the world through the new paradigm. Yet for such a small group, it was thoughtful indeed. As Mead says, these tiny circles of courageous thinkers are the only thing that have ever changed the world.

If they can do it, so can you.

Will we collectively have the same courage to stand up and take the steps required to change our world? There is much we will have to do and sacrifice to make it all happen. Granted, society could reach a tipping point and wake up to the mechanistic dilemma that is all around it. Yet how does that become a new stable civilization, rather than a sudden descent into the resource-starved, totalitarian-techno nightmares—so often depicted in movies and science fiction?

Simple. Knowing how to use the tools that the Organic Shift has given us, we can find the courage and fight for the planet, for our children's future. History is not preordained. If we ourselves are not active enough in creating a better world, the nightmare future will arrive. The American Revolutionaries, without a free press, and under pain of death, met in secret and spread the truth through a democratic underground to the patriotic third of the colonial population. They had the will, the courage, to persevere—as do today's young millennials.

If we understand our own power as change agents by seeing the incredible utility of the regenerative sciences—and how those could be deployed systematically to quickly change the world—then little can stop a better future from coming about. So it's actually not up to the government, nor to the many institutions of the world, it's up to us. It's up to me. It's up to you.

CLUSTER PROBLEMS AND CLUSTER SOLUTIONS

The realm of the idea is called the noosphere. We stand on the geosphere. We live in the biosphere. Earth is covered with an atmosphere, but we think in the noosphere. It is in this noosphere where all our collective thoughts reside. When the noosphere becomes connected, the global brain will wake up to a new level of consciousness and being. The global brain is an interconnected environment. In an interdependent environment, things tend to affect other things. So almost anything we think and do has spin-off effects—either positive or negative.

All the present crises in the world reinforce each other and produce a synergistic effect, making problems even worse. These crises and problems are called cluster problems because they interact with and overlap each other. As we look at the big picture, we see that all problems are interconnected.

There is a relationship between the problem and the solution. Hidden within the problem, we often find a surprising solution—great opportunities for the regeneration of environment, civilization, and culture to sustainable peace, abundance, and continued life on Earth. Together, solutions to the present network of crises can be seen as the creation of a network of opportunity and change. In this network, opportunity realized in one area improves the chances for finding solutions to problems in all other areas. The power of synergy then creates—both individually and globally—the great transformation.

When we take the time to look at the main crises of the world and investigate them with a holistic worldview, we can find regenerative solutions. Mechanistic science is based on a limited linear view of cause and effect and often does not plan for unintended consequences, for interactive synergies, or for the long term. So many solutions and technologies that appeared to be good end up having unintended consequences. We look beyond the linear cause and effect and search for possible unintended consequences and synergistic interactions. If we find a drawback to a particular solution, we may find a way to redesign the solution so that the drawback is avoided. In addition, we find that the *best* solutions solve more than one problem. These are called cluster solutions.

Regenerative solutions are cluster solutions that have beneficial spin-off effects and positive interactions with other problems or combine with other solutions for a more powerful and far-reaching change. Interactivity, synthesis, and synergy are characteristics of regenerative science and solutions. In the global regeneration network, cluster solutions are placed within the regenerative network model and then broken down into steps or the components of the solutions. The new solutions are built up and implemented from combining the specific solutions and the people who are working on them.

We find and connect with people who have invented various things that solve specific problems in a regenerative way. As new solutions are found in each field, they are evaluated for the many ways they interconnect and interact with other areas. We look for people who have been successful at doing something or creating something or who are interested in doing something to regenerate the planet.

GLOBAL REGENERATION
THROUGH CLUSTER SOLUTIONS

Life and nature exemplify the power of synergy. It is synergy that has the power to regenerate. It's the ability of nature to regenerate after disaster or damage. Global regeneration is based on the organic development and interdependence of life processes—the sum total of all purposeful activity that produces synergy and regeneration. The first step is to become educated about what global regeneration is and how it works so that you know how to plug yourself in and interact not only on a personal and community level but on a *global scale*.

Global regeneration is a way to transform the world. It is the way to rebuild entire environments and cultures from the ground up. Whether a region has been decimated or "degenerated" by political causes, economic causes, or environmental causes, global regeneration uses interconnected solutions to restore the people and the environment and everything about our planet, even the atmosphere.

Global regeneration looks at the big picture to see the many different interconnected causes of global degeneration. Among them are war, tyranny, colonialism, terrorism, poor or corrupt government, corporate and government exploitation of resources and people, economic depression, lack of education, overcrowding, and environmental causes such as drought, hurricane, and, of course, global warming or global climate disruption.

Many of the problems we are faced with are interconnected and require a multifaceted holistic approach to analyze the situation and arrive at appropriate solutions. It helps to look at things as a network. For example, major areas of crisis are in agriculture, environment, economics, education, foreign policy, and health. These areas or nodes in the network can be considered critical drivers or attractor points of change in the world today. If we look at just these attractor points we will be looking at major areas where people can plug into the network and make a difference. Other major nodes in the network are energy, government, global climate change, and species extinction. Connecting to these major nodes or attractor points are these major connections: agriculture, education, environment, health, human resources, foreign policy, technology, and peace—and there are many other nodes and connections in the network. You will see how some areas are listed in both the major nodes and the connecting nodes. This happens because when one moves to a connecting node a new crisis or solution might present itself that you hadn't thought of before when thinking of just the larger overarching crisis. There is a link back or a looping back connection to consider.

For instance, perhaps the field you are working in is agriculture. By knowing the connections between agriculture and the environment, you might discover the connection of, say, neurotoxic pesticides used in agriculture or on lawns to various deleterious effects on the environment. As you become aware of a crisis and educate yourself, you then understand it enough to arrive at a cluster solution.

COLONY COLLAPSE DISORDER AND HOLISTIC THINKING

As an example of holistic thinking and problem solving, let's examine the honeybee die-off or the problem of colony collapse disorder. In my book *A Spring without Bees: How Colony Collapse Disorder Has Endangered Our Food Supply*, I used holistic thinking to investigate the crisis of disappearing and dying honeybees and in the process discovered many interconnections.

First, we define the crisis and think of possible connections:

Crisis: Honeybees are disappearing and dying off and this threatens our food supply.

Connections: Agriculture, economics, education, environment, government policy, health, and technology.

Next, we begin the inquiry. We look at apiculture and note what has changed over the past few years in the way people keep bees and what the effects are of the bee die-off. This brings us to economics: we see that this bee die-off will have costly effects on agriculture and even endanger the food supply. Next, we discover a possible cause of bee death: the use of pesticides. Then we look at recent agriculture practices, which brings us to technology and the development of new pesticides and another economic link—these pesticides improve crop yield. Then we link to the environment and discover that neurotoxic pesticides are killing not only bees, but also frogs, birds, and bats. We look at health and find that neurotoxic pesticides also endanger human health: some human diseases and birth defects are being linked to these neurotoxins. Then we discover these neurotoxins have not been tested and approved in the proper way, and that in turn leads us to government policy.

We now have several links: technology is used to develop chemicals that destroy pests that endanger crops but these pesticides have unintended consequences of killing honeybees and other pollinators. The loss of pollinators is costly and, worse, will endanger the food supply.

Then we discover that neurotoxic pesticides, as of 2007, are the fastest growing market in the United States. So we are back to economics again! Wait a minute—what was that about government policy? We look into it and find that the manufacturer is a lobbyist and is also funding scientific research in the colleges.

You connect all these things and begin to see how this problem has wide effects: it's a cluster problem that, if solved, will have far-reaching beneficial effects. After connecting everything up, it looks like we need to suspend the use of neurotoxic pesticides in order to save the bees and our food supply and our own health. If we examine the interrelated or *synergistic* causes of the problems, we find that we will need to work in each one of these areas—government, education, agriculture, and technology—to be successful.

This type of thinking will save the world—or at least make it a better place. It is holistic thinking, global regenerative thinking. Regeneration is not a utopian view. It is rather a way to make a positive difference without overinflating ourselves or adopting overidealistic or fantastic views that will always fall short. Regenerative thinking remains realistic. The organic paradigm and the way of global regeneration is not egocentric nor is it anthropocentric. It is Earth-centric. The life of our planet and our relationship to it is the most important thing of all. This is the type of thinking that we must learn and must teach to our children for future generations.

Colony collapse disorder is a microcosm of our world today. We are now seeing the beginnings of "civilization collapse disorder," where combined environmental degradation and poisoning suddenly reaches a critical mass—causing a massive die-off of species, including our own. The difference between the bees and us is that *we,* in our ignorance of how global regeneration works, are the ones *causing* the collapse.

AN INTEGRATED FRAMEWORK
FOR GLOBAL CHANGE

The Global Regeneration Chart (see appendix 1, pages 400–414) specifies the crises and opportunities for change and shows how one solution reinforces the other solutions. The core areas of crises are: the environment, agriculture, economics, foreign policy, energy, education, and health. In the chart, each solution is offered after an analysis of each crisis. This analysis is based on a philosophy of finding the true cause or causes for each of the crises. The Global Regeneration Chart is available at the Global Regeneration Network at www.GlobalRegen.net.

Most political solutions just treat the symptoms of a problem, not the true cause. Treating only the symptoms creates an ever-more expensive treadmill effect on the budget as the root cause of the crisis keeps on churning out problems. Treat the cause, however, and there is hope of a cure. Together, the proposed solutions create a social network—the global regeneration network that can be a solutions generator for the planet. New alternative technologies are expanding at a rapid rate and will be connected in the network. Energy and education solutions and other main nodes of the regeneration network, such as foreign policy and economics, will develop in this ongoing process.

GLOBAL REGENERATION ZONES

The global regeneration network will offer resources on global regeneration planning and development of regenerative zones in the world. Each regenerative zone will get an online center or portal on a global regeneration website. The global regeneration plan will help the world by boosting agricultural and economic self-sufficiency on a region-by-region basis. Each region will have a volunteer or paid regional planner as well as a regenerative agriculture expert assigned to support it online. (See appendix 1, "Global Regeneration: The Real Solution to War and Poverty," for a more thorough explanation of this plan and the potential outcomes.)

Each regenerative zone development plan is thus a comprehensive effort to develop a region according to its own unique resources and opportunities, creating a powerful synergy between improved education, leadership training, agriculture, economic development, health, and energy production.

Agriculture and basic economics are key to regional regeneration. If the soil has been depleted and can no longer yield large crops for local food consumption, usually regional money circulation has plunged to low levels. After teaching farmers how to restore and conserve soil moisture and build the soil, regional farmers' markets are often established or expanded to help increase money circulation. Farming and the economy are thus inextricably linked in most areas. The regenerative zone development plan systematically restores the natural health of the soil, while raising the economic self-sufficiency of a region, increasing money circulation and overall well-being. Rather than exporting one crop or resource to far-off lands, with most receiving pay not much above that of colonial times, farmers are taught how to raise different kinds of crops, which makes the region more self-sufficient.

While the soil and natural ecosystem are being restored, each regenerative zone development plan has a second focus: making these simple economies more complex and more self-sufficient by creating products the region or the world needs. Care is taken to grow the businesses in such a way that money circulation is promoted rather than lowered. Rather than large loans to established businesses that often lower regional money circulation, the initiative concentrates on starting many small, local businesses. Microcredit or microgrant programs from the World Bank and other institutions will finance small businesspeople and farmers with needed equipment and seed. These grants will be used to start a variety of businesses, including direct sale of goods, music, and art over the Internet through the Regenerative Zone Center.

There are future regeneration programs for urban areas as well as rural, in which neighborhoods are approached and organized like the villages, but with differences for the urban environment. Regeneration also helps reverse the trend from rural to urban, which threatens the future in so many countries.

The point in regional regeneration is to create multiple forces for transformation—all focused in one place so that natural synergies are released. This is far more cost-effective than top-down, which releases little or no synergy. It is a bottom-up approach that leaves the control in the hands of the people. It starts with the basic need to sustain life on Earth, a true grassroots movement for rapid change to restore and reinvent a healthy sustainable culture and environment for the present and for future generations.

GLOBAL CURRICULUM

Education is a key factor. The core of the global regeneration website will be the global curriculum (see appendix 2, "Progressive Global Education for the World"), online courses with materials that can be printed out. Rather than a nation-based structure, which tends to emphasize differences and old prejudices, the global curriculum emphasizes common human values and accomplishments from every culture around the world. Students are taught to not only appreciate their own culture but to appreciate and love other cultures—so important for world peace. All subjects are taught through history, with students advancing from time horizon to time horizon, understanding all the connections within and between cultures and learning all math and science through the history of those subjects. Discussion groups let students around the world talk to each and learn together for free.

Satellite links between the villages can provide ongoing online dialogues or support groups that connect the different parts of the region. Experts and workshop facilitators can hold specially scheduled online seminars and conferences and then follow up with their participating villages over the Internet/satellite link to the villages of the region or to the world. Successful farmers can be filmed and then train other farmers online on specially designated days, who in turn show their village the new methods. Graphics and video clips will be posted to help educate one and all around the world. Everyone gets to talk to each other—for free—about their similar problems and solutions. Since the needed services in each region can be accessed through the same zone center portal, the multiple regenerative sources within the zone are easily synergized, region-wide community actions undertaken. People can sign up online for composting day, for example.

The online dialogues will also connect the new mini-entrepreneurs and give them support, resources, and guidance. As with agriculture, successful small businesspeople tell others how to do it and train regional trainers, while inexpensive online global workshops on how to produce and market goods are given by leading regeneration or NGO experts. The Internet/satellite link makes it economical for all involved.

The recent recession in the United States has created economic dead zones and whole areas and communities in need of regeneration. The same method can be applied to developed countries in economic degradation. The schools would adopt the global curriculum as a core studies course and link math, literature, art, science, and other regular subjects to the core. It could be taught as a global studies course in high school or adapted into the gifted and talented programs for junior high students. Independent study programs, home schooling, and charter schools can adopt the curriculum.

DEVELOPING A SUSTAINABLE REGENERATIVE FUTURE: HOW TO DO REGENERATION PLANNING

Using the organic-paradigm thinking to develop a global regeneration plan will help you adopt a big picture view and break down the crisis into interconnected problems along with their solutions. Look at the present condition or crisis and then pose what possible positive changes and interactions can be made as the beginning framework of regenerative planning. Using the global regeneration chart, place the crisis segments in the first column and the regenerative solutions or changes in the second column. An action plan, chart, and whole "Global Regeneration Vision" map, each with specific crises and solutions, can be found at the Global Regeneration Network at GlobalRegen.net.

You can start with any problem or topic and plot out the interconnections to produce an integrated vision of what can and should be done to solve a crisis or regenerate a region or field of study. The topic could be a broad field like education, overpopulation, democracy, or the environment or a smaller specific field like homelessness, parenting, or local health services. This method can be used for businesses, nonprofit organizations, and government agencies or for personal development and self-improvement. Though the framework is laid out in chart form, imagine the content displayed as nodes in a neural network that are connected to each other. Add these connections in your chart or map. Then imagine this cluster to be connected to everything else in the global brain—a vast ever-expanding organism that creates the overall conditions of the planet.

Global regeneration and regenerative sciences can shape our future. It takes imagination, a vast knowledge base, and courage to create a picture of a positive view of the future. There are amazing people in the world who are coming up with astounding solutions—regenerative solutions. The mainstream media tells little if anything about them. Strangely, we seem to not be very interested in optimistic plans or visions of the future. Currently the media—movies and the news—mostly focus on the negative. Negative futures abound in movies and science-fiction novels. These negative images reflect our fears of what the world might turn into if the mechanistic paradigm is not superseded by a more holistic organic paradigm. These negative future scenarios do not reflect human hope and the essential goodness of an uncorrupted human spirit.

We are given the false impression that there aren't any answers, that there aren't solutions to our problems—but there are! In fact, amazing new projects, organizations, innovations, and inventions have surfaced just in the last two years. Spain built the largest solar plant in the world. It stores solar energy in molten salt, allowing it to be the first solar generating plant to produce power twenty-four hours a day and seven days a week. Idaho-based Solar Roadways founder Scott Brusaw has invented something truly brilliant: every roadway, parking lot, and driveway in the country could be turned into a massive solar generating system. Solar power generated by the system could be delivered directly into homes or offices. Instead of asphalt or cement, Solar Roadways uses a glass material that is stronger than steel, embedded with solar power cells. In addition to generating electricity, this technology would eliminate the need for snowplows in winter by warming the surface to melt the snow and ice. It would also embed electric programmable road signs warning drivers of detours, accidents ahead, or other hazardous road conditions as well as running street lights and traffic lights.

When multiple Solar Road Panels are interconnected, the intelligent Solar Roadway is formed. These panels replace current driveways, parking lots, and all road systems, be they interstate highways, state routes, downtown streets, residential streets, or even plain dirt or gravel country roads. Panels can also be used in amusement parks, raceways, bike paths, parking garage rooftops, remote military locations, etc. Any home or business connected to the Solar Roadway (via a Solar Road Panel driveway or parking lot) receives the power and data signals that the Solar Roadway provides. The Solar Roadway becomes an intelligent, self-healing, decentralized (secure) power grid. The west can power the east in the evening and the east can power the west in the morning hours. Everyone has power. No more power shortages, no more roaming power outages, no more need to burn coal (50 percent of greenhouse gases). Less need for fossil fuels and less dependency upon foreign oil. Much less pollution. How about this for a long term advantage: an electric road allows all-electric vehicles to recharge anywhere: rest stops, parking lots, etc. They would then have the same range as a gasoline-powered vehicle. In the 48 contiguous states alone, pavements and other impervious surfaces cover 112,610 square kilometers—an area nearly the size of Ohio—according to research published in the 15 June 2004 issue of *Eos,* the newsletter of the American Geophysical Union. Continuing development adds another quarter of a million acres each year. Internal combustion engines would become obsolete. Our dependency on oil would come to an abrupt end.[1]

The Department of Energy has taken this invention very seriously and has already funded the working prototype. The technology also creates enormous job potential and the company plans to export it to other countries. It is a *regenerative* solution, solving the greenhouse gas and pollution problems and working synergistically with electric cars. This one invention could transform the world.

These are just two examples of the tremendous potential and real solutions that are now being discovered. The general public is unaware, but there are geniuses and activists that are already regenerating Earth. The Global Regeneration Network is designed to inform and announce new innovations and developments in the Organic Shift and connect as many as possible. You will find links to the many and most important organizations that you can plug into. GlobalRegen.net is the network of change for Regen solutions and Global Awakening. A free media channel, you can publicize your solution and find answers there.

Here are possible events that could effect positive change in the future as well as comments on the trends and shifts occurring in the timeline. This scenario may not make complete sense to you until you have educated yourself with the global curriculum and started work or volunteering for regenerative projects. I urge you to take up life-long learning and achieving and I invite you to expand your consciousness and action in the world through the Global Regeneration Network and other social networks and the collective genius it offers to you.

THE MOST LIKELY FUTURE SCENARIO

2012–2020: Causes, Trends, Events

- Paradigm flip to holistic worldview probably after 2011. The new worldview affects all aspects of civilization and life.
- Holistic curriculum gains acceptance in the West and other areas. Global human rights movement intensifies.
- Terrorism continues but weakens. World grows closer as tensions ease.
- The people assert their rights and regain control. Peaceful demonstrations and revolutions.
- Green technologies and innovations are invented and deployed at an accelerating rate. These new technologies raise the standard of living.
- Effects of regenerative science and regeneration zones begin and then continue to bring positive change to the world. Nuclear power plants shut down. Atmospheric carbon dioxide levels—if we can make the necessary changes—begin to drop by 2015, reversing global warming.
- A feeling of hope begins to return as progressive, holistic movements and solutions begin to change the world and become more and more successful.

2021–2050: Causes, Trends, Events

- Shift to holistic worldview East and West.
- New time sense, environmental ethic, international cooperation, near-global holistic education and peacemaking. Peace council and peacemaking, conflict resolution, and control of violence are part of every level of society. Universal health care preserves the money-making human resource base of world economies.
- Small pockets of fundamentalism remain. Old liberalism and conservatism nearly disappear. Progressive, democratic, and cooperative ideologies prevail.
- New world political structures. Nations of the world converge along holistic lines, yet each region continues to retain its own unique character.
- Human and species rights protected; population begins to level off.
- Nuclear bombs outlawed by midcentury. Nuclear waste cleanup and sequestering is a top priority.
- Global warming continues to be reversed and the planet is saved from total environmental destruction and civilization collapse.
- Some economic, political, cultural differences remain. As competitive alternative energy come online, old economic structures are transformed.

A NEW DEMOCRATIC REVOLUTION

Our democracy has been limited by the mechanistic paradigm. Mechanists create bureaucracies not democratic management systems. They don't like democratic processes because then the leader, the CEO, the president, the head of the board, the wealthiest, the biggest army, the deadliest weapon, the monopoly, the corporation loses absolute power and that power is given to the people. They no longer have control of everything. A true democracy demands that all heads serve the people and that all people are equal.

We say we have a democracy, but then why are so few of our

institutions, government departments and agencies, and businesses run democratically? How can we expect democracy to work if all or most of what we have is bureaucracy? Why isn't democracy taught in schools? What if democracy permeated the entire culture? What if we decided to take democracy seriously and participate in it? Why do we have to submit a résumé to prove our qualifications and experience in order to get a job when our presidential candidates and elected officials do not? Why don't we have a national holiday for national elections so everyone can vote like almost every other democratic country in the world? Why don't we stagger the hours the polling booths open and close so that all of them close simultaneously, preventing the East Coast from "throwing the election" before the West Coast has a chance to vote? Why don't we take charge of our own destinies and restore self-determination? Why not go around the system and create our own solutions? Asking questions is the single most important way we can shift our thinking and find solutions. One of the first meaningful sentences I said as I started to recover my speech was "The answer is . . . *questions.*" Now it's your turn.

EXPANDING FROM THE PERSONAL TO THE GLOBAL

What do you want to do with your life now that you know your planet will not survive without your help? Think about what you can do in the context of whatever profession you are in or whatever your job is. Use the skills you already have and apply them. But don't wait until you think you are skilled or educated enough. *There isn't time.* Start working on it *now.* Movements in the past have failed to effect change because people felt they weren't ready or had nothing to offer. Once you plug yourself into global regeneration, you will find yourself becoming more skilled and educated. Through the synergy released, you will be reinventing yourself and regenerating right along with it.

Choose a part of the problem you want to focus on and then find out what you can do about it. Get involved in groups, demonstrations, events, school boards, social networks, and organizations that are working for change. Try joining projects like GlobalRegen.net, Bill McKibben's 350.org, Lester Brown's Plan B, Al Gore's Climate Reality Project, Vandana Shiva's projects at Vandanya.org, and ThriveMovement.com. Socially conscious corporations and small green businesses and corporations will also be found in the Global Regeneration Network. By the time this book is published there will be many more.

Learn how to work as a team. You must join with others who are working on specific problems with the awareness that your actions are either directly or indirectly a part of a collective effort. Follow ethical practices. Don't let your ego or someone else's ego get in the way. Be kind and only use constructive criticism. Don't be overly critical or jump to conclusions. Let everyone speak and don't put down an idea just because "it's never worked in the past" or "it's too difficult" or "we can't afford it," or "that's just part of it." Constant analysis or overanalyzing creates "analysis paralysis." Using only critical thinking also doesn't produce truly innovative solutions. Being overly critical won't work because you are making an assumption or jumping to a conclusion. You are not investigating *why* didn't it work in the past, *why* it's too difficult, *how* we can find an easier way to do the same thing, *how* we can find a way to afford it, *what* other parts need to be added to it, and so forth. Allow the process to flow.

Answers come only when you have questions, plenty of questions. The essential questions we might ask are: What haven't we done before? What has been overlooked? Ingenious solutions are created out of inquiry and making connections—combining this part of the solution with that solution while always seeing how or *asking* how it fits into the whole—the big picture. See it as a process with immediate steps: What can we do now? And at the same time, keep in mind the long term. We need to document things that didn't work (and why), as well as document things that did work and were successful (and why). Create a knowledge base. Figure out how successful actions can be duplicated. Put this in the knowledge base and then share this with other people and groups. Use your imagination! Think outside the box. Don't get distracted. Stay positive. Dare to dream, to envision something new. This is how real transformation will be *generated*. We know that real change will never happen with our current way of trying to find solutions and solve problems—our current way of thinking.

But beware of just talking about a problem. Talking can give the false sense that you are involved, but talking is not doing something about the problem. We have to develop a plan with goals, strategies, and tactics; work through consensus; and then take collective action. And we've got to think and act on a global scale. The only way to make this happen is through global regeneration. Although the transformation begins with each person awakening and taking action, the global crisis is too big for just isolated efforts and individual planning and action.

To create a positive future requires a whole shift in worldview, an awakening of consciousness. It requires cooperation, rationality and direct democracy, a participatory democracy. It will take every man, woman, and child to contribute their time, money, and energy to make a better world and ensure that there will *be* future generations that will not just *survive* but *thrive*. As we travel this path of awakening, we will find the courage to go out into the world and apply the solutions that are already here and to find the solutions that we have yet to discover.

We are at the threshold of just that global shift in worldview. This turning point in history and in consciousness is a great adventure—perhaps the greatest adventure of all time. It is time to awaken from the nightmare and take our place in the new story of the world—the great transformation.

It is time.

GLOBAL REGENERATION— THE REAL SOLUTION TO WAR AND POVERTY

A NEW MARSHALL PLAN

The Marshall Plan, also called the European Recovery Program (ERP), was created by the United States to rebuild the countries of Western Europe damaged by World War II. The initiative was named for George Marshall, who was secretary of state at the time. It is cited here as the prime example of an enormously successful regenerative project and was the first cooperative regenerative plan ever to be implemented. We look to the dynamics of this plan as a model in regenerative planning and cooperation because it laid new foundations for the future that have lasted to this day.

It is now accepted that part of the solution to terrorism is to alleviate poverty and oppression in the Third World, the breeding ground of extremism. If we don't address this major cause of the problem, we will be fighting the symptom of terrorism far into the future. That is why donor countries have already agreed in principle to fund an extra $50 billion a year in a New Marshall Plan; aid from the United States alone has increased $10 billion. The goal: raise the standard of living throughout the world for below and at subsistence level up through low income level populations and cut the worst poverty possibly in half by 2015 or sometime thereafter. All details of this plan apply to nations such as the United States, Europe, India, Russia, and many other countries—that are not considered "developing countries" but now have a large disparity of wealth with endangered middle-classes and serious poverty issues.

There are two billion people on Earth living in abject poverty in six hundred thousand villages around the world. For most, their local soil is poor and incapable of supporting a high-yield crop under traditional farming methods. Most cannot feed their families properly. Millions starve every year. As a rule, children are malnourished. The soil of these people has been depleted by years of farming, colonial monocropping, and herbicide use, as well as drought, wind, and sun erosion. The first real success of a New Marshall Plan would be to create an agrarian/information age society that really works, allowing villagers to become self-sufficient in food and raise their income at the same time.

Regenerative methods have worked in the past to completely restore a region's soil, agriculture, water, energy, social fabric, and economy. In contrast, earlier attempts to help developing countries often involved top-down funding of large projects designed to bring them into the industrialized world. These large projects provided foreign investors with profits and created jobs but failed to create real change. In this regard, the United States has the highest percentage of the donor nations, with 70 percent of its foreign aid being used to purchase goods and services manufactured in the United States. Often foreign aid is given only to buy the goods of that country, the opposite of what people on the ground say is needed: food self-sufficiency and increased local money circulation.

This policy must be aligned with the New Marshall Plan by allocating a large part of the new funding to go to new strategies like sustainable development and the global regeneration plan. We need to see that by creating self-sufficient regional and national economies, developing countries can lift themselves up to a new level of well-being.

Yet how should this new money be spent? How can we be sure it is not stolen—as has often been the case in the past? Rather than the top-down strategies of the past, the approach of working on the village level and sustainable development have provided a new bottom-up approach that works—according to all the data of the last few years.

The money goes directly to the villagers through NGOs, and it is based primarily on microloans and agricultural assistance. This bottom-up approach to aid fosters responsibility. Even though there's no credit rating system, in most of the well-run systems, people pay their micro-credit loans back at 98 percent—the same as American mortgage holders.

Unfortunately, the implementation of the bottom-up approach on the village level has been piecemeal and has been stuck in the pilot stage. We thus have a top-down model on one hand that has been mostly ineffective and a successful but poorly funded bottom-up model on the other. The new funds must not be spent with the old approach.

Policy makers must be brought into the bottom-up paradigm so that the needed changes in aid funding can be made. This history, this situation, raises two critical questions:

1. How will we measure the success of the New Marshall Plan?
2. How do we reach that goal with the new bottom-up approach?

How should we measure the success of a New Marshall Plan? Are the old methods of using foreign aid as a way to sell American goods or watching indicators like an increase in the average wage enough to ensure success? These are the failed, top-down methods. We need a new way to measure the transformation of these economies, an economics to match the complexity of the new bottom-up strategy. Many areas of the world need transformation—a true transformation, a profound lift upward to a new agrarian/information age society—and we need a new science that understands how to help them make that leap. It is here today—studied, proven, and developed over the last twenty to thirty years by researchers like Robert Rodale. The knowledge is now in our hands: we need only apply it. GlobalRegen.net does just that (see www .globalcurriculum.com).

REGENERATION:
OUR MOST PRECIOUS RESOURCE

Restoring Earth has, in recent years, become a science, or rather many sciences. There are several main branches in the regenerative sciences. These include:

- The philosophy of regeneration
- Regenerative economics
- Regenerative development and planning
- Regenerative soil science
- Regenerative agriculture
- Regenerative agro-forestry
- Urban regeneration

Regeneration is the ability of nature to restore that which has been lost or diminished. There are many examples of this near-miraculous phenomenon, but most important of all for our planet is the ability of soil to regenerate itself. Regenerative soil science teaches us how to restore the soil health of poor and drought-ridden farmlands. Something is regenerative if it helps to restore Earth's environment or a region's economy or culture in an ecologically beneficial way. Conventional farming and its use of chemicals is degenerative, while organic farming, which is designed to restore the environment, is regenerative. We now know that ecosystems are living things and that chemical agriculture and pollution disrupt and can even kill that living entity.

Soil science understands soil as a living organism, one that can be restored to health and fertility within a matter of months to a few years. By inputting organic matter in a various number of ways, the natural bioindicators of healthy soil return, restoring the natural nutrient cycles. There is then no need for chemical fertilizers.

Regenerative agriculture and the raising of organic food and livestock are thus the foundation of regenerative development. Beyond sustainable agriculture, regenerative farming is designed to *actually restore* the local soil and environment.

The first step in sustainable development in the agrarian areas and cultures of the developing world is to improve farming methods and raise yields without using expensive, polluting pesticides and fertilizers. Farmers in developing countries cannot afford chemicals, and those that can often use them in an unsafe manner.

There are many different variations of regenerative agriculture, depending on the local region. Agro-forestry is another component in regenerative agriculture. Biodiversity on the farm, by planting or leaving forested areas, makes room for beneficial birds and insects that help manage pests. Planting trees for future generations and reclaiming and preserving old trees are worthy projects. Intercropping is another important technique, especially for tropical climates. In Rwanda, maize is grown in between east–west rows of trees, while soy and sweet potatoes are raised in the shade. In Kenya, maize and millet are grown together, reducing labor for weed control by two-thirds.

Agriculture is the foundation of all cultures. So regeneration is centered on it. By teaching farmers how to invest in their soil by feeding it, within three to five years the region's yields and diversification of foods at local markets and stores will soar, raising money circulation and the health index, making a vast improvement in people's lives. Women, especially, are taught how to grow organic home vegetable gardens, greatly increasing their ability to feed their children. Malnourishment would become a thing of the past in a regenerated region.

REGENERATIVE DEVELOPMENT: BOTTOM-UP DEVELOPMENT BASED ON A NEW ECONOMICS

Regenerative development is the next step beyond sustainable development. As discussed in chapter 11, it is based on the work of Robert Rodale in the 1980s, who outlined a new regenerative economics. Regenerative economics is designed to track and restore economic strength from the bottom up rather than the old trickle-down or top-down strategies.

There's a whole system to regional regeneration, and since the 1980s, Rodale International and other NGOs have developed and fine-tuned this model. Rather than dissipating funding, the New Marshall Plan should focus with the bottom-up strategy on the most likely regions for early success. The Global Regeneration Chart presents some flashpoints as well as some opportunities of this bottom-up system. A success story follows the chart to help inspire the rest of the world and serve as model for global regeneration in practice.

GLOBAL REGENERATION CHART

REGENERATIVE INDEX

A regenerative index first establishes a baseline and set of goals to be established. In short, the regenerative index describes an ideal regenerated state, such as full employment (100 percent employment) compared to current levels of local employment. The same goes for health, energy, food, and so on.

The regeneration table explains the measurements contained in the regenerative index and how much money the regenerative zone loses when index levels are low. The whole idea here is to develop each region's regenerative sources in a single zone, unleashing multiple beneficial synergies across the index.

Once the regenerative zone is chosen, a development plan is created with as many people from the region as possible. The idea is to match resources with local needs—as well as possible overseas needs for foodstuffs and other products. At the same time, regenerative agriculture and microcredit programs unlock the potential of the land and the villagers themselves. Rather than depleting natural resources, regenerative agriculture and microcredit build them up.

Based on the regeneration of the land, the regenerative zone progressively opens up more and more regenerative forces within the one region. This allows the natural, untapped synergy of the region to gather together and make the needed leap to a new agrarian/information age. Here the Internet brings farmers and whole communities together, educating them, healing them, transforming their use of energy, and increasing their income directly and indirectly.

The whole aim of the regenerative zone is to raise regional money circulation and self-sufficiency. Farmers suddenly start raising high-yield crops, while microcredit funds village businesses that often double or triple family income. Simultaneously, a regeneration team helps provide solar power, an Internet connection, cultural and community development, and clean water. The team also arranges for lagoons to be dug that serve as natural sewage systems.

ENVIRONMENT

ENVIRONMENT · FOREIGN POLICY

CRISIS	OPPORTUNITY FOR CHANGE
Degradation of the ozone layer, greenhouse effect, burning of Amazonia—prevention is crucial to avoiding global disaster. Not being effectively addressed by UN or world governments. Environmental disaster would destabilize the world.	Stop ozone depletion, dramatically reduce; eliminate burning of all fossil fuels. Healthy biosphere helps to stabilize the planet. Must educate that fossil fuels kill in three ways: climate change, air pollution, and acid rain. We need the electric commuter car.
No computer model of the environment. Little cooperation internationally.	Full cooperation, computer model of collective action. Ozone agreement is a start but not enough.
Misuse of Third World environment by present economic establishment.	Stop bio-rape, restore, reforest, make healthy economic development. Regional regeneration of Third World crucial, especially Latin America and Africa.
United States losing lead in development of alternative energy.	United States could lead the world in electric cars, photo-voltaics, wind machines, and other alternative energies.
United States not analyzing and putting to use the massive wind and solar power available in the Third World.	Analyses could discover the largest wind-farms in the world. The Third World is practically all in the high solar efficiency belt between the Tropics.
No global education here, no globally required course on environment.	Environment course in global, holistic education.

ENVIRONMENT · AGRICULTURE

CRISIS	OPPORTUNITY FOR CHANGE
Pesticide and herbicide poisoning of land and water causing high cancer rate among farmers and, many times, high toxic residue levels in foodstuffs.	IPM (integrated pest management) and organic methods plus regeneration can drastically reduce pollution while increasing profit.
Monoculture environmental disaster from time to time, such as the year grasshoppers ate the Dakotas.	Diversified organic agriculture healthier. Amish farmers have been economically fine all these years.
Unused land left unforested.	Reforest unused farmland. Subsidize drip irrigation.
Large agribusiness wasting land, wiping out small farmers.	Do not promote or subsidize agribusiness, encourage and subsidize family farms.
Old mechanistic agriculture taught to farmers.	Reeducate all farm advisors, entire Department of Agriculture to non-chemical agriculture, regenerative agriculture community.

ENVIRONMENT · EDUCATION

CRISIS	OPPORTUNITY FOR CHANGE
Environmental protection taught poorly, if at all, in global and national classrooms.	Expansion of teaching environmental awareness. "Environmental Awareness" could be a four-year required course in high school, making everyone environmentally literate.
Time, and our knowledge of it, taught poorly, if at all, in global and national classrooms. Poor time-sense promotes short-term actions, degrading the environment.	Evolution of time major strand of four-year expansion of Environmental Awareness course, making everyone aware of a much broader time sense, encouraging long-term action.
Mechanistic education turning out new mechanists; holistic ecological thinking suppressed. Environment can only take so many years of mechanist society. Overspecialization of knowledge major cause of environmental blindness.	Holistic education in this century will create millions of holistically educated scientists. New holistic science/technology paradigm replaces mechanistic.

Mechanistic pessimism promoting negativism and discouraging proactive approach to solving environmental crisis.	Holistic meliorism promotes activism and encourages proactive solutions.

ENVIRONMENT • TECHNOLOGY

CRISIS	OPPORTUNITY FOR CHANGE
Government subsidizes environmentally degrading power solutions, nuclear power plants, and oil. Export of nuclear technology to Third World, increasing chances of Chernobyl-type disasters and nuclear bomb-making capabilities.	Government subsidizes power companies to go alternative in their centralized systems. In 2012, German solar power plants produced a world record twenty-two gigawatts of electricity per hour—equal to twenty nuclear power stations at full capacity. Environment improved. (See Global Regeneration Network for the newest solar technology and alternative technology solutions and updates.)
Gasoline car: 85 percent of national passenger miles are for commuting. Pollution threatens planetary survival with greenhouse effect, acid rain, and lung disease and cancer.	More mass transit and grants for efficient electric commuter car. (Hydrogen fuel cell technology and hybrid cars; electric cars running on computer batteries—see Global Regeneration Network for updated information.)
Non-super-conducting technology terrific waste of energy.	Super-conducting technology will waste zero energy, meaning less power needed.
CFUs and other gases destroying ozone. Rocket flights degrade upper atmosphere.	Global space elevator should be next step for NASA, not space station without elevator.
Corporate abuses of government influence, like the space station plans for global security. "Star Wars" Plan in 1983. Huge expenditure of money for something that will quickly become obsolete and have to be continually upgraded. Thinking only of profit, the station would have been nothing more than a huge target and would have created a space war. (Public opposition defeated the plan.)	Space stations—now many have been converted to peaceful uses such as tracking environmental data to solve global problems. What are the regenerative scientific innovations?

ENVIRONMENT • TECHONOLOGY (continued)

CRISIS	OPPORTUNITY FOR CHANGE
Space programs take away from funding global regeneration of the planet. Earth's self-organizing systems will not allow us to successfully go beyond our limits to get off planet's surface into space unless we evolve higher-level consciousness. "The Dinosaur Principle" means we must evolve our consciousness first.	Invest in global regeneration plans for "Spaceship Earth" instead of, or in addition to, space stations and other space programs. If we foolishly squander the planet's finite resources and continue global wars we will not get very far with space exploration— the "self-organizing," "fail safe system" of Earth will not allow us to leave the solar system until we have reached a certain level of consciousness.
Global models needed but none really exist. Reactive planning builds in insecurity.	Global models, global space agency to create space elevators and space hybrid jets running on solar energy for commercial travel. Plan for space tower in this century.
Technology tends to remove human from nature.	Use technology to bring nature to people more, organize nature walks, programs, and societies.
Little public understanding of the "illusion of technique" or the ramifications of mechanistic dogma in environmental science. Absolutions of various kinds guide many technological projects.	Holistic educational curriculum teaches difference between holism and mechanism. Mechanistic dogma exposed. All absolutisms in technology undermined.
Little public understanding of the "principal of unintended consequences," the finite quality of the Earth and its resources, or the long-term effects of technology. Government agencies approve technologies without fully testing them, as in toxic synergies of chemicals, for example. The mechanistic model fails to take into account human error and limits.	By looking at the interconnectedness of biological systems, the curriculum should teach the interconnected causes and effects of everything man-made. By using the organic or biologic model, a new science, an "organic science," will surpass the previous mechanistic paradigm. This organic science will use mechanisms, but will take a long-term view. The possibility of human error will always remain a factor.

ENVIRONMENT • HUMAN RESOURCES

CRISIS	OPPORTUNITY FOR CHANGE
Children learn to exploit and litter environment.	School children should help clean up community as learning task. Holistic education and organic, regenerative science.
Third World and poor western and Communist use of resources help degrade environment, waste and use up the finite resources of Earth.	Renewable economy puts society more in balance with nature.
Holistic analysis of human resources and environment, region by region.	System analysis of many human resources and environment can greatly increase regional environmental health as well as economic well-being (regeneration).

ENVIRONMENT • PEACE

CRISIS	OPPORTUNITY FOR CHANGE
Environmental dislocations have, through history, created needs and then many of the wars.	Healthier environment promotes less dislocation and thus peace.
Poor knowledge of fragile nature of biosphere encourages selfishness and exploitation. No values are taught in school.	Holistic education teaches fragility of the biosphere, promoting whole Earth, person/planet identification, greatly suppressing war.
Lack of holistic history curriculum keeps everyone separate and ignorant of other cultures and the events that changed the world and the environment to create peace.	Global curriculum teaches the history of change, and the most important people who change the world for the better. The mistakes of history need not be repeated.
Competitive view of resources and human concept of creation of wealth and success. Either/or thinking keeps groups polarized and in conflict and win-lose scenarios.	Cooperation is taught in the schools and in training in the work force and government systems. The both/and thinking brings people together and promotes win-win solutions.

ENVIRONMENT • PEACE (continued)	
CRISIS	**OPPORTUNITY FOR CHANGE**
Universal values are not taught in school. Unless values are taught, children adopt negative, regressive, and destructive patterns and worldviews in regard to the environment, culture, and the future.	Holistic, universal value system is explained early on in schools. Respect and conservation are universal values and constructive worldviews in regard to the environment.
Ultra-nationalism still preventing or retarding international environmental and political cooperation needed to protect it.	Globalism can knock down artificial barriers of many kinds and allow everyone to expand their awareness to encompass the whole needed for cooperation.

AGRICULTURE

AGRICULTURE • EDUCATION	
CRISIS	**OPPORTUNITY FOR CHANGE**
Improving agricultural knowledge but agriculture colleges not teaching IPM, organic agriculture, or family farming.	Holistic education reveals "corruption of science" and damage done by chemical agriculture.
Poor public knowledge of regeneration. Poor rural education in general.	Educate public to regeneration projects through education and media.
No knowledge in agriculture about sixty-year commodities cycles.	To switch over to IPM, need national plan and program, organic education. Need organic co-ops and services set up. Farmers can create organizations and convert themselves over.

AGRICULTURE • TECHNOLOGY	
CRISIS	**OPPORTUNITY FOR CHANGE**
Farmers drained by chemical agriculture and fossil fuel technology. Study fuel and pesticide cost rises in the 1970s, '80s, '90s and on. These can be correlated with economic decline.	Switch over to IPM and transition into organic agriculture. National plan and program is needed.

AGRICULTURE • PEACE

CRISIS	OPPORTUNITY FOR CHANGE
Inadequate, chemical agriculture degrading economies, and environments causing instability around the globe.	Better agriculture reinforces network, promotes stability and peace.
Not enough American farmers and agricultural graduates going overseas to teach agriculture.	Government program to send out-of-work farmers to the Third World. Farmers work overseas for two to four years, then given farm credit loans to stake them in return to farming.

AGRICULTURE • HUMAN RESOURCES

CRISIS	OPPORTUNITY FOR CHANGE
Misdirection of human resources in agriculture. Agribusiness, monoculture farming increasing.	Regenerate America. Regional analyses and action to crate renewable economic self-sufficiency. (No area is 100 percent self-sufficient.)

AGRICULTURE • HEALTH

CRISIS	OPPORTUNITY FOR CHANGE
Poisoning of farm regions is seen in the high rate of unusual cancers among farmers and those living in farm areas. Studies obviously link this to pesticide and herbicide use. Further undermining of economic structure.	Dramatic improvement in quality of life. Drastic reduction in healthcare costs, suffering and death from cancer, neurological diseases, heart disease, stroke, birth defects, and many other conditions and illnesses. Improved economy.

ECONOMICS

ECONOMICS • ENVIRONMENT

CRISIS	OPPORTUNITY FOR CHANGE
Nuclear power plants, traditional economics, and oil dependence delivering big blows to the economy.	Alternatives boost economy and are reliable.
Economics not based on ecology.	Listen to ecology-based economists.
Not based on renewable energies.	Super-conductive photovoltaic utilities, Concentrated photovoltaic technologies, electric cars, mass transit.

ECONOMICS • ENVIRONMENT (continued)

CRISIS	OPPORTUNITY FOR CHANGE
Regional environmental/economic crises like Chesapeake, Long Island Sound.	Restore and understand marine ecology with cleanup and research.
Non-ecology-based economics leads to much greater instability in economic cycles.	Ecology-based economics create a more stable, sustainable economic cycle.
Non-renewable energy leads to greenhouse effect or global climate disruption.	Renewable energies reduce CO_2 emissions, which in turn reduce global warming and climate disruption.
Non-renewable energy produces acid rain and air pollution.	Electric cars, hydrogen fuel cell technology, and solar power stored in solar battery or molten salt reduces and then eliminates man-made acid rain and pollution.

ECONOMICS • FOREIGN POLICY

CRISIS	OPPORTUNITY FOR CHANGE
A "Third Industrial Revolution" has de-industrialized large regions, causing unemployment in America, while the Pacific Rim and China an economic explosion.	Understand and work with this Third Industrial Revolution rather than be damaged by it.
Budget deficit cuts back foreign aid.	Help undeveloped countries become developed on renewables, becoming stable markets for U.S. exports.
Debt crisis undermines banking stability.	Relieving debt crisis. Trade regulations to favor dept-ridden countries.
Centralized approach failing.	Change to successful village approach.
Instability leads to defense arms race.	Stability cuts defense needs, freeing more funds, lowering deficit and taxes.
Starting with the Reagan/Shumway (1983) and through all Bush Administrations, foreign policy only reacts, is short-term not like Carter/Camp David or Clinton's and Obama's efforts that started to build stability. Economic instability.	Promote long-range planning, goodwill, and integrity in global affairs. Democracy must provide the best example and lead the world in reforms. Must keep trying to make peace. Peace brings stability.

ECONOMICS • TECHNOLOGY

CRISIS	OPPORTUNITY FOR CHANGE
Current economic set-up pours technological resources wastefully into complex weapons systems that are unreliable, creating economic instability.	Opportunity to convert war plants into peaceful production, avoiding waste of taxpayer's money, freeing researchers, better investment of funds for the future.
Current economic establishment pours money into polluting, wasteful and dangerous technology.	New, renewable super-efficient and earth-friendly technologies are not wasteful or dangerous and actually generate new wealth.
Current economic establishment promotes gasoline-powered ICE and acid rain. Dependence on foreign oil brings instability.	Electric commuter cars, recharged at night during low-use hours. Good for air pollution, acid rain, economic independence, and electric companies. (No extra power used by electric companies.)
According to Barry Commoner, we currently lose over half of energy to waste.	Much greater efficiency in the future. Computer model of global economy will help prevent new depression or reoccuring recessions. We are still in a general global depression.
Oil subsidies further drain the economy. The money flows out of the country to foreign-owned oil companies.	Invest and subsidize renewable energy technologies based in this country such as Pyron Solar and others. Money recycles directly back into our economy.

ECONOMICS • PEACE

CRISIS	OPPORTUNITY FOR CHANGE
Economics tied strongly to peace/war Cycle. In the Cold War, our armies never "stood down" as they did throughout history. Now the practice of keeping multi-million men armies deployed in terrorist and oil wars keeps defense industry strong, other industries weakened or on the brink.	Ending wars and unnecessary occupations while working with allies to promote and keep the peace will free great economic resources to help all nations achieve a whole new, strong peacetime economy.

ECONOMICS • PEACE (continued)

CRISIS	OPPORTUNITY FOR CHANGE
Maintaining defense spending or increasing defense spending in times of relative peace sends money into a "black hole." The money never recycles back into the economy. Money and resources are lost forever.	Completely reform armies and military to train for peaceful regenerative activities when not actively engaged in defense. Empty barracks and military bases should be converted into housing for the homeless and jobs for the unemployed. The concept of the Samurai.
Increasing global economic interdependence usually baffling to most politicians and businessmen inadvertently blocks a healthy global interdependence, which in turn helps to block true peace.	Recognizing and promoting a healthy global village and global interdependence will help make war unthinkable. Regeneration projects heal wounds and bring diverse peoples together.

ECONOMICS • HEALTH

CRISIS	OPPORTUNITY FOR CHANGE
Diet, pollution, technology all directly connected to disease, high health care costs.	Education and government policy can reduce this environmental crisis and lower health care costs.
Poor health education on preventive medicine. (Vitamins, occupational hazards, etc.) High costs to taxpayers and in unintended consequences.	Good health education and preventive medicine save billions and prevent loss of wages, wealth, and tax revenues.

ECONOMICS • EDUCATION

CRISIS	OPPORTUNITY FOR CHANGE
Economic instability leads to cutbacks on needed school funds and services, reduces the number of subjects taught. It leads to more inequality between people, erodes the middle class, and increases poverty.	More stable, renewable-energy economics.
Mechanistic economics taught. Basic economics never taught in high school.	Holistic ecology-based economics taught in high school.
Teachers not paid as professionals. Degraded and undervalued.	Education should be seen as the "core industry" of the nation.

ECONOMICS • AGRICULTURE

CRISIS	OPPORTUNITY FOR CHANGE
Old oil-based economic establishment draining farmers dry.	New ecology-based economics protect and replenish agriculture resources.
Ignorance of sixty-year commodities cycle leads to misguided business decisions.	Understanding of sixty-year cycle will prevent farmers from buying land on loans at top of cycle.
Old agri/economics structure cheats farmers.	Regeneration solutions such as in Lehigh Valley, Pennsylvania.
Current agriculture policy promotes monoculture, soil depletion, and chemical contamination and creates economic dislocations as small farmers go bankrupt.	Diversification, IPM, and organic agriculture leads to fewer dislocations, more production, and more wealth.

FOREIGN POLICY

FOREIGN POLICY • AGRICULTURE

CRISIS	OPPORTUNITY FOR CHANGE
Third World starved to death by "seven sisters of grain."	Regenerate Third World through holistic analysis of regions and community/regional regeneration plans.
U.S. agriculture monoculture system subject to world price and demand swings.	Diversity in U.S. agriculture and other nations.
Sahara is moving south.	Stop the Sahara with barbed wire, wills, panning, and enforcement to prevent overgrazing.
Countries decimated by famine, environmental diasters, and drought.	Work with super powers on terracing irrigation and waterworks and rebuilding regeneration.
Amazon being burned up for ranching by hungry farmers.	Change trade regulations to favor Latin America instead of Pacific Rim to create manufacturing in Brazil.

FOREIGN POLICY • EDUCATION

CRISIS	OPPORTUNITY FOR CHANGE
Many people all over the world don't go to school; most can't go past sixth grade. Ignorance breeds war, intolerance, and poverty.	Global, holistic education with holistic history curriculum teaching the compression and convergence of humanity.
Students learn U.S. history three times from K–12. Little world history or culture taught.	Holistic high school education based on "horizontal" curriculum using history of consciousness to teach everything except foreign language and math. Students know how to synthesize as well as analyze.
Political favors and positions given to ambassadors. No education or training in global studies, awareness, or regeneration.	Ambassadors should come up through ranks from foreign service or be specially trained.
Oil and nuclear technology wreaking economic and environmental havoc in the Third World and elsewhere. Education geared toward chemical and mechanical technology.	With the proper holistic education communities turn to renewable energy and technologies and an ecology-based economics.
No inexpensive computer network to educate the Third World professionals, like doctors and paramedics, to help make diagnosis.	Global network in every aid station in the Third World, every hospital.
Global education not computerized.	Holistic education software distributed globally by UNESCO in over a hundred languages.

FOREIGN POLICY • HEALTH

CRISIS	OPPORTUNITY FOR CHANGE
Shortage of hypodermic needles and poor or no sterilization helped the rapid spread of AIDS in Africa and elsewhere.	Radically improve medical knowledge in the Third World. Global hookup through computers and translation software to plug Third World into medical diagnostic databases.

Poverty breeds disease, creating economic drain and even more poverty.	Economic regeneration projects reduce poverty. Global health education and more education in general will reduce disease.
Water shortage and contamination due to little or no planning, conservation, and water systems.	Global campaign to improve water supply, planning, and quality.
Lack of education and available contraceptive promotion create too many pregnancies and population explosion in the Third World.	More education proven to reduce pregnancy rate. Best way to prevent a tragic, dangerous situation in the Third World in this century is to educate children so the pregnancy rate goes down.

FOREIGN POLICY • HUMAN RESOURCES

CRISIS	OPPORTUNITY FOR CHANGE
Misdirected human resources create and perpetuate foreign policy crises (economic and military), result of extremism: fundamentalism, mechanistic, or not enough education.	Holistic global education and regeneration/regional projects promote shift to holistic, organic paradigm in this century. This will bring convergence of the global culture, redirecting human resources toward the Regeneration Network. Holistic analysis of history can reveal many solutions and point out the wrong directions.
Foreign Service content to perpetuate status quo or worsen the crisis.	Foreign Service should help other countries realize Regeneration Network and accomplish it.

FOREIGN POLICY • PEACE

CRISIS	OPPORTUNITY FOR CHANGE
"Peace through strength" makes debilitating arms race, promoting military-industrial-complex influence.	"Peace through convergence" promotes cultural convergence, making war unthinkable. Already underway naturally.
Centralized planning for foreign aid is huge disaster, degraded by poor or no planning and gross corruption.	Reward democracy not dictatorships. Regeneration village plan for goods, services, and technology rather than money whenever possible to avoid corruption and hidden exploitation.

FOREIGN POLICY • PEACE (continued)	
CRISIS	**OPPORTUNITY FOR CHANGE**
Third World debt bomb will eventually explode, causing a worldwide depression, a climate for revolution and war. Bank's paper band-aids make problems worse.	Defuse debt bomb with trade regulation changes to favor Africa and Latin American at expense of Pacific. Marshall Plan–type action for Latin America.
Bank loans to dictatorships, arms exchanges with non-democratic countries (some actual enemies); money to pseudo-democratic revolutionaries.	Arms industry is not a commodity or export item, ethical policies (ethical investment); support in non-money ways.

SENEGAL:
A REGENERATIVE SUCCESS STORY

In Senegal, colonial peanut farming based on chemical methods had severely depleted the soil, and the natural ecosystem had broken down. After years of regeneration—by Rodale teaching local farmers—soil health was restored, crop yields improved, and the natural wildlife returned. With Rodale International's regenerative training, successful regenerative farmers in Senegal have, on the average, doubled their income, typically selling their new bounty at local farmers' markets. The region has become more self-sufficient. In Senegal, Rodale has also:

- Trained 50,000 farmers in various regenerative agricultural techniques with an adoption rate of 60 percent.
- Taught use of natural pesticides as well as improved planting techniques to 340 farmers.
- Introduced improved varieties, plant protection, and the establishment of agro-forestry plantations in seven villages in Senegal.
- Improved an estimated 50,000 hectares by Rodale activities.

- Planted 11,000 trees as windbreaks and live fences around home vegetable gardens.
- Increased revenue in pilots due to improved production and microcredit.

According to the Rodale Institute Research Center, "Simply put, every management step is focused primarily on improving the soil ecosystem in favor of moisture and nutrient retention and availability. This regenerative strategy represents a major shift in emphasis from a short-term, production-oriented strategy to a long-term, rehabilitative strategy in which farmers invest in the soil resource as a first priority, and subsequently receive the long-term benefit of increased crop yields and sustained production."[1]

The most powerful factor in the reeducation of Senegal turned out to be word-of-mouth communication. Many times villagers had heard of the new farming before the Rodale people could even get to them and had implemented the organic methods by themselves. Increased use of the Internet in the Third World will leverage development plans, as online dialogues and workshops link villages within a region. The Internet is the new word-of-mouth media.

This is but a small part of the work done by just one NGO in Senegal. Their experience there and in other parts of the world—and the experience of other NGOs—shows that the ideal small organic farm is five acres, with ten to fifteen field and cover crops per year, five to ten tree crops, and three to five types of livestock. More crowded lands require the teaching of biointensive gardening—the science of using raised beds to get the most out of vegetable and fruit production out of maybe half an acre.

THE REGENERATIVE ZONE DEVELOPMENT METHOD

Each regenerative zone has a planner assigned to create the zone development plan and then oversee the regeneration process. This person must be a facilitator or guide to the transformation; they must not see themselves as the leader or authority. A development plan task force is set up, made up of many different elements within the region. One measure of success would be the number of people the planner is able to bring into the planning process. What follows are general guidelines for regenerative development, as first identified by Robert Rodale and updated for the global regeneration plan.

Step 1: Set Up Zone Team and Development Task Force
Project Development
Train a regenerative zone development team with several subregional regeneration teams.

Zone Actions and Implementation
Local people are trained in regeneration. They help bring together a zone development task force, which operates by consensus.

Step 2: Define, Evaluate, and Inventory Zone
Project Development
Define the geographical limits of the zone. Formulate a baseline regenerative index for zone, which becomes the baseline data.

Zone Actions and Implementation
Make an inventory of land and the potential of soil, climate, water, and natural resources. Also make an inventory of industrial energy, materials, food, water, and capital in and out of zone.

Step 3: Market Search Establishes Resources and the Potential of the Zone to Regenerate
Project Development
- AgMarket Search. Survey all in-zone bulk buyers of food. Replace their buys with local products; usually a large market exists.
- Farmer Search. A zone survey locates all farmers producing commodities. The survey is made available to buyers and consumers.

The regeneration team begins signing up farmers who want to learn regenerative agriculture.

- Business Market Search. Survey all current businesses as well as potential business opportunities, no matter how small.

Zone Actions and Implementation

- Energy Search. Make a market search of all in-zone users of energy: homes, businesses, and government. Alternative energy businesses are financed. Potential for solar home systems and GreenStar Community Centers are identified.
- Wellness Search. Make a survey of needed wellness programs. Find the potential to expand programs; put clinics in place and direct toward prevention.
- Shelter Search. Make a survey of needed shelter in the zone as well as property owners to identify the potential for renovation by in-zone workers. Identify potential subsoils and villages where compressed earth block housing can best be built.

Step 4: Outreach and Implementation

Project Development

- Farmers' Market Day Shows and Sign-ups. High-powered video-driven shows on market days educate and sign up farmers from several villages who receive free cover crop seed, tree seedlings, and even upgraded livestock for signing up and pledging to go regenerative. Follow-up visits to farms produce success and income doubles and triples. (Just home vegetable gardening taught to women is an enormous boon.)
- Meetings with Local Groups. The zone development plan is presented and every village in the region is invited to join and receive benefits.
- Microcredit Groups. Microcredit groups in villages start microenterprises, doubling and tripling income. Farmers can also join and get equipment, seed, and so on.

Zone Actions and Implementation

- Progress Monitored by Regenerative Index. Progress is measured by how much the index moves against its original numbers.
- Wellness Outreach. Expansion of wellness programs; clinics are put in place, creating a strong push toward prevention.
- Shelter Businesses, Sweat Equity. Local villagers are encouraged to form businesses to build earth-block housing, or villages can create co-ops. Those who join help others get housing, and then they themselves receive an earth-block home.
- Education. Put into place Market Day shows, GreenStar Community Centers, and web mobiles to provide agriculture, health, business, and basic literacy and education over the Internet. Make video clips in the region's main native language. Global online curriculum for local high schools has been put into place.

Step 5: Follow-up, Evaluation, and Modification

Project Development

- Farmer Follow-up. Villages are visited once a month to make sure they are doing well with the organic farming techniques. Study circles of farmers go over methods and teach problem solving. Further web video talk ties region's farmers together in support group.
- Microloan Follow-up. Regeneration team leads circles of fifteen or so participants on starting and running a business, how to market, and how to make a profit.
- Market Development. Market development director searches out and develops any and all in-zone and out-of-zone markets for region's products. Cultural arts are sold in and out of zone.
- Securing New Funding. Regenerate the rest of villages or communities in the region. Regional money circulation soars.

Zone Actions and Implementation

- Regenerative Index of Zone Monitored. Close monitoring of the zone's regenerative index allows effective evaluation of development process. As components of plan fall behind or exceed expectations, the plan is adjusted accordingly.

- Report Three to Five Years Out. Soil is fully regenerated and able to produce decent yields even in times of drought. Microloan revolving fund has gone through several microcredit cycles and expanded to rest of region. Money circulation approaches 25 to 30 percent rise.
- Light Industrial and Exports Grow. After leaping to the new level, light industry and exports develop.

This is the step-by-step method of the bottom-up approach. Within two to three years, money circulation within the regenerative zone should arise, starting the land and the people on their way to full regeneration and a new chance at the future. Within three to five years, full regeneration of the soil and the impact of the multiple regenerative sources within the one zone should lift that region up to a 15 to 25 percent money circulation increase and a corresponding rise in self-sufficiency.

Pilot regions in areas of the world where it is most needed, such as famine and drought areas in India, Asia, and Africa, would begin the regeneration process immediately. Areas devastated by natural and man-made environmental disasters, such as Haiti's and Eastern Japan's quakes and tsunamis, the Fukashima nuclear power plant disaster area, hurricanes such as Katrina in the United States, and man-made disasters like the Gulf of Mexico's BP oil spill, are areas that need regeneration projects. The deforestation and devastation of the Amazon jungle would use regenerative projects to restore the jungle in an organic way, saving the world's largest CO_2 sequestering region (and atmospheric oxygen generator) while saving innumerable plant and wildlife species from extinction.

War zones need regeneration. Areas where genocide has destroyed populations and infrastructure and places where an ecological crisis combined with violent conflict, like the drought and genocide in Darfur, are still in need of regeneration. The holistic approach of global regeneration will create partnerships with many different humanitarian organizations and government efforts. Regenerative indexes will be tracked online for each zone, allowing donors to see how their contributions are producing results. This will also allow for fine-tuning of the regenerative zone development method before bringing it to other areas of the world.

Current regenerative-type organizations, such as Still Require Aid (SRA), have projects in various parts of the world. SRA creates partnerships, coordinating funds, and workforce from entities such as UN relief, Hollywood philanthropists, Christie Communications, the Village Reforestation and Advancement Initiative (VRAI), Garden of Eatin', Common Ground Relief, Comprehensive Development Program (CODEP), and charter schools, to name just a few. The Global Regeneration Network's mission is to connect all of the world's projects, foundations, NGOs, charity groups, solutions, and resources. A fairly extensive listing of these is available at www.GlobalRegen.net. You can explore this ongoing network to find a project you might support or get involved in. Projects using the new regenerative model and all the organic methods do not really exist yet. In the future, www .GlobalRegen.net will be working to help create "true regenerative" projects, which will be self-replicating in order to build a new interconnected regenerative planetary culture—a truly awakened humanity.

PROGRESSIVE GLOBAL EDUCATION FOR THE WORLD

THE CURRENT STATE OF EDUCATION

The first step in understanding how education can be changed at its core is to deconstruct the recent history of education and see how it can be improved. Today's curriculum is centered on passing the SATs and other standardized tests. In the academic classes, we now mostly teach how to pass the test, how to get by in the school bureaucracy, not how to reach one's full potential or to succeed in the new economy world. Although many schools do their best with the old structure, for others it is a complete fiasco, beset by a 30 percent dropout rate, plummeting test scores, and the wasting of billions of dollars. Recent test scores have gone up in many cases only because dropouts and underachievers were dropped from the statistics.

Our current curriculum structure developed from the early 1800s to the early 1920s and then "devolved" into the problematic system we have today. Education needs reform at its core—the outdated curriculum. Because there is no integrated narrative to the curriculum, students are left to fend for themselves in a great sea of detail. Sorting out all that detail in order to find meaningful information becomes a frustrating and boring task. Students are thus demotivated because they do not see the value in the outdated and irrelevant information they are learning—this when the world around them is lurching through multiple global dilemmas.

Fig. A2.1. The sea of detail currently taught in our educational curriculums must be replaced by an integrated narrative that helps us move toward a progressive future.

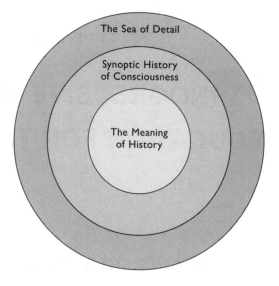

High school does not even properly prepare young people for making their way in the new realm of computers, the Internet, and new sustainable technologies and is not focused on teaching skills and information in preparation for meaningful and long-lasting careers related to students' individual abilities, talents, and interests. With the world in crisis, most high school graduates emerge from twelve years of formal education with no idea what they want to do. The students who do know what they want to do are unable to find their way into higher education due to lack of financial resources. All students are unprepared to be fully functioning adults. They leave high school and are faced with a pitifully deficient job market. In short, traditional education is in deep crisis.

There has not been a recent reshaping of the high school curriculum, based on the twentieth century and all its changes. Nor has there been a teaching of new-paradigm science and thinking—now quite substantial—which leads the way out of the modern dilemma of war, pollution, poverty, and slavery to a system that doesn't work anymore.

HOLISTIC AND
MOTIVATIONAL INSTRUCTION

Our current public high school curriculum is a watered-down version of the concepts of just a few original thinkers; in fact, there are many original thinkers who have created extraordinary lives and careers that we know nothing, or very little, about. These thinkers and innovators would be the best instructional models for all students. The global curriculum teaches the stories of men and women throughout history who made a difference and changed the world, along with global awareness— a new way of thinking. By using this model, the education becomes meaningful and inspiring and motivates students to follow their hearts. The new education would show them how to find their path, their purpose in life. Learning stories from around the world about how change happened and who made it happen develops the awareness and respect of other cultures and lays the foundation for cooperative and collaborative global economies and peacemaking.

In contrast to the current inadequate and fragmented curriculum, the global curriculum is a multidiscipline framework. It provides a foundation of thought for the rest of the curriculum to connect to, a map to navigate the sea of detail. It is synoptic, giving the most important information in all areas of study and leaving details to be researched and discovered by students when they decide what field they want to study in depth. It is global-studies based, linking high school subjects with history, through a presentation method that goes from time horizon to time horizon. It is thus a *horizontal* rather than a *vertical* presentation. Such a restructuring emphasizes periods of change in the past and reveals interconnections between disciplines—such as between science and culture, philosophy and art, East and West. Beyond making multidisciplinary and cultural connections, a number of beneficial things occur when this horizontal approach is taken.

Students perform better academically because the curriculum has a single narrative, starting with the simple human societies and progressing to the more complex. Younger students study the earlier, simpler times, building their understanding one global change at a time. Complex concepts and events are taught through human stories and the dynamics of social transformation, rather than through rote learning and the accumulation of meaningless facts. In this way, students can better understand modern history and current events. Because they have a context and deeper understanding, they can focus on what is truly important and retention is vastly increased.

The whole process is designed to create self-directed learners, who are motivated enough to teach themselves. This global curriculum and holistic approach to learning puts within students' reach a level of sophistication and interest not possible with the current high school curriculum.

The main question, however, still remains: What are the specific topics to be taught? What is most meaningful? What must one know to be culturally literate in the twenty-first century? What would a global curriculum entail?

SYNTROPY: THE KEY TO WHAT TO LEARN AND WHAT TO DO

How do we know what to emphasize and what should be saved for later—for specialized study in high school and college? The global curriculum is based on time horizons, history, and global awareness. Since human history developed out of the evolution of life, understanding how that occurred holds insights for our study of the history of everything. The new curriculum therefore first studies evolution then compares the development of human culture to the biological model. This comparison helps identify the key events, trends, and people to study.

There is a new-paradigm understanding of evolution. Darwin thought it to be gradual and constant, yet the fossil record indicates something far different. There were long periods of genetic equilibrium that were suddenly upset by rapid leaps, by sudden transformations, creating a staircaselike chart of change. It is called the punctuated equilibrium theory, put forth by Stephen Jay Gould and his colleagues.

Change accumulates around genetic attractor points, which suddenly explode into new species and families of organisms.

Today, the field understands evolution and its ever-increasing complexity of life as the opposite of entropy, the seemingly inexorable force that is slowly winding down the universe. In what could eventually be seen as the greatest discovery of the twentieth century, Teilhard de Chardin was the first to realize that the underlying movement of evolution is in the direction of reverse entropy.

Evolution is a movement toward ever more complex arrangements of molecules—the opposite of the entropic breaking of the bonds between molecules. The entropy charts have been flipped over to point upward, as evolution is now explained by this reverse entropy or, as Buckminster Fuller called it, syntropy—the movement toward greater and greater syntheses. Rather than a clockwork universe whose spring is slowly winding down, we live in a self-organizing universe, which is climbing upward toward syntheses. Here science begins to tread on ground formerly reserved for spiritual beliefs; the new science of chaos theory and evolution touch on the spiritual realm.

Syntropy explains all of biological evolution and also uncovers the central meaning of history. Think of all the sudden leaps in social complexity and sophistication such as the farming revolution, Classic Greece, or the Industrial Revolution—as well as the long, dark age equilibriums in between. Syntropic attractor points in history are explained as springboards for change that power the transformational leaps to the future.

These syntropic attractors are the key. Rather than smothering students in the seemingly irrelevant details, an integrated lesson concentrates on teaching the attractor points of our past, the most significant periods of innovation and the most important thinkers. Knowing these facts and people is the core curriculum that the world needs. Syntropy therefore is key to understanding life evolution and human history and helping us to decide what should be learned.

What do we teach and what do we leave for later? That is the main question in creating a new curriculum based on history. Who and what led humanity to leap forward? We can actually use Prigogine's chaos theory to measure the significance of all information, permitting the creation of a far more meaningful curriculum. Events, trends, and thinkers that led to great syntropic social change—or prevented it—are thus presented in the global curriculum. The usual details of the history of war and politics are secondary to this core learning.

Some detail is taught, yet it is hung on a solid framework in the horizontal curriculum, rather than being part of the "data smog" of the vertical structure. Instead of learning science, students live science through projects and experiential simulations. Many of the vertical strands remain, but they are woven into the horizontal framework, presented at their appropriate place within the time horizon project of the day.

We can divide all information into three layers. The outermost layer is the vast sea of detail, which includes every bit of information ever produced by humankind and nature. Then there is the middle layer, the synoptic history of consciousness and change, how society was transformed at various points in the past. Finally, the innermost layer is the syntropic meaning of history itself, how syntropy worked in each period of change to lift humanity up to new levels of awareness and social structure. In this center is the series of core transformations that make up human history. These inner two layers are what every student (and adult) should know. It is the foundation for the rest of knowledge.

With the new syntropic model of learning, we then can apply this learning style to a way of living and acting in the world. This is how we can tackle the gargantuan task of global regeneration. At first it may seem impossible because we are used to learning and seeing everything in unconnected and fragmented details. To plan for global change, we must apply what we have learned in the global curriculum to the here and now.

The global curriculum asks questions. It recognizes natural curiosity as both the *starting point* and the *destination* of learning. It teaches through nurturing curiosity: We have to *want* to know. Teaching intelligent inquiry will develop minds that can think creatively and critically—minds that will *want* to *know*. The global curriculum starts with the questions: Where have we been? Where are we now? Where are we going? What is *our* story? What is *your* story? How can we change the story that we are in? What are the answers to the problems we face?

Most of what you've just read about the global curriculum is a higher-level curriculum framework, something on the level of high school or college. These ideas and methods can be broken down into simpler concepts and "chunks" to prepare young students for their future roles in a sustainable future as individuals acting in the community and on a global scale—and *it can be great fun!*

Such a curriculum already exists. The global curriculum is available as an online progressive core curriculum at GlobalCurriculum.com. It can be used as a high school core curriculum program or an introductory college level course—even a junior high gifted program or enrichment program. A lifelong learning course is also available for adults.

THE NEW EDUCATION: VALUING ALL SKILLS AND INTELLIGENCES

During the nineteenth century, new directions in education were brought to the fore. In John Dewey and William James's progressive education, the idea emerged that acquiring skills and doing creative exploration in one domain could overlap into all other domains. Art, music, physical education, home economics, and mechanical skills could improve skills in reading, writing, and arithmetic—the types of learning so prized in conventional education, both then and now.

Fifty years later, the work of Howard Gardner on multiple intelligences confirmed this idea and expanded on it. Not only is there no abstract or innate hierarchy between the skills that make a successful academic or political leader higher than those of, say, a skilled mechanic or farmer, there is now evidence that skill sets acquired in one domain do indeed transfer to another. A hedge-fund trader or high-level politician who has worked as a plumber or carpenter develops an intrinsic respect for the overlay of skills required by those professions. He or she is therefore less likely to demand that such persons work for low wages especially when those too-easily-demeaned skill sets keep your water flowing or your house standing.

Currently, the world is rent by skewed values, by making the accumulation of wealth an end in itself. Being rich is a kind of fulfillment of the American Dream. We all admire, and perhaps rightly so, people who have overcome humble economic origins to found wealthy dynasties. But is trading on the stock market, or acquiring other corporations, or divesting oneself of assets because of insider information a higher skill set than that of a committed teacher, construction worker, policeman, or fireman? These are the kinds of unconscious hierarchical assumptions that are being tested in today's political and economic climate. Is there a kind of arrogance or "inflation" that comes with wealth? Conversely is there a kind of automatic self-devaluation and inferiority complex that comes with poverty?

A school teaching the new education with a global curriculum should endeavor to redress such imbalances at the starting gate. Hard work in the trades certainly should be rewarded, as should education in a profession (which always involves hard work). Mature citizens whose work included not only a stint in the army, but also in the Peace Corps, the Longshoremen's Union, a hospital emergency room, or an inner city classroom are likely to be not only more broadly versed but more tolerant of those who make their life professions in one of these occupations.

And what of elementary survival skills? What educated person could start a fire without matches, build a survival shelter, or find food in the wild? Who learns in elementary school to read the night sky, use a compass, or sense the approach of storms by proprioceptive signals or animal behavior? Why are the works of Tom Brown Jr., a naturalist and wilderness tracker, not included in elementary school curricula, and what of Lyall Watson, Rupert Sheldrake, and others, who teach us about the vast intelligence of the complex living systems all about us? We all have been exposed to foreign language instruction in the schools, but what about courses that teach us how to learn a language generically, rather than a particular one? Where in the early elementary curricula, for that matter, are courses that teach us how to learn?

This leads us to the core of this appendix, which is about the "how to" of cultural and global transformation. Which of us was given proprioceptive training as children? This would include how to monitor our own subjective states and conditions in a variety of circumstances. Does this setting (group, situation) make me comfortable or uncomfortable? Why? How can I learn self-soothing (especially if I am high-strung or reactive)? Are there healthful and less-healthful ways of doing something as everyday as breathing, eating, exercising, or making use of free time? What constitutes good sleep? What is dreaming for? Do dreams (or daydreams) have a meaning? How do I learn to monitor the vagaries and fluctuations of my own consciousness? How do I manage my attention deficit or ADD? Or even decide if that is a problem? These questions address the elemental preconditions of our very being in the world and the very basic functioning of our consciousness.

Corollary questions that come to mind are: Why do we wait till college to study psychology? Why aren't health and self-management skills taught in the very elementary years? Could kindergarteners *learn how to learn* before they start learning anything else? Why are physical education courses geared only toward competitive sports instead of teaching every boy and every girl techniques of skillful use of the human body, such as how to relax or how to use the natural human urge for excitement or "pushing the limits" without damaging ourselves permanently in the process? Is there sex education beyond gross anatomical charts, dismal facts, and dire warnings? How do we help children accept the inevitabilities of their sexual natures without acquiring sexually transmitted diseases or becoming pregnant? And if a young woman becomes pregnant, what better opportunity for planning parenthood or realizing one is in no way equipped for it? Could there be any more important zone for a skillful and sensitive educational process than this? What an opportunity for adults to transcend their own flawed and skewed educations and really rise to the occasion of setting the next generation off to a wiser and sounder experience.

In the last few decades, incredible tools for health management and self-skills have become available; in fact they have entered the popular lexicon: yoga, meditation, qi gong, martial arts, shiatsu, reiki, jin shin jutsu, Alexander and Feldenkrais techniques, Matrix Energetics, Brain Gym, positive psychology, positive parenting. Classes in the above arts should not be restricted to adults, nor weekend spas or growth centers. They should be there in the primary schools and high schools and available on every college campus, along with opportunities for personal coaching or counseling, so students can find their ways in experiential encounters with available methods.

The global curriculum would have students read the seminal works from famous authors instead of using the standard textbooks. Curriculum development committees in the schools along with the school librarian would develop and spend their budgets on alternative progressive sources of knowledge. Individual teachers might use textbooks in a limited way—sifting the information, discarding useless details, and presenting the essential information. Teachers would focus mainly on authoritative works that could be summarized for the students.

Currently, all standard textbooks are controlled by a group of fifteen people who are part of the Texas Department of Education Board. The mandated guidelines for content in U.S. schools were amended in 2010 by the conservative Texas Textbook Board to conform to the right-wing agenda. All publishers of textbooks drafts must be submitted and approved by this board. The board has ten Republican members and five Democratic members. Thomas Jefferson was removed from the list of figures whose writings inspired the revolutions of the eighteenth and nineteenth centuries and replaced by John Calvin. Martin Luther King Jr. was marginalized to make room for Phyllis Schlafly, Contract with America, the Heritage Foundation, and the National Rifle Association. Evolution is now presented as an unproven theory, and the separation of church and state is overthrown by false interpretations of the Constitution and the founding fathers. Standard textbooks are useless and can't give our children the information and thinking skills that they need to survive in the real world.

If we keep it relevant and down to Earth, inspiring and creative, it will be fun! Education and the chance to learn and develop is a *gift* we give our children and future generations. To do this we need to *learn* from our children, teach them by example to be self-educating and life-long learners, and then teach our children to be teachers to the next generation—to pass it forward. First, *we* have to learn and *be* the joy of living and learning: asking questions, making inquiries, finding the answers and then *acting* on them.

A NEW EDUCATION
DELIVERED WORLDWIDE

At the same time regional economies are being lifted up by the bottom-up microcredit and regenerative agriculture programs, another part of any regenerative zone development plan is using the global curriculum. Due to a lack of resources, 125 million children do not go to school in the world today. In many Islamic countries, fundamentalist schools have filled the void, offering the only education available. For a mere $6 billion to $9 billion a year, donor countries and foundations can educate the Third World. Education is key to a workable future in so many ways, but since 9/11 it has also become a matter of national and global security.

The core curriculum for world education is based on global studies and history. East, West, and everywhere in between are covered in this new horizontal educational structure. Whereas the old Herbartian curriculum of the West stifled multidisciplinary and creative thought and presented the world through Eurocentric lenses, the new system looks at the world as a whole, time horizon by time horizon. Its interactive design also promotes creativity and writing.

By introducing people to the world as a whole, the global curriculum promotes cooperation and trust among nations, as each is taught the constructive parallel strands within neighboring cultures, not just the negative stories and half truths. In particular, the role of Islamic science will stand out in the global curriculum, as it was Islam that taught the West the science and math of the Greeks and Romans, leading directly to the Renaissance. In the sixteenth century, Islamic religious doctrine shut down learning, weakening the position of Muslim countries everywhere.

Whereas the West used to be well behind Islam, by the eighteenth century, the armies of Europe began to regularly defeat those of the Turks and Arabs. This rise of the West is what drives some fundamentalist sects to call for violence, to overthrow the infidel and restore the supremacy of Islam. Rather than the fundamentalist call for violence, however, the best way for Islam to reach parity with the West is to restore the greatness of Islamic science and learning. This begins by teaching one generation after another through a structure like the global curriculum. Using the history of science to change Islamic attitudes toward modernity will, in the end, appeal to the man and woman in the street and the young.

The global curriculum will be delivered online as well as on computer software disks. A new, updated universal curriculum would be delivered in a single simplified package over the Internet and satellite in the local language or translated by local teachers from English to the native tongue. Instructors could receive a password and be given their entire year's lesson plan to customize for their classes. Paper materials for student binders can be printed straight from the browser. Web mobiles, already in use in Malaysia to bridge the digital divide, can deliver twenty computers by bus to all the villages within a region, enabling them to connect via the Internet.

Rockefeller Technology Grants and the Gates Foundation have already pledged to contribute the computer connections and web hookups for many remote villages. This must be expanded exponentially. The global curriculum, on the web and in the classroom, is a practical solution for world education and the rise of a strong global movement toward cultural understanding. This core curriculum is thus the foundation of the global regeneration plan—for without global education, the future is far more problematic.

LOW-IMPACT TECHNOLOGY AND RENEWABLE ENERGY

QUICK IMPLEMENTATION OF ALTERNATIVE ENERGY CRUCIAL

All the development of the Third World—and the industrialized countries—must be powered by alternative energy and low-impact technology to be sustainable. Atmospheric CO_2 levels already stand at 395 parts per million (ppm) throughout the world and 400 ppm at the poles,[1] up from 315 ppm in the 1950s—and it will be the Third World that gets hit the hardest by the early greenhouse effects. The UN Environmental Programme projects that at 450 ppm, CO_2 damage will become irreversible, as climate change and rising seas overwhelm the ability of the global economy to repair the damage—creating an unstoppable downward spiral.

As mentioned earlier, Bill McKibben, of 350.org has created a global project so that people and governments all over the globe become aware of the need to reduce carbon emissions to 350 ppm and then reduce to 300 ppm before it is too late. This is the amount that we must now make all possible efforts to achieve to stabilize and start reversing the course of global climate change.

PROMISING TECHNOLOGIES

Some of the most promising new technologies for the world include the following:

- **Biodiesel, Methane Fuel, and Air Fuel Synthesis Petrol**
  Biodiesel fuel can be created out of vegetable oil, especially soy and hemp oil, through a simple refining and cracking process. Biodiesel can be used directly in nonmodified diesel engines, emitting 86 percent less black smoke and pollution. At present, soy diesel costs seventy-five cents more per gallon than conventional diesel. Mass production would clearly bring this gap down. In China, the International Center for Sustainable Development is creating one-hundred-person villages powered by methane made from animal and human manure, with the remainder of the manure turned into compost. "Air Fuel Synthetic Petrol, the first 'petrol from air,' is produced by extracting CO_2 from the air, and hydrogen from water. It is a carbon neutral synthetic liquid fuel when renewable energy is used to produce it."[2]

- **Solar Power** Photovoltaic (PV) panels have 15 percent efficiency now: $4 per watts peak (Wp) down from $20 per Wp. Solar PVs produce 1,200 megawatts (MW) today in four hundred thousand to eight hundred thousand systems worldwide, a 200 MW per year growth, up from 40 MW per year in 1990. A typical 50W PV solar home system in the Third World offsets the emissions of about 400 kilograms of CO_2 on a yearly basis. Solar panels, solar ovens, passive solar, and solar hot water have all become more efficient and less expensive to install over the last ten years, with brilliant new technology and applications that have come about in 2010, 2011, and 2012, and will emerge in 2013 and beyond.

  Developing countries situated near the equator have far more solar insulation than the developed world, thus solar is a natural low-impact energy source. In 2005, fifty thousand solar water pumps in developing countries have been a great boon already. Many households run TVs on car batteries they charge once a week, transporting them miles to battery-charging services. These families are the first to buy the $350 solar home system. The World Bank has pilot programs and has done much in this area but should now scale up and go global by financing PV pumps, village charging stations, and solar home systems. A program using microcredit to start and then expand PV sales and service companies should be a major World Bank program in the New Marshall Plan. Solar PV building materials will be a large future source of energy. The simple construction of solar ovens, replacing precious trees used as firewood, can have a big beneficial impact in a village.

- **Wind** With 18 calories for every 1 calorie put in, wind has *the best efficiency of any renewable.* In 2005, 17,500 MW were installed in thirty countries. The average generator size is now 750 kilowatts (kW), up from 250 kW. Small wind generators are the best option in many areas for battery-charging stations and farmers selling energy. Wind farms are possible!
- **Biomass** 14,000 MW today.
- **Geothermal** Nine thousand MW of geothermal power are already produced. Geothermal heat pumps provide heat in winter and cool in summer by using ground temperatures.
- **Microhydro** A five-foot drop and 550 gallons per minute (GPM) will produce nearly 250 watts of continuous power (650 watts with the medium unit needing 1,100 GPM, 1,000 watts with the large model needing 2,150 GPM). A continuous 250 watts is enough to keep batteries charged for several 12-volt lighting systems in a village.
- **Bamboo** Bamboo, a miraculous plant, will be a major factor in creating a sustainable future. Bamboo can now be processed into wood products, meaning we can dramatically reduce timber harvesting and start using a quickly growing species of grass to fill the world's need for wood. In addition, one-fifth of the world already lives in bamboo houses. Not only can some species of bamboo be eaten, other species actually soak up toxins from soil and water,

cleaning polluted soil. Still another species is ideal for power generation. A biomass plant in Asia runs on a surrounding ring of bamboo. Once the entire circle is harvested and burned for power, the harvesting crew find themselves back where they started six years earlier, ready to start a new six-year cycle of growth and cutting.

- **Compressed Earth Block Housing** Shelter is another major area of concern. New compressed earth block (CEB) technology, using a TerraBuilt machine, allows local subsoil to be mixed with concrete and create tongue-and-groove bricks. A medium-size middle-class house can be built in three weeks. Smaller structures can be built in a week. No mortar is required, and the resulting structure is far stronger than a typical American wooden home. People can build their own houses, or communities can work together to improve village housing. Most of the building and construction, however, would ideally be done by local contractors. The basic building material—soil—is nonpolluting and incurs no transportation costs, and it raises local money circulation by providing employment for unskilled labor.

- **Inexpensive Water Filtration** Most of the Third World suffers from unsafe drinking water; only 2 percent of the world's drinking water is now considered safe. New technology, using a series of sophisticated filters, now allows a village of six thousand to have safe drinking water for a cost of only $65,000, plus $15,000 worth of filters every six months. Funding new technology for villages will improve life dramatically while increasing economic development. There are also solar-powered UV water purifiers available. There is even a $90 portable solar water filter.

- **Solar Aquatic Sewage Treatment** New technology invented by John Todd allows natural sewage treatment of larger villages, even in colder climates. This is a low-cost alternative for developing countries, where sewage is a tremendous problem. The raw sewage is routed through a series of cylinders containing plants, snails, bacteria, and fish, all of it kept warm in a regulated greenhouse.

The end result is clear, clean water, ready to return to the river. A pilot project in Bear River, Nova Scotia, attracts thousands of visitors a year. These sewage plants, called Living Machines, are sold by John Todd Ecological Design around the world. For small villages, there is also a cost-effective, organic lagoon system, as well as composting toilets for smaller villages. The lagoons become fish farms while the composting toilets produce safe organic matter for building soil. What used to be polluting waste can be recycled through these organic sewage systems and become valuable to the community.

All of these low-impact technologies—and there are many more under study—offer practical ways to transform global energy use and technology. It does require a different way of thinking. Rather than building a transmission grid and one enormous power plant in a region, planners must now facilitate large-scale distribution of solar home systems to each household, or it might be microhydro energy or solar ovens—whatever is most appropriate to the region.

John Spears of the International Center for Sustainable Development (www.solarcities.org) has six principles of rural and urban construction, allowing local planners to construct their own regenerative communities, ones that use no fossil fuels while they regenerate the region as a whole. These principles will be taught as an e-course on the Global Regeneration Network.

By combining all of the above with health workshops, community development, village leadership development, and other programs already in place, this bottom-up approach applied to several thousand key regions would certainly raise standards of living around the world, "draining the swamp" of the conditions that breed terrorism. And by doing so on a low-emission technology base, there is the second prize of helping to avert the greenhouse effect and global climate change.

NOTES

CHAPTER 1. THE COMING CHANGE IN WORLDVIEWS

1. Chang, "Globe Grows Darker as Sunshine Diminishes 10% to 37%."
2. NOAA, "Carbon dioxide levels reach milestones at Arctic sites."
3. Anderegg et al., "Expert credibility in climate change."
4. "The Pebble Bed Modular Reactor (PBMR)."
5. Earthjustice, "Pennsylvania and Fracking."
6. "Dead zone may boost shark attacks."
7. Barstow and Stein, "Under Bush, a New Age of Prepackaged News."
8. Cullen, "Biggest Study of GMOs Finds Impact on Birds, Bees."
9. "Kazakhstan Country Profile."
10. Kuhn, *The Structure of Scientific Revolutions*.
11. Ibid.
12. Gay, *The Enlightenment: The Rise of Modern Paganism*.
13. Kuhn, *The Structure of Scientific Revolutions*.
14. Kant, "The End of All Things (1794)," in *Perpetual Peace and Other Essays*, 93–106.
15. Kant, "Preface to Second Edition," in *Critique of Pure Reason*, 17.
16. Steiner, *Nature's Open Secret*.

CHAPTER 2. KANT: THE ORGANIC COPERNICUS

1. Johann Gottfried Herder, *Letters on the Advancement of Humanity*, letter 79, quoted in Cassirer, *The Philosophy of the Enlightened*, 84.
2. Kant, *Kant: On History*.
3. Kant, "Preface to Second Edition," in *Critique of Pure Reason*, 17.
4. Kant, *Kant: On History*, 79.
5. Kant, *Critique of Pure Reason*, 82.

6. Fung, *A New Treatise on the Methodology of Metaphysics.*

7. Fung, *A Short History of Chinese Philosophy,* 341.

8. Ibid.

9. Kant, *Kant: On History.*

10. Kant, "Preface to Second Edition," in *Critique of Pure Reason,* 17.

11. Ibid.

12. Kant, *Metaphysics of Morals.*

13. Kant, *Kant: On History,* 3–5.

14. Kant, *Kant: On History.*

15. Ibid., 20, 23.

16. Ibid., 24, 25.

17. Ibid., 21, 25.

18. Ibid, 25.

19. Kant, *Perpetual Peace and Other Essays,* 111.

CHAPTER 3. GOETHE: THE ORGANIC GALILEO

1. Barnes, review of *The Wholeness of Nature.*

2. Merchant, *The Death of Nature,* 171.

3. Bortoft, *The Wholeness of Nature.*

4. Hanson, *Perception and Discovery.*

5. Goethe quoted in Steiner, *Goethe's Conception of the World,* 57.

6. Goethe quoted in Steiner, *Nature's Open Secret,* 33.

7. Goethe quoted from "Winckelmann," in Nisbet, *German Aesthetic and Literary Criticism,* 237.

8. Steiner, *Nature's Open Secret.*

9. Barnes, *The Third Culture.*

CHAPTER 4. THE ORGANIC SHIFT
IN SCIENCE AND TECHNOLOGY

1. Kevles, *In the Name of Eugenics,* ix–xiii.

2. Bergson, *Creative Evolution,* 108.

3. Ibid., 107.

4. Ibid., 285–86.

5. Haeckel, *Generelle Morphologie der Organismen.*

6. Haeckel, *The Riddle of the Universe,* 103.

7. Haeckel quoted in Gasman, *The Scientific Origins of National Socialism,* 41.

8. Ricketts et al., *Between Pacific Tides.*

9. Leopold, "The Land Ethic," in *A Sand County Almanac,* 201.

10. Ibid., 221–22.
11. Susskind, *Heinrich Hertz: A Short Life.*
12. Kuhn, *Structure of Scientific Revolutions,* 88.
13. Bohr quoted in Capra, *The Tao of Physics,* 132.
14. Smuts, *Holism and Evolution.*
15. Kant, "Idea for a Universal History from a Cosmopolitan Point of View," in *Kant: On History.*
16. Jung, *Letters,* vol. 1, 360.
17. Jung, *Letters,* vol. 2, 567.
18. Witt, "The in vitro evidence for an effect of high homeopathic potencies—a systematic review of the literature."
19. King, *History of Homoeopathy,* 346.
20. Makover, "Doctor of Medicine Profession."
21. Twain, "A Majestic Literary Fossil," 444.
22. Starfield, "Is US Health Really the Best in the World?"
23. Null, et al., *Death by Medicine,* 12.
24. Ibid., 97.
25. Ibid., 3.
26. Harris, "F.D.A. Official Admits 'Lapses' on Vioxx."
27. "ALEC Exposed."
28. Wolf, "How Independent Is the FDA?"
29. Mouchot quoted in Butti and Perlin, *A Golden Thread,* 63.
30. Ibid., 73.
31. Ibid.
32. Ibid., 84–87.
33. Ibid, 100–111.
34. Carver, "Being Kind to the Soil."
35. Carwell, *Blacks in Science,* 18.
36. Carver quoted in Clark, *The Man Who Talks with the Flowers,* 39.
37. Quoted in Lewis, *The Public Image of Henry Ford,* 283.
38. Ibid., 284.
39. Bird and Ikerd, "Sustainable Agriculture: A Twenty-First-Century System," 92–100.

CHAPTER 5. A SOCIAL/CIVIL RIGHTS REVOLUTION

1. Murray, "On the Equality of the Sexes."
2. de Gouges, "The Declaration of the Rights of Woman."
3. Wollstonecraft, *A Vindication of the Rights of Woman.*

4. Nelson, *Thomas Paine: Enlightenment, Revolution, and the Birth of Modern Nations*, 221.

5. Walker, *David Walker's Appeal, in Four Articles*, 25.

6. Ibid., 70.

7. Stanton, "The Declaration of Sentiments."

8. Norton, "Walt Whitman, Prophet of Gay Liberation."

9. Carpenter quoted in Norton, "Walt Whitman, Prophet of Gay Liberation."

10. Shaw, *The Lincoln Encyclopedia.*

11. Josephson, *The Robber Barons,* 354.

12. Brecher, *Strike! Boston.*

13. Illinois Labor History Society, www.illinoislaborhistory.org/articles.html.

14. "Preamble," Populist Party Platform (1892), www.wwnorton.com/college/history/eamerica/media/ch22/resources/documents/populist.htm.

15. Laurie, "The United States Army and the Return to Normalcy in Labor Dispute Interventions."

16. Bancroft, ed., "Platform of the American Anti-Imperialist League," in *Speeches, Correspondence, ard Political Papers of Carl Schurz,* 77.

17. James, "The Philippine Tangle."

18. Bancroft, ed., *Speeches, Correspondence, and Political Papers of Carl Schurz.*

19. "Mark Twain Home, An Anti-Imperialist."

20. Twain, *The Greatest American Humorist, Returning Home.*

21. Roosevelt, "Free Silver, Trusts, and the Philippines."

22. Arrhenius, *World in the Making,* 63.

23. "U.S. Public Health Service, Proceedings of a Conference to Determine Whether or Not There Is a Public Health Question in the Manufacture, Distribution or Use of Tetraethyl Lead Gasoline."

24. Trotsky quoted in Irvine, "The Prophet Misarmed: Trotsky, Ecology and Sustainability."

25. Stauber and Rampton, *Toxic Sludge Is Good for You.*

26. Bernays, *Propaganda.*

27. Higham, *Trading with the Enemy: An Exposé of the Nazi-American Money Plot,* xv, 20, 97.

28. Perfect, *Camera dei Fasci e delle Corporazioni.*

29. Higham, *Trading with the Enemy,* xv, 20, 97.

30. Lasby, *Operation Paperclip*; Hunt, "U.S. Coverup of Nazi Scientists."

31. Bell, Davis, and Fletcher, "A Retrospective Assessment of Mortality from the London Smog Episode of 1952."

32. Landerman quoted in "Consumers and Environmentalists Call on State Dental Board to 'Open Wide and Say the M Word.'"

33. McKinley, "Bill Seeks to Ban Use of Mercury in a Variety of Common Products."

34. Geier and Geier, "Early Downward Trends in Neurodevelopmental Disorders Following Removal of Thimerosal-Containing Vaccines."

35. Kennedy quoted in "Review of Manned Aircraft Nuclear Propulsion Program."

36. *Deseret News,* April 24, 1979.

37. Carter, "H-Bomb off Georgia Coast: Is It a Danger?"

38. Einstein, "Atomic Education Urged by Einstein: Scientist in Plea for $200,000 to Promote New Type of Essential Thinking."

39. Einstein, "The Menace of Mass Destruction," in *Out of My Later Years,* 205.

CHAPTER 6. THE ORGANIC
COUNTERCULTURES AND THE NEW ARTIST

1. Blake, "Songs of Innocence and Experience," in *The Complete Poetry and Prose of William Blake.*

2. Blake, "The Emanation of the Giant Albion," in *The Complete Poetry and Prose of William Blake,* 146.

3. Ackroyd, *Blake,* 340.

4. Shelley, *Prometheus Unbound,* 115

5. Emerson, *The Collected Works of Ralph Waldo Emerson.*

6. Richardson, *Emerson: The Mind on Fire.*

7. Thoreau, *Walden,* 119.

8. Ibid., 427.

9. Thoreau, *On the Duty of Civil Disobedience,* 19.

10. "The Literature of Bohemia," 17, 18.

11. Houssaye, *Man about Paris,* 30.

12. Uzanne, *Fashion in Paris,* 195.

13. Whitman, preface to *Leaves of Grass.*

14. Huneker, *Steeplejack,* 10.

15. Vorse quoted in Stansell, *American Moderns,* 13.

16. James, *The Varieties of Religious Experience,* 76.

17. Muir, *My First Summer in the Sierra,* 211.

18. Muir, *Our National Parks,* 56.

19. Muir, notes for after-dinner speech at Harvard University.

20. Swett, "John Muir," 120–23.

21. Roosevelt, "A Layman's Views of an Art Exhibition."

22. Luhan, *Movers and Shakers.*

23. Cox quoted in "Cubists of All Sorts."

24. Altshuler, *The Avant Garde in Exhibition,* 73.

25. Luhan, *Movers and Shakers,* 269.

26. Hapgood quoted in Luhan, *Movers and Shakers.*

CHAPTER 7. THE FUTURE ACCORDING
TO TEILHARD DE CHARDIN

1. Teilhard, *The Future of Man,* 284.

2. Ibid., 91.

3. Ibid., 173–74, 180.

4. Ibid., 95.

5. Ibid., 124.

6. Ibid., 238–39.

7. Ibid., 296–97.

8. Ibid., 269, 271.

CHAPTER 8. THE EARLY ENLIGHTENMENT: 1950–1975

1. Eisenhower, Farewell address to the nation.

2. DeGroot, *Student Protest;* Sale, *SDS.*

3. "Hillbilly Bopster in Show Here."

4. McClure quoted in "City Lights Bookstore 50th Anniversary: The Birth of Cool."

5. Snyder quoted in "City Lights Bookstore 50th Anniversary: The Birth of Cool."

6. McClure quoted in Schumacher, *Dharma Lion.*

7. King, "I Have a Dream."

8. Snyder, "Buddhist Anarchism."

9. Clarke and Hemphill, "The Santa Barbabra Oil Spill, A Retrospective," 157–62.

10. "Vietnam war protests."

11. Goldsmith, *The Beatles Come to America.*

12. Lisker, "Homo Nest Raided—Queen Bees Are Stinging Mad."

13. Clendinenm and Nagourney, *Out for Good.*

14. Fuller, *Operating Manual for Spaceship Earth.*

15. Bertalanffy, *General System Theory,* 37.

16. Gorton and Jarvis, "The effectiveness of vitamin C in preventing and relieving the symptoms of virus-induced respiratory infections," 530–33.

17. Lilly, *The Center of the Cyclone,* 39.

18. Ornstein, *The Psychology of Consciousness,* 2.

19. Schumacher, *Small Is Beautiful,* 52.
20. Roszak, "Introduction," in Schumacher, *Small Is Beautiful,* 7.
21. Schumacher, *Small Is Beautiful,* 137.
22. Ibid., 78, 177.
23. Grossman, "Radioactivity in the Ocean: Diluted, But Far from Harmless."
24. "Sunken Nuke Subs Decay Toward Catastrophes."
25. Naess quoted in Fox, *Toward a Transpersonal Ecology,* 93–94.

CHAPTER 9. THE CONSERVATIVE BACKLASH: 1976–1990

1. Hume, *A Treatise of Human Nature,* 264.
2. Calavita, Pontell, and Tillman, *Big Money Crime, Fraud and Politics in the Savings and Loan Crisis.*
3. "The Iran-Contra Affair."
4. Solomon, "Rumsfeld's Handshake Deal with Saddam," 2273.
5. Cole, "US Ambassador to Iraq April Glaspie."
6. Cockburn, *Corruptions of Empire.*
7. Watt, "Ours Is the Earth," 74–75; Wolf, "God, James Watt, and the Public Land," 65.
8. Wetstone quoted in "A look back at Reagan's environmental record."
9. Quoted in Amador, "Fairness Doctrine, R.I.P."
10. Fairlie and Sumner, "The Other Report on Chernobyl."
11. Ehrlich, *The Machinery of Nature,* 17.
12. Capra, *The Web of Life,* 7.
13. Snyder, *The Gary Snyder Reader,* 193.

CHAPTER 10. THE INTENSIVE PHASE AND THE COMING PARADIGM FLIP: 1991–2010

1. Smith, "10 Reasons to Avoid GMOs."
2. Augustin, "Eight glacial cycles from an Antarctic ice core," 623–28.
3. United Nations Foundation, "World Is Nearing Climate Point of No Return."
4. Hansen et al., "Target Atmospheric CO_2: Where Should Humanity Aim?," 217.
5. 350 Science, 350.org, 2011, www.350.org/en/node/26.
6. Hardt and Safina, "How Acidification Threatens Oceans from the Inside Out."
7. Gagosian, "Abrupt Climate Change: Should We Be Worried?"
8. *Declaration of the Occupation of New York City,* www.nycga.net/resources/declaration.
9. Ibid.
10. The Peacemaker, www.iroquoismuseum.org/PEACEMAKER.htm.

11. Archer and Jacobson, "Evaluation of global wind power."

12. "In 2011, Wind Power Reaches New Heights."

13. De Gamo, *Herbart and the Herbartians.*

14. Kant, "Idea for a Universal History with a Cosmopolitan Purpose," in Brown and Held, *The Cosmopolitanism Reader,* 23.

15. Yoffe, "Doctors Are Reminded, 'Wash Up!'"

16. Beijer and de Blaey, "Hospitalisations caused by adverse drug reactions (ADR): A meta-analysis of observational studies," 46–54.

17. Madland and Teixeira, *New Progressive America,* 5.

CHAPTER 11. AFTER THE PARADIGM FLIP THE TRANSFORMATIONAL PHASE, 2011–2050

1. Strauss and Howe, *Millennials Rising.*

2. Laszlo, *The Sustainable Company.*

3. Lampkin and Padel, *The Economics of Organic Farming.*

4. McNeely and Scherr, *Ecoagriculture.*

5. Magdoff and van Es, *Building Soils for Better Crops.*

6. Gershon, *The Livable Neighborhood Workbook.*

7. Brown, "Wind Power Set to Become World's Leading Energy Source."

8. "Harnessing the wind: Clipper energizes turbine industry."

9. Marty, *Search for a Usable Future.*

CHAPTER 12. THE NEXT STEP: WHAT YOU CAN DO

1. "Overview," in "Introduction," Solar Roadways, http://solarroadways.com.

APPENDIX 1. GLOBAL REGENERATION— THE REAL SOLUTION TO WAR AND POVERTY

1. Westley, www.africa.upenn.edu/Org_Institutes/Rodale_Institute_14582 .html.

APPENDIX 3. LOW-IMPACT TECHNOLOGY AND RENEWABLE ENERGY

1. NOAA, "Carbon dioxide levels reach milestones at Arctic sites."

2. Connor, www.independent.co.uk., search on "scientists turn fresh air into petrol."

BIBLIOGRAPHY

Ackroyd, Peter. *Blake.* New York: Alfred A. Knopf, 1995.

"A look back at Reagan's environmental record." http://grist.org/article/griscom reagan (article dated June 11, 2004; accessed September 4, 2012).

"ALEC Exposed. Talk: Government-industry revolving door." www.sourcewatch.org/index.php?title=Talk:Government-industry_revolving_door (article dated October 20, 2005; accessed August 31, 2012).

Altshuler, Bruce. *The Avant Garde in Exhibition: New Art in the 20th Century.* New York: Harry N. Abrams, 1994.

Amador, Jorge. "Fairness Doctrine, R.I.P." www.thefreemanonline.org/columns/fairness-doctrine-rip (article dated June 1989; accessed September 4, 2012).

Anderegg, William R. L., et al. "Expert credibility in climate change." Proceedings of the National Academy of Sciences of the United States of America, April 9, 2010, www.pnas.org/content/early/2010/06/04/1003187107.abstract (article dated April 9, 2010; accessed September 4, 2012).

Archer, Cristina L., and Mark Z. Jacobson. "Evaluation of global wind power." www.stanford.edu/group/efmh/winds/global_winds.html (article dated 2005; accessed August 31, 2012).

Arnold, Edwin. *Light of Asia, or The Great Renunciation, Being the Life and Teachings of Gautama, Prince of India and Founder of Buddhism.* Boston: Roberts Brothers, 1892.

Arrhenius, Svante. *World in the Making: The Evolution of the Universe.* New York/London: Harper & Brothers, 1908.

Augustin, Laurent, et al. "Eight glacial cycles from an Antarctic ice core." *Nature* 429, no. 6992 (June 24, 2004): 623–28.

Bancroft, Frederic, ed. *Speeches, Correspondence and Political Papers of Carl Schurz.* Volume VI. New York: G. P. Putnam's Sons, 1913.

Barnes, John Michael. Review of *The Wholeness of Nature. Waldorf Library Journal,* Research Bulletin, vol. 2, no. 1 (January 1997).

———. *The Third Culture: Participatory Science as the Basis for a Healing Culture.* Hillsdale, N.Y.: Adonis Press, 2009.

Barrett, William. *The Illusion of Technique: A Search for Meaning in a Technological Civilization.* New York: Anchor Books, 1978.

Barstow, David, and Robin Stein. "Under Bush, a New Age of Prepackaged News." *New York Times,* March 13, 2005.

Bateson, Gregory. *Steps to an Ecology of Mind.* Chicago: University of Chicago Press, 1972.

Beijer, H. J., and C. J. de Blaey. "Hospitalisations caused by adverse drug reactions (ADR): A meta-analysis of observational studies." *Pharmacy World and Science* 24 (April 2002).

Bell, Michele L., Devra L. Davis, and Tony Fletcher. "A Retrospective Assessment of Mortality from the London Smog Episode of 1952." *Environmental Health Perspectives* (October 15, 2003).

Bergson, Henri. *Creative Evolution.* Translated by Arthur Mitchell. New York: Macmillan, 1920. First published 1911 by Henry Holt and Co.

———. *The Creative Mind: An Introduction to Metaphysics.* New York: Carol Publishing, 1992. First published 1946 by Philosophical Library.

Berkeley, George. *Treatise Concerning the Principles of Human Knowledge.* Dublin, Scotland: Printed by Aaron Rhames, for Jeremy Pepyat, bookseller, 1710.

Bernays, Edward. *Propaganda.* Brooklyn, N.Y.: Ig Publishing, 2004.

Berry, Thomas. *The Dream of the Earth.* San Francisco: Sierra Club Books, 1990.

Bertalanffy, Ludwig von. *General System Theory: Foundations, Development, Applications.* New York: George Braziller, 1969.

Bird, G. W., and J. Ikerd. "Sustainable Agriculture: A Twenty-First-Century System." *Annals of the American Academy of Political and Social Science* 529 (1993): 92–100.

Blake, William. *The Complete Poetry and Prose of William Blake.* Edited by David V. Erdman with an introduction by Harold Bloom. Berkeley: University of California Press, 2008. First published 1965.

Blum, Linda. *Between Feminism and Labor: The Significance of the Comparable Worth Movement.* Berkeley: University of California Press, 1991.

Bohm, David. *Wholeness and the Implicate Order.* New York: Routledge, 2002.

Bortoft, Henri. *The Wholeness of Nature: Goethe's Way toward a Science of Conscious Participation in Nature.* Great Barrington, Mass.: Lindisfarne Press, 1996.

Brecher, Jeremy. *Strike! Boston.* Boston: South End Press, 1997.

Brown, Garrett Wallace, and David Held, eds. *The Cosmopolitanism Reader.* Malden, Mass.: Polity Press, 2010.

Brown, Lester R. "Wind Power Set to Become World's Leading Energy Source." Earth Policy Institute. www.earth-policy.org/plan_b_updates/2003/update24 (article dated June 25, 2003; accessed September 4, 2012).

———. *Plan B 4.0: Mobilizing to Save Civilization.* New York: W. W. Norton, 2009.

Butti, Ken, and John Perlin. *A Golden Thread: 2500 Years of Solar Architecture and Technology.* New York: Cheshire/Van Nostrand Reinhold Co., 1980.

Calavita, Kitty, Henry Pontell, and Robert Tillman. *Big Money Crime, Fraud and Politics in the Savings and Loan Crisis.* Berkeley/Los Angeles: University of California Press, 1997.

Capra, Fritjof. *The Tao of Physics: An Exploration of the Parallels between Modern Physics and Eastern Mysticism.* Boston: Shambhala Publications, 1999. First published 1975.

———. *The Turning Point: Science, Society, and the Rising Culture.* New York: Simon & Schuster, 1982.

———. *The Web of Life: A New Scientific Understanding of Living Systems.* New York: Anchor Books, 1997.

Carson, Rachel. *Silent Spring.* New York: Houghton Mifflin, 1962.

Carter, Chelsea. "H-Bomb off Georgia Coast: Is It a Danger?" Associated Press, May 2, 2004.

Carver, George Washington. "Being Kind to the Soil," *The Negro Farmer,* www .scribd.com/doc/51547121/Special-History-Study-George-Washington-Carver (article dated January 31, 1914; accessed September 4, 2012).

Carwell, Hattie. *Blacks in Science: Astrophysicist to Zoologist.* Hicksville, N.Y.: Exposition Press, 1977.

Cassirer, Ernst. *The Philosophy of the Enlightenment.* Princeton, N.J.: Princeton University Press, 1979.

Chang, Kenneth. "Globe Grows Darker as Sunshine Diminishes 10% to 37%." *New York Times,* May 13, 2004.

Chopra, Deepak. *Quantum Healing.* New York: Bantam, 1990.

"City Lights Bookstore 50th Anniversary: The Birth of Cool," *San Francisco Chronicle,* June 8, 2003.

Clark, Glenn. *The Man Who Talks with the Flowers: The Intimate Life Story of Dr. George Washington Carver.* Shakopee, Minn.: Park Publishing, 1939. Reprinted 1994.

Clarke, K. C., and Jeffrey J. Hemphill. "The Santa Barbara Oil Spill, a Retrospective." *Yearbook of the Association of Pacific Coast Geographers* 64 (2002): 157–62.

Clendinenm, Dudley, and Adam Nagourney. *Out for Good: The Struggle to Build a Gay Rights Movement in America.* New York: Simon & Schuster, 1999.

Cockburn, Alexander. *Corruptions of Empire.* London/New York: Verso, 1988.

Cole, Carleton. "US Ambassador to Iraq April Glaspie." *Christian Science Monitor,* May 27, 1999.

Connor, Steve. "Exclusive: Pioneering Scientists Turn Fresh Air Into Petrol in Massive Boost in Fight Against the Energy Crisis." www.independent.co.uk and search on "scientists turn fresh air into petrol" (article dated October 19, 2012; accessed October 26, 2012).

"Consumers and Environmentalists Call on State Dental Board to 'Open Wide and Say the M Word.'" Campaign for Mercury Free Dentistry: A Project of Consumers for Dental Choice. www.toxicteeth.org/pressRoom_recentNews/ August-2001/Say-NO-to-Corrupt-Mercury-Amalgam-Fact-Sheet.aspx (article dated August 3, 2001; accessed September 4, 2012).

"Cubists of All Sorts." *New York Times,* March 16, 1913.

Cullen, David. "Biggest Study of GMOs Finds Impact on Birds, Bees." Reuters, March 21, 2005.

Davids, Caroline Rhys. *Buddhist Psychology, an Inquiry into the Analysis and Theory of Mind in Pali Literature.* Glouchestershire, U.K.: G. Bell, 1914.

"Dead zone may boost shark attacks." BBC News. http://news.bbc.co.uk/2/hi/ science/nature/3534658.stm (article dated August 4, 2004; accessed September 4, 2012).

Declaration of the Occupation of New York City, www.nycga.net/resources/ declaration (article dated September 29, 2011; accessed August 31, 2012).

De Gamo, Charles. *Herbart and the Herbartians.* Charleston, S.C.: Nabu Press, 2010.

de Gouges, Olympe. "The Declaration of the Rights of Woman." www.library .csi.cuny.edu/dept/americanstudies/lavender/decwom2.html (article dated September 1791; accessed September 4, 2012).

DeGroot, Gerald J., ed. *Student Protest: The Sixties and After.* London/New York: Longman, 1998.

Earthjustice. "Pennsylvania and Fracking." http://earthjustice.org/features/ campaigns/pennsylvania-and-fracking (article dated 2011; accessed September 4, 2012).

Ehrlich, Paul. *The Machinery of Nature.* New York: Simon & Schuster, 1987.

Einstein, Albert. "Atomic Education Urged by Einstein: Scientist in Plea for $200,000 to Promote New Type of Essential Thinking." Telegram sent to *New York Times,* May 25, 1946.

———. *Out of My Later Years.* New York: Carol Publishing, 1995.

Eisenhower, Dwight D. Farewell address to the nation, January 17, 1961. http:// mcadams.posc.mu.edu/ike.htm (accessed September 4, 2012).

Elgin, Duane. *Voluntary Simplicity.* New York: William Morrow & Company, 1981.

Emerson, Ralph Waldo. *The Collected Works of Ralph Waldo Emerson.* Edited by Robert Spiller et al. Cambridge, Mass: Harvard University Press, 1971.

Fairlie, Ian, and David Sumner. "The Other Report on Chernobyl." www.chernobylreport.org/?p=summary (article dated April 4, 2006; accessed August 31, 2012).

Ferguson, Marilyn. *The Aquarian Conspiracy.* New York: Jeremy P. Tarcher, 1980.

Fox, Warwick. *Toward a Transpersonal Ecology: Developing New Foundations for Environmentalism.* Albany: State University of New York, 1995.

Friedan, Betty. *The Feminine Mystique.* New York: W. W. Horton and Co., 1963.

Fuller, R. Buckminster. *Operating Manual for Spaceship Earth.* New York: E. P. Dutton & Co., 1963.

Fung Yu-Lan. *A New Treatise on the Methodology of Metaphysics.* Edited by Derk Bodde. New York: Free Press/Simon & Schuster, 1966.

———. *A Short History of Chinese Philosophy: A Systematic Account of Chinese Thought from Its Origins to the Present Day.* Edited by Derk Bodde. New York: Free Press/Simon & Schuster, 1976.

Gagosian, Robert B. "Abrupt Climate Change: Should We Be Worried?" Woods Hole Oceanographic Institute. www.whoi.edu/page.do?pid=83339&tid=3622&cid=9986 (article dated January 27, 2003; accessed September 4, 2012).

Gasman, Daniel. *The Scientific Origins of National Socialism.* New York: American Elsevier, 1971.

Gay, Peter. *The Enlightenment: The Rise of Modern Paganism.* New York: W. W. Norton & Company, 1995.

Geier, D. A., and M. R. Geier. "Early Downward Trends in Neurodevelopmental Disorders Following Removal of Thimerosal-Containing Vaccines." *Journal of American Physicians and Surgeons* 11, no. 1 (Spring 2006).

Gershon, David. *The Livable Neighborhood Workbook: About Making Life Better on the Street Where You Live.* Ann Arbor, Mich.: Aatec Publications, 2002.

Goldsmith, Martin. *The Beatles Come to America.* Hoboken, N.J.: John Wiley & Sons, 2004.

Goleman, Daniel. *Emotional Intelligence.* New York: Bantam, 1995.

Gorton, H. C., and K. Jarvos. "The effectiveness of vitamin C in preventing and relieving the symptoms of virus-induced respiratory infections." *Journal of Manipulative and Physiological Therapeutics* 8 (October 22, 1999).

Grossman, Elizabeth. "Radioactivity in the Ocean: Diluted, But Far from Harmless." *Environment 360: Opinion, Analysis, Reporting & Debate.* http://e360.yale.edu/search (article dated April 7, 2011; accessed August 31, 2012).

Haeckel, Ernst. *Generelle Morphologie der Organismen*. Berlin: G. Reimer Verlag, 1866.

———. *The Riddle of the Universe*. Translated by Joseph McCabe. New York: Buffalo Books, 1992.

Hahnemann, Samuel. *The Organon of the Medical Art*. Edited and annotated by Wendy Brewster O'Reilly. Palo Alto, Calif.: Birdcage Books, 1996. First published 1842.

Hanley, Charles J. "CO_2 Hits Record Levels, Researchers Find." Associated Press, March 20, 2004.

Hansen, James, et al. "Target Atmospheric CO_2: Where Should Humanity Aim?" *Open Atmosphere Science Journal* (2008): 217.

Hanson, Norwood Russell. *Perception and Discovery*. San Francisco: W. H. Freeman, 1969.

Hardt, J., and C. Safina. "How Acidification Threatens Oceans from the Inside Out." *Scientific American* (August 9, 2010).

"Harnessing the wind: Clipper energizes turbine industry." *Pacific Coast Business Times,* December 10–16, 2004.

Harris, Gardiner. "F.D.A. Official Admits 'Lapses' on Vioxx." *New York Times,* March 2, 2005.

Hartmann, Thom. *Unequal Protection: How Corporations Became "People"—and How You Can Fight Back*. 2nd ed. San Francisco: Berrett-Koehler Publishers, 2010. First published 2002.

Herder, Johann Gottfried. *Auch eine Philosophie der Gerschichte zur Bildung der Menschheit*. Germany: n.p., 1774.

———. *Outlines in the Philosophy of the History of Man*. Translated by T. Churchill. London: n.p., 1800.

Higham, Charles. *Trading with the Enemy: An Exposé of the Nazi-American Money Plot*. New York: Delacorte Press, 1983.

"Hillbilly Bopster in Show Here." *San Antonio Express,* January 11, 1956.

Houssaye, Arsene. *Man About Paris: Confessions*. West Sussex, U.K.: Littlehampton Book Services Ltd., 1972.

Howard, Sir Albert. *An Agricultural Testament*. Oxford, UK: Benediction Classics, 2010. First published 1940.

Hume, David. *A Treatise of Human Nature: Being an Attempt to Introduce the Experimental Method of Reasoning into Moral Subjects. Book I: Of the Understanding*. Edited by L. A. Selby-Bigge. Oxford, U.K.: Clarendon Press, 1896.

Huneker, James Gibbons. *Steeplejack*. New York: Charles Scribner's Sons, 1922.

Hunt, Linda. "U.S. Coverup of Nazi Scientists." *Bulletin of Atomic Scientist* 41, no. 4 (April 1985).

"In 2011, Wind Power Reaches New Heights." *Renewable Power News.* www
.renewablepowernews.com/archives/3005 (article dated March 18, 2012; accessed August 31, 2012).

"Iran-Contra Affair, The." www.jewishvirtuallibrary.org/jsource/US-Israel/Iran_
Contra_Affair.html (article dated November 1987; accessed August 31, 2102).

Irvine, Sandy. "The Prophet Misarmed: Trotsky, Ecology and Sustainability."
What's Next: Marxist Discussion Journal, www.whatnextjournal.co.uk/Pages/
Latest/Misarmed.html.

James, William. *The Principles of Psychology.* 2 vols. New York: Henry Holt and
Co., 1890.

———. "The Philippine Tangle." *Boston Evening Transcript,* March 1, 1899.

———. *The Varieties of Religious Experience: A Study in Human Nature.* Rockville,
Md.: Arc Manor, 2008. First published 1901 by Longmans, Green, and Co.

Jantsch, Eric. *The Self-Organizing Universe.* Oxford, U.K.: Pergamon, 1980.

Jaynes, Julian. *The Origin of Consciousness in the Breakdown of the Bicameral Mind.*
Boston: Hougton Mifflin Harcourt, 1976.

Josephson, Matthew. *The Robber Barons: The Great American Capitalists, 1861–
1901.* New York: Harcourt Brace and Co., 1962. First published 1934.

Jung, Carl Gustav. *The Archetypes and the Collective Unconscious.* Translated by R.
F. C. Hull. Princeton, N.J.: Princeton University Press, 1990. First published
1959.

———. *Letters.* 2 vols. Princeton, N.J.: Princeton University Press, 1973.

Kant, Immanuel. *Metaphysics of Morals.* Germany, n.p.: 1797.

———. *Critique of Pure Reason.* Translated by Norman Kemp Smith. London:
Macmillan and Co., 1929.

———. *Kant: On History.* Edited by Lewis White Beck. Indianapolis: Bobbs-
Merrill, 1963.

———. *Perpetual Peace and Other Essays.* Edited by Ted Humphrey. Indianapolis:
Hackett Publishing Co., 1983.

"Kazakhstan Country Profile," http://news.bbc.co.uk/2/hi/europe/country_
profiles/1298071.stm (article updated January 31, 2012; accessed August 31, 2012).

Kevles, Daniel J. *In the Name of Eugenics: Genetics and the Uses of Human Heredity.*
Cambridge, Mass.: Harvard University Press, 1995.

King, Martin Luther, Jr. "I Have a Dream." Washington, D.C., August 28, 1963.
http://abcnews.go.com/Politics/martin-luther-kings-speech-dream-full-text/
story?id=14358231#.T-4YQd25Dms (accessed September 4, 2012).

———. *Stride toward Freedom.* New York: Joanna Cotler Books, 1979.

King, William Harvey. *History of Homoeopathy.* Vol. 2. New York: Lewis, 1905.

Kübler-Ross, Elisabeth. *On Death and Dying.* New York: Charles Scribner's Sons, 1997.

Kuhn, Thomas S. *The Structure of Scientific Revolutions*. 3rd ed. Chicago/London: University of Chicago Press, 1996. First published 1962.

Lampkin, N. H., and S. Padel, eds. *The Economics of Organic Farming: An International Perspective*. Wallingford, UK: Centre for Biological Science International, 1994.

Larsen, Stephen. *The Healing Power of Neurofeedback: The Revolutionary LENS Technique for Restoring Optimal Brain Function*. Rochester, Vt.: Healing Arts Press, 2006.

Lasby, Clare. *Operation Paperclip*. New York: Athenaeum, 1975.

Laszlo, Chris. *The Sustainable Company: How to Create Lasting Value through Social and Environmental Performance*. Washington, D.C.: Island Press, 2003.

Laszlo, Ervin. *Introduction to Systems Philosophy: Toward a New Paradigm of Contemporary Thought*. New York: Harper Torchbooks, 1973.

Laurie, Clayton D. "The United States Army and the Return to Normalcy in Labor Dispute Interventions: The Case of the West Virginia Coal Mines, 1920–21." *West Virginia History* 50 (1991): 1–24. www.wvculture.org/history/journal_wvh/wvh50-1.html (accessed September 4, 2012).

Leopold, Aldo. *A Sand County Almanac: And Sketches Here and There*. New York: Oxford University Press, 1989. First published 1949.

Lewis, David L. *The Public Image of Henry Ford*. Detroit, Mich.: Wayne State University Press, 1987.

Lilly, John C. *The Center of the Cyclone: Looking into Inner Space*. Oakland, Calif.: Ronin, 1972.

Lisker, Jerry. "Homo Nest Raided—Queen Bees Are Stinging Mad." *New York Daily News,* July 6, 1969.

"The Literature of Bohemia." *The Westminster Review* 78 (July–October 1862).

Lovelock, James. *Gaia: A New Look at Life on Earth*. Oxford, U.K.: Oxford University Press, 2000. First published 1979.

Luhan, Mabel Dodge. *Movers and Shakers*. New York: Harcourt, Brace and Co., 1936.

Madland, David, and Ruy Teixeira. *New Progressive America: The Millennial Generation*. Progressive Studies Program, Center for American Progress, May 2009.

Magdoff, Fred, and Harold van Es. *Building Soils for Better Crops*. 2nd ed. Beltsville, Md.: Sustainable Agricultural Network, 2000.

Makover, Michael. "Doctor of Medicine Profession," www.nlm.nih.gov/medlineplus/ency/article/001936.htm (article updated February 3, 2011; accessed August 31, 2012).

"Mark Twain Home, An Anti-Imperialist." *New York Herald,* October 15, 1900. www.nlm.nih.gov/medlineplus/ency/article/001936.htm (accessed September 4, 2102).

Marty, Martin. *Search for a Usable Future.* New York: Harper & Row, 1969.

McKinley, James C., Jr. "Bill Seeks to Ban Use of Mercury in a Variety of Common Products." *New York Times,* February 7, 2001.

McNeely, Jeffrey A., and Sara J. Scherr. *Ecoagriculture: Strategies to Feed the World and Save Wild Biodiversity.* Washington, D.C.: Island Press, 2002.

Merchant, Carolyn. *The Death of Nature: Women, Ecology, and the Scientific Revolution.* New York: HarperCollins, 1980.

Mouchot, Augustin. *La Chaleur Solaire et Ses Applications Industrielles* (Solar Heat and its Industrial Applications). Paris: A. Blanchard, 1980. First published 1879 by Gauthier-Villars.

Muir, John. *My First Summer in the Sierra.* New York: Houghton, Mifflin and Co., 1911.

———. Notes for after-dinner speech at Harvard University, June 24, 1896. Muir Papers, Holt-Atherton Special Collections, University of the Pacific Library, Stockton, California.

———. *Our National Parks.* Boston/New York: Houghton, Mifflin and Co., 1901.

Murray, Judith Sargent. "On the Equality of the Sexes." *Massachusetts Magazine, or, Monthly Museum Concerning the Literature, History, Politics, Arts, Manners, Amusements of the Age* 2 (1790). http://digital.library.upenn.edu/women/murray/equality/equality.html (accessed September 4, 2012).

Nelson, Craig. *Thomas Paine: Enlightenment, Revolution, and the Birth of Modern Nations.* New York: Viking Penguin, 2006.

Nisbet, H. B., ed. *German Aesthetic and Literary Criticism: Winckelmann, Lessing, Haman, Herder, Schiller and Goethe.* Cambridge, UK: Cambridge University Press, 1985.

NOAA. "Carbon dioxide levels reach milestones at Artic sites." http://researchmatters.noaa.gov/news/Pages/arcticCO2.aspx (article dated May 31, 2012; accessed September 4, 2012).

Northbourne, Lord. *Look to the Land.* San Rafael, Calif.: Angelico Press/Sophia Perennis, 2011. First published 1940 by J. M. Dent & Sons.

Norton, Rictor. "Walt Whitman, Prophet of Gay Liberation." *Gay History and Literature.* http://rictornorton.co.uk/whitman.htm (article dated November 18, 1999, updated June 20, 2008; accessed September 4, 2012).

Null, Gary, et. al. *Death by Medicine.* Mount Jackson, Va.: Praktikos Books, 2010.

Ornstein, Robert. *The Psychology of Consciousness.* San Francisco: Harcourt Brace Jovanovich, 1977. First published 1972 by W. H. Freeman.

Osborn, Fairfield. *Our Plundered Planet.* London: Faber and Faber, 1948.

Otto, Rudolf. *Mysticism East and West.* Whitefish, Mont.: Kessinger Publishing, 2003.

Palmer, R. R. *The Age of the Democratic Revolutions: The Challenge.* Vol. 1. Princeton, N.J.: Princeton University Press, 1959.

"Overview," in "Introduction," *Solar Roadways.* http://solarroadways.com (undated article; accessed August 31, 2012).

Patanjali, Vyasa. *The Yoga-System of Patanjali.* Translated by James Haughton Woods. Cambridge, Mass.: Harvard University Press, 1914.

"The Pebble Bed Modular Reactor (PBMR)." Nuclear Information and Resource Service, www.nirs.org/factsheets/pbmrfactsheet.htm (undated article; accessed September 4, 2012).

Perfect, Francisco. *Camera dei Fasci e delle Corporazioni.* Formello, Italy: Bonacci Editore, 1991.

"Platform of the American AntiImperialist League." In *Speeches, Correspondence, and Political Papers of Carl Schurz,* vol. 6. New York: G. P. Putnam's Sons, 1913.

Prigogine, Ilya. *From Being to Becoming.* London: W. H. Freeman & Co., 1981.

"Review of Manned Aircraft Nuclear Propulsion Program." Government report from the Atomic Energy Commission and the Department of Defense. Washington, D.C.: Government Printing Office, 1963.

Richardson, Robert D., Jr. *Emerson: The Mind on Fire.* Berkeley/Los Angeles: University of California Press, 1995.

Ricketts, Edward F., Jack Calvin, and Joel W. Hedgpeth. *Between Pacific Tides.* 5th ed. Stanford, Calif.: Stanford University Press, 1985. First published 1939.

Rodale, Maria. *Organic Manifesto, How Organic Farming Can Heal Our Planet, Feed the World and Keep Us Safe.* New York: Rodale Books, 2010.

Roosevelt, Theodore. "Free Silver, Trusts, and the Philippines." Speech of Governor Roosevelt at Grand Rapids, Michigan, September 7, 1900. http://voicesofdemocracy .umd.edu/theodore-roosevelt-free-silver-trusts-and-the-philippines-speech-text (accessed September 4, 2012).

———. "A Layman's Views of an Art Exhibition." *Outlook* 103 (March 29, 1913): 718–20.

Roszak, Theodore. *Person/Planet: The Creative Disintegration of Industrial Society.* New York: Doubleday & Co., 1978.

Russell, Peter. *The Global Brain.* New York: Jeremy P. Tarcher, 1983.

Sale, Kirkpatrick. *SDS: The Rise and Development of the Students for a Democratic Society.* New York: Vintage Books, 1973.

Schacker, Michael. *A Spring Without Bees: How Colony Collapse Disorder Has Endangered Our Food Supply.* Guilford, Conn.: Lyons Press, 2008.

Schumacher, E. F. *Small Is Beautiful: Economics as if People Mattered.* London: Blond & Briggs, 1973.

Schumacher, Michael. *Dharma Lion: A Biography of Allen Ginsberg.* New York: St. Martin's Press, 1994.

Senge, Peter. *Fifth Discipline Guide.* New York: Doubleday Business, 1994.

Shaw, Archer H. *The Lincoln Encyclopedia.* New York: Macmillan, 1950.

Sheldrake Rupert. *A New Science of Life.* Rochester, Vt.: Park Street Press, 1995.

Shelley, Percy Bysshe. *Prometheus Unbound: A Lyrical Drama in Four Acts.* London: M. Dent and Co., 1898.

Smith, Jeffrey. "10 Reasons to Avoid GMOs." Institute for Responsible Technology. www.responsibletechnology.org/10-Reasons-to-Avoid-GMOs (article dated August 2, 2011; accessed September 4, 2012).

Smuts, Jan Christiaan. *Holism and Evolution.* New York: Macmillan Co., 1926.

Snyder, Gary. "Buddhist Anarchism." *Journal for the Protection of All Beings* 1 (1961). San Francisco: City Lights.

———. *The Gary Snyder Reader: Prose, Poetry, and Translations, 1952–1998.* New York: Counterpoint, 1999.

Solomon, Norman. "Rumsfeld's Handshake Deal with Saddam." www.fair.org/index.php?page=2773 (article dated December 8, 2005; accessed August 31, 2012).

Spretnak, Charlene, and Fritjof Capra. *Green Politics.* Rochester, Vt.: Bear & Co., 1986.

Stansell, Christine. *American Moderns: Bohemian New York and the Creation of a New Century.* New York: Henry Holt and Co., 2000.

Stanton, Elizabeth Cady. "The Declaration of Sentiments." Seneca Falls Conference, 1848. www.fordham.edu/halsall/mod/senecafalls.asp (accessed September 4, 2012).

Starfield, Barbara. "Is US Health Really the Best in the World?" *Journal of the American Medical Association* 284, no. 4 (July 26, 2000).

Stauber, John, and Sheldon Rampton. *Toxic Sludge Is Good for You: Lies, Damn Lies and the Public Relations Industry.* Monroe, Maine: Common Courage Press, 1995.

Steiner, Rudolf. *Goethe's Conception of the World.* Somerset, U.K.: Anthroposophical Publishing Co., 1928.

———. *Nature's Open Secret: Introductions to Goethe's Scientific Writings.* Great Barrington, Mass.: Anthroposophic Press, 2000. First published 1950.

———. *The Story of My Life.* Whitefish, Mont.: Kessinger Publishing, 2010.

Strauss, William, and Neil Howe. *Generations: The History of America's Future, 1584 to 2069.* New York: William Morrow and Co., 1991.

———. *Millennials Rising: The Next Great Generation.* New York: Vintage Books, 2000.

"Sunken Nuke Subs Decay Toward Catastrophes." www.rense.com/general41/ suken.htm (article dated September 5, 2003; accessed September 4, 2012).

Susskind, Charles. *Heinrich Hertz: A Short Life.* San Francisco: San Francisco Press, 1995.

Suzuki, D. T. *Outlines of Mahayana Buddhism.* Chicago: Open Court Publishing Co., 1908.

———. *Essays in Zen Buddhism.* N.p.: Souvenir Publishing Company, 2010.

Swett, John. "John Muir." *The Century Magazine* (new series 24) 46, no. 1 (May 1893).

Swimme, Brian, and Thomas Berry. *The Universe Story: From the Primordial Flaring Forth to the Ecozoic Era.* New York: HarperOne, 1994.

Tart, Charles. *Altered States of Consciousness.* New York: Grove Press, 1994.

Teilhard de Chardin, Pierre. *The Future of Man.* New York: Harper & Row, 1964. First published 1959 by Editions de Seuil.

Tellier, Charles. *Pacific and African Conquest by the Sun.* New York: Harper, 1990.

Thompson, William Irwin. *At the Edge of History: Passages about Earth.* New York: Harper & Row, 1971.

Thoreau, Henry David. *On the Duty of Civil Disobedience.* Rockville, Md.: Arc Manor, 2008. First published 1849 as *Resistance to Civil Government.*

———. *Walden.* New York: Thomas Y. Crowell & Co., 1910.

Trewartha, Glenn Thomas. *An Introduction to Weather and Climate.* New York: McGraw-Hill, 1937.

Twain, Mark. "A Majestic Literary Fossil." *Harper's Magazine,* February 1890.

———. *The Greatest American Humorist, Returning Home.* London: New York World, 1900. *UN Wire.* www.historywiz.com/primarysources/marktwain-imperialism.htm (accessed August 31, 2012).

United Nations Foundation. "World is Nearing Climate Point of No Return." www.smartbrief.com/servlet/ArchiveServlet?issueid=B52BA8C5-547A-4E2D-923F-D05C177DDF83&lmid=archives (article dated November 9, 2011; accessed August 31, 2012).

"Uranium and Nuclear Power in Kazakhstan." World Nuclear Association. www .world-nuclear.org/info/inf89.html (undated article; updated May 2012; accessed September 4, 2012).

"U.S. Public Health Service, Proceedings of a Conference to Determine Whether or Not There Is a Public Health Question in the Manufacture, Distribution or Use of Tetraethyl Lead Gasoline." PHS Bulletin No. 158. Washington, D.C.: U.S. Treasury Department, August 1925.

Uzanne, Octave. *Fashion in Paris: The Various Phases of Feminine Taste and Aesthetics from 1797 to 1897.* London: W. Heinemann; New York: C. Scribner's Sons, 1898.

"Vietnam war protests." www.history.com/topics/vietnam-war-protests (undated article; accessed August 31, 2012).

Walker, David. *David Walker's Appeal, in Four Articles: Together with a Preamble to the Coloured Citizens of the World, but in Particular, and Very Expressly, to Those of the United States of America.* New York: Hill and Wang, 1995.

Watt, James. "Ours Is the Earth." *Saturday Evening Post,* January/February 1982, 74–75.

Watts, Alan. *The Way of Zen.* New York: Pantheon Books, 1957.

Westley, Karen. www.africa.upenn.edu/Org_Institutes/Rodale_Institute_ 14582 .html (article dated October 29, 1993; accessed August 31, 2012).

Wheatley, Margaret. *Leadership and the New Science.* San Francisco: Berret-Koehler Publishers, 2006.

Whitman, Walt. *Leaves of Grass.* New York: Simon and Brown, 2011.

Wilber, Ken. *A Brief History of Everything.* Boston: Shambhala Publications, 1996.

———. *Up from Eden.* Wheaton, Ill.: Quest Books, 2007.

Witt, C. M., et al. "The in vitro evidence for an effect of high homeopathic potencies—a systematic review of the literature." *Complementary Therapies in Medicine* 2 (2007): 128–38.

Wolf, Ron. "God, James Watt, and the Public Land." *Audubon* 83, no. 3 (1981): 65.

Wolf, Sidney. "How Independent Is the FDA?" www.pbs.org/wgbh/pages/ frontline/shows/prescription/hazard/independent.html. (November 2003; accessed August 31, 2012).

Wollstonecraft, Mary. *Thoughts on the Education of Daughters: With Reflections on Female Conduct in the Important Duties of Life.* London: Joseph Johnson, 1786.

———. *Original Stories from Real Life.* Illustrations by William Blake. London: Joseph Johnson, 1791.

———. *A Vindication of the Rights of Woman: With Strictures on Political and Moral Subjects.* London: T. Fisher Unwin, 1792.

Yoffe, Emily. "Doctors Are Reminded, 'Wash Up!'" *New York Times,* November 9, 1999.

INDEX